THE LAST REPUBLICANS

The
LAST
REPUBLICANS

Inside the Extraordinary
Relationship Between
George H. W. Bush
and
George W. Bush

MARK K. UPDEGROVE

HARPER

NEW YORK • LONDON • TORONTO • SYDNEY

HARPER

A hardcover edition of this book was published
in 2017 by HarperCollins Publishers.

HarperCollins books may be purchased for educational, business, or
sales promotional use. For information, please email the Special Markets
Department at SPsales@harpercollins.com.

FIRST HARPER PAPERBACKS EDITION PUBLISHED 2018.

Designed by Bonni Leon-Berman

The Library of Congress has catalogued the hardcover edition as follows:
Names: Updegrove, Mark K., author.
Title: The last Republicans: inside the extraordinary relationship
between George H. W. Bush and George W. Bush / Mark K. Updegrove.
Description: First edition. | New York, NY: Harper, 2017. | Includes
bibliographical references.
Identifiers: LCCN 2017034575 (print) | LCCN 2017035636 (ebook) |
ISBN 9780062654144 (ebook) | ISBN 9780062695109 (digaudio) | ISBN
9780062654120 (hc: alk. paper) | ISBN 9780062688156 (largeprint: alk.
paper) | ISBN 9780062695116 (audio: alk. paper)
Subjects: LCSH: Bush, George, 1924– | Bush, George W. (George
Walker), 1946– | Presidents—United States—Biography. | United
States—Politics and government—1989– | Fathers and sons—United
States—Biography. | Bush family.
Classification: LCC E882 (ebook) | LCC E882 .U63 2017 (print) | DDC
973.928092/2—dc23
LC record available at https://lccn.loc.gov/2017034575

ISBN 978-0-06-265413-7 (pbk.)

18 19 20 21 22   LSC   10 9 8 7 6 5 4 3 2 1

For my wife, Amy—the love of my life

CONTENTS

THE LAST REPUBLICANS

INTRODUCTION

SURELY THE NEWS GAVE the old man a feeling of redemption as well as pride. Gnarled and decrepit at eighty-nine years of age, but still sharp of mind, John Adams learned on February 14, 1825, that his eldest son, John Quincy, would become the sixth president of the United States.

Nearly a quarter of a century had passed since Adams had bitterly yielded the presidency to Thomas Jefferson, his revolutionary protégé turned political rival, when the election of 1800 saw candidates from opposing political parties contend for the office for the first time. On March 4, 1801, in the waning hours of his single White House term, the second president, and the last Federalist to hold the office, stole away from Washington rather than bear witness to the noon inauguration of Jefferson, the founder of the Democratic-Republican Party and, effectively, the first Republican. Under the light of a crescent moon, Adams boarded a 4:00 a.m. stagecoach bound for Baltimore on the first leg of a five-day, 450-mile journey that would take him back to his native Quincy, Massachusetts, and into political exile.

It took just as long—five days—for word of his son's election to reach Adams in Quincy. The news had been considerably longer in the making. It had taken over three months for the outcome of the election of 1824 to be decided. At a time of growing factionalism, the Democratic-Republican Party, the only political party during the time from 1801 to 1825, put up four presidential candidates representing regional distinctions: John Quincy Adams of Massachusetts, Andrew Jackson of Tennessee, William Crawford of Georgia, and Henry Clay of Kentucky. While Andrew Jackson had handily won the popular vote—41 percent overall—he lacked the majority of electoral votes needed to win the office. The presidency hung in the balance.

The victor would be determined by the U.S. House of Representatives in a runoff vote of the three top candidates: Adams, Jackson, and Crawford. But it was the fourth candidate, Henry Clay, who controversially put the matter to rest by convincing his supporters to pledge their electoral votes to the candidate from Massachusetts; Adams emerged as the winner. In return, Clay exacted assurances that he would be appointed Adams's secretary of state and thus, by historical proclivity in those years, the presidential heir apparent.

The elder Adams considered his son's ascent to the country's highest office to be one of the most fulfilling moments of his long life. Among the congratulations he received was a warm letter from Jefferson, whose enmity, along with Adams's, had long ago faded with the recession of political passions and the burnishing of their intertwining legacies as essential founding fathers. The third president wrote from Monticello: "It must excite ineffable feelings in the breast of a father to have lived to see a son to whose educ[atio]n and happiness has been so devotedly so eminently distinguished by the voice of his country."

One hundred and seventy-six years would pass before the son of another president would become the nation's commander in chief. George Herbert Walker Bush's feelings of pride in his son George Walker Bush's ascent to the presidency at the dawn of the twenty-first century undoubtedly stirred in his breast just as deeply as had Adams's for his own son. "It was all about family loyalty and pride for a father in his son," Bush reflected later. But like Adams, he may have felt in it some absolution. After a life devoted largely to public service, Bush had climbed the ranks of the Republican Party through elections and appointments, serving as vice president to Ronald Reagan for two terms before succeeding him and becoming the first incumbent vice president to be elected to the presidency since Martin Van Buren had done it in 1836, over a century and a half earlier.

Bush skillfully presided over a peaceful end to the Cold War, reunited Germany despite European resistance, and established an unprecedented international coalition around the liberation of Kuwait from Iraqi forces in the Gulf War. But his waterloo came on the domestic front when he broke a "no new taxes" campaign pledge to facilitate a 1990 budget deal while conservatives, always wary of Bush as a closet moderate, cried foul. He left the White House unseated by his Democratic challenger, Bill Clinton, after a single term. Though Bush handled defeat with graciousness that eluded the irascible Adams, he left Washington with an unfinished agenda and an ambiguous legacy, a disappointment Adams would not have found unfamiliar.

Just as the fathers stood on common ground, there were similarities in circumstance between their oldest sons. Like John Quincy Adams's, George W. Bush's election had come hard. After a mutable election night grounded in uncertainty, Bush's contest against Democratic challenger Al Gore came down to the state of Florida, where voting irregularities held up the final results. Gore had won the popular vote nationwide, but Florida would give either candidate the necessary electoral votes to swing the election overall. It fell to the U.S. Supreme Court to resolve the dispute. In a divisive judgment reminiscent of the vote rendered by the U.S. House of Representatives in 1825, the high court made its ruling in *Bush v. Gore* on December 12, 2000, a month and five days after the election: George W. Bush would be going to the White House.

So, in effect, would his father.

Few fathers have lived to see their sons become president. It had happened only twice in the twentieth century: Calvin Coolidge Sr., a modestly successful politician and businessman—and a notary public—swore his son into office at his home in Plymouth, Vermont, in 1923, after Vice President Calvin Coolidge Jr. assumed the presidency upon the death of Warren Harding. Thirty-seven years later, in 1960, Joseph Kennedy's wealth and power were instrumental in propelling his son Jack into the White House. After an earlier race for the U.S. Senate, the younger Kennedy joked that his father had sent a wire instructing him to "buy no more votes than necessary. I'll be damned if I'm going to pay for a landslide." When John F. Kennedy won the presidency "in a squeaker," he realized not only his own ambitions, but also those of the imperious Kennedy patriarch. Coolidge Sr. died in 1926, midway through his son's tenure as president, and the elder Kennedy suffered a massive stroke less than a year into his own son's presidency, rendering him partially paralyzed and unable to speak. Regardless, whatever paternal guidance they could offer their sons in their journeys at the nation's helm, neither had been there before them.

Bush *had* been there. And unlike John Adams, who died in Quincy (providentially on July 4, 1826), having seen his son only once for a few short days in Quincy during the course of the latter's presidency, Bush was readily accessible to his son in a communications age the Adamses could only have imagined. Twenty-four years had elapsed between the Adams presidencies, but only eight years separated the Bushes' terms in office—interrupted only by Clinton's administration—and the elder Bush, a nimble seventy-six upon his son's 2001 inauguration, could be there for *him*.

And he *was* there: not only in the White House, where he spent many nights during his son's tenure and where much of the staff had remained unchanged since his presidency, but also at Camp David, the presidential mountain retreat, on his son's Texas ranch outside Waco, and at the Bush family compound in Kennebunkport, Maine. When he wasn't there physically, he could be on the other end of the phone whenever the president called. He could provide not only succor but also the counsel of one who knew firsthand the burden his son carried. Father and son would come to refer to themselves numerically as "41" and "43," a nod to their places in the presidential continuum and to their mutual attainment.

Together they became known to insiders as 84, the sum of 41 and 43. But what transpired between them has mostly remained a mystery. Family, close friends, and staff respectfully kept their distance, giving them privacy and space. As 43 explained in a 2010 interview, "What people can't possibly imagine is what it's like to have two presidents who have a relationship as father and son—they envision us sitting around the table endlessly analyzing the different issues and strategies and tactics. It's much simpler than that, and in many ways more profound."

Though 43 wrote *41: A Portrait of My Father*, he admitted the work was a "love letter" devoid, understandably, of objectivity. Initially due to be released upon his father's death, 43 moved up the book's publication date so that his father could see the work. Though it reflects 43's deep love and respect for the family patriarch, it reveals nothing of the jumble of emotions and behavior inherent in any father-son relationship, and aside from cursory glimpses behind the curtain, sheds precious little light on how the relationship came to bear during 43's presidency.

Nearly any relationship between a father and son contains between the two parties not only love but also pride and shame, patience and frustration, expectation and resentment, obedience and defiance, and loyalty and betrayal. But the outsize tracks of an ambitious, successful father can be particularly daunting for his oldest son, especially if he's consciously—or even subconsciously—following his progenitor's path.

In this regard the Bush 41-43 relationship is particularly complex, compounded by their natural personality differences: 41, a blend of prim, patrician New England, which bred him, and rough-and-tumble Texas, where he came into his own in the oil business and in Republican politics. A comer since his earliest days, he was modest but ambitious, exuding quiet competence and judgment beyond his years. Forty-three, by contrast, is a cultural product of Midland, Texas, where he spent most of his formative years, exuding Lone Star swagger and unvarnished charm. Quick, decisive, cocksure—and inauspicious. While Bush the son arrived at the White House at age fifty-four, eleven years younger than Bush the father, who attained the office at sixty-five, 43's fate defied prediction. Though often overstated (most egregiously in Oliver Stone's bloated 2008 film, *W.*), 43's life from young adulthood through his thirties had been defined largely by mild rebelliousness; underachievement, at least by Bush standards; and a patina of booziness. No one knew more than he that "people have certain expectations from the son of a president—particularly the oldest one." He further allowed in a 1988 *Houston Chronicle* interview that if he were to follow his father in the family business of politics, he would have to "work hard at establishing my own identity."

And he did, famously putting down the bottle and finding God at the threshold of middle age, forging business success as co-managing general partner of the Texas Rangers, and going on to a string of spectacular and unexpected political triumphs. But as he emerged from his father's shadow, he must have wondered, *Could I have done it without Dad's name?* For that matter, would he have found the need to prove himself if not for the expectations—self-imposed or otherwise—that his father's name heaped upon him?

Real or imagined, those psychological undercurrents came into play during the course of 43's administration. Pundits, reporters, and armchair psychologists had a field day raising questions about the Shakespearean machinations and impulses between father and son, some based on love and loyalty, others on emulation and rivalry: Was 43's rise to the presidency a Prodigal Son proving himself to his father? Or was restoring the White House in the Bush name after his father's 1992 defeat the greatest tribute he could pay to the man he worshipped, and to whom he had tried to measure up throughout his life?

Much of the speculation revolved around 43's decision to wage war on Iraq, his most fateful as president. When U.S. troops marched into Baghdad taking out Saddam Hussein with the aim of building a democratic government, was 43 remedying what his father had failed to do after he drove Iraqi troops out of occupied Kuwait in 1991? Or was 43 avenging Hussein's attempt on the life of his father shortly after he left the presidency in 1993?

Abundant theories about the father and son were rooted in psychobabble, and the Bushes as a breed aren't ones to wallow on the couch. As far back as 1989, when an editor of *Texas Monthly* asked him about his father, George W. snapped, "Are you going to write that kind of article? One of those pseudo-psychological me-and-my-dad stories?" As Jeb Bush put it, self-absorbtion was anathema in the Bush-Walker household. "[I]t's part of our problem in this country," he said. "You should keep your head up, help others, and be well versed in the world around you." For what it was worth, 43 downplayed his father's influence. When asked by *Washington Post* reporter Bob Woodward in a 2004 interview if he sought his father's counsel, the born-again Bush replied that there was "a higher father I appeal to." After innocently stumbling during W.'s 2000 campaign by praising "this boy, this son of ours," 41 mostly relegated himself mutely to the sidelines. "Any nuance of difference" detected between him and 43, he recognized, "would be a big story and complicate his life."

Absent any clear read on their interplay, musings about them often became subtext during the course of 43's administration before crystallizing into conventional wisdom. *New York Times* columnist Maureen Dowd echoed mainstream sentiments when she wrote in her 2004 book *Bushworld: Enter at Your Own Risk*, "W. avenged his dad, replaced his dad, made his dad proud and rebelled against his dad, all in the same war."

Forty-three left the White House after two tempestuous terms in office—besting John Quincy Adams by earning reelection, which had eluded Adams just as it had the elder Adams and Bush—but with a pallid outgoing presidential approval rating of 22 percent, the lowest in modern presidential history. He and his father watched on January 20, 2009, along with a record inaugural crowd of 1.8 million, as Barack Obama became the forty-fourth president, and the Bush presidential dynasty, ostensibly, receded into history.

Then came a development unique to the American canon. As attention turned inevitably during the latter part of Obama's tenure on who would succeed him, Jeb Bush, George H. W. Bush's second son, emerged as the leading Republican candidate. The Bush family—emblems of decency and public service but devoid of the radiant passion and ebullience of the Roosevelts and the megawatt charisma, glamor, and splayed ambition of the Kennedys—was once again poised for another turn at the White House. As the Bush machine revved up for another presidential bid with big money pouring into the campaign, Jeb, possessed of an impressive conservative record and policy profile, and the proven Bush name, quickly emerged as the presumptive nominee in a field of also-rans. The second son of the forty-first president, and the brother of the forty-third, had a good shot at being the forty-fifth . . .

Until the Republican Party of the Bushes was hijacked by insurgents with little interest in politics as usual. Jeb Bush, whose family had had a knack for adapting successfully to their political surroundings and finding a way to win without yielding their family ethos, was of no material interest to a party that now placed ideological purity over pragmatism and compromise in governance. Prudence, a favorite word of George H. W. Bush's during his administration to the point of *Saturday Night Live* parody, had given way to populist rage. Civility, a Bush family hallmark, became a quaint notion of the past. George H. W. Bush had campaigned for the presidency in 1988 with the hope and promise of leading a "kinder, gentler nation," while George W. Bush had championed "compassionate conservatism" in his own presidential bid in 2000.

But the prevailing GOP campaign rhetoric in 2015 and the election season of 2016, echoed in tweets, was sharp and cutting, devoid of kindness, compassion, and optimism. Jeb Bush, sanguine, knowledgeable on the issues, and ready for battle on the merits, was blindsided and dazed by an onslaught of damn-the-consequences barbs dished out by candidates callow and crude. Donald Trump, boastful, provocative, and openly narcissistic, dominated the public consciousness, the news cycle, and the polls while effectively mocking and marginalizing Jeb as "low energy." The party that the Bushes had helped to define as dusk set on the twentieth century and at the dawn of the twenty-first was all but unrecognizable to them. And the predominant common denominator among its members was not belief in a binding party platform but searing anger toward the establishment—not just Barack Obama and the Washington status quo but the *Republican* establishment epitomized by the name Bush.

As Jeb was relegated to the edge of the national political stage and GOP debate platforms, the forty-first and forty-third presidents, bound by blood and party tradition, became symbols of a bygone era. And yet Americans—Republicans and Democrats alike—*revered* them. Forty-one had lived long enough to see a new appreciation of his one-term presidency. *Newsweek*, which had published a 1987 cover story on Bush blaring the infamous headline "Fighting the 'Wimp Factor,'" reflected a retrospective change of heart many felt was due with a 2011 story titled "A Wimp He Wasn't." Forty-three saw resurgence, too, with polls showing his approval rating at over 50 percent in 2015—nine points higher than that of Barack Obama— and over 75 percent among Republicans, prompting Jeb to boast of his brother, "He's probably the most popular president among Republicans in this country."

But despite the anticlimax of Jeb's failed campaign and a party that has strayed decidedly and discernibly from the Bushes' principles, including their antipathy toward isolationism, protectionism, and nativism, the singularity of the Bushes puts them in the pantheon of American political families along with the Adamses, Tafts, Roosevelts, and Kennedys. The narrative of their two standard-bearers warrants reexamination beyond the mist of supposition and begs for clarity that can come only from the principals themselves.

This, based on myriad exclusive interviews over the years with the forty-first and forty-third presidents, is the story of the relationship between George Herbert Walker Bush and George Walker Bush, the last Republicans. Theirs is the most consequential father-son relationship in America's history—two men, in many ways as different as their generations, who stood astride the world as commanders in chief in times that mattered: one at the close of the Cold War as the Berlin Wall fell into rubble out of which a new world order emerged; the other after Manhattan's World Trade Center towers tumbled into heaps of ruin that marked a war on terror against a diffuse, insidious enemy. It is not the story that has been told; until now, there are depths that have not been plumbed.

At heart, theirs is a love story. But it is far more.

PART I

"I Never Looked Back"

1

"A FLASH OF LIGHT"

IN THE FALL OF 2012, nearly four years after leaving the White House, six months before the opening of his presidential library, George W. Bush sat in his sleek high-rise Dallas office devoid of presidential trappings—no presidential seals or flags, no oil paintings, no Frederic Remington bronzes—and considered his father, the man he had revered since his earliest memories. His sneakered feet and khakied legs propped up on a generic executive desk, he leaned back in his black leather chair in a gray golf shirt as he fingered an unlit cigar, looking uncannily like his father at times, particularly when his face turned upright and to the side. The resemblance between the two had deepened through the years, as the younger Bush aged. One could see the old man clearly in his bushy eyebrows, arched and slightly askew, the eyes narrow and squinting at times, and the amorphous opened mouth, though he lacked the patrician refinement in his features; his face was craggier, his hair wirier, and his forehead less expansive.

Whatever pressures he had toward his father in his life—to prove himself to him, or surpass him, or rebel against him—had long ago fallen gently away. The stuff about him being the Prodigal Son was "bullshit," he said in his curt, matter-of-fact Texas twang. Sure, in his late teens and early twenties he had "chased a lot of pussy and drank a lot of whiskey," but he added, "I was never the Prodigal Son because I never left my family." He and his dad had gone through some rocky patches—what father-and-son relationship hasn't?—but his heart was with his father at every step.

Also in the "bullshit" category was media speculation that he and his dad were locked in a competition that intensified during Bush 43's White House years. George W. never trusted the media, an instinct going back to his days serving his father during the latter's bid for the presidency in 1988, when he would aggressively demand of reporters seeking interviews, "Why should I talk to you? How do I know you're not out to get my dad?" The *New York Times* and the *Washington Post*! They never let up—and they never got it right. He and his father had always been close, he insisted. Always.

Yet after sixty-six years, much of his father remained a mystery to him. George Herbert Walker Bush wasn't one to go on about himself, and despite all they had shared, some things—many things—had gone unsaid, adding to his mystique. "All these questions people assumed my family knew about—my parents had been brought up to be completely reserved and not to talk about those things," Jeb Bush had observed. And with his life's shadow growing longer at eighty-eight, the elder Bush had gotten quieter, becoming, in the words of Jean Becker, his longtime chief of staff, "a man of few words." As George himself put it, "I've run out of things to say."

George's early adult years in particular were a source of wonder for his oldest son. The latter chapters of his life George W. could better understand because he had gone down similar roads himself: Both found success in business, embarked on ascendant political careers after early losses, and ultimately achieved the pinnacle of power in the White House. But the crowded, storied years of his father's late teens and twenties were as awesome to him as they were elusive. He marveled at what it was for his father to enlist in the navy on his eighteenth birthday, go to war at nineteen, and become a husband at twenty-one and a father at twenty-two. He could only imagine his father's excruciating pain after losing his three-year-old daughter, Robin, to leukemia before he was thirty. And he wondered what it was that prompted him to forgo a family pass to the riches of Wall Street after graduating from Yale in 1948 and break away toward adventure and uncertainty in the desolate oil fields of West Texas. Those years were as formative for his father as they were propitious. They shaped him and ushered him early into manhood.

For George W., finding himself and establishing himself in the world would come much later in his own life. It was one of the differences between him and his dad. Indeed, lateness was a hallmark in George W.'s early years, just as it had been when he entered the world on a hot Saturday morning in July, when his father first cast eyes on him.

Barbara Bush, standing five feet eight, had put on sixty pounds awaiting the birth of her first child, who was three weeks past due, as her husband of eighteen months, Yale University freshman George Herbert Walker Bush, pursued his studies. By her own telling, she weighed "more than a Yale linebacker." During a short summer break, the couple drove from New Haven fifty miles southwest to the Bush family home on Grove Lane in Greenwich, Connecticut, to visit his parents, Prescott and Dorothy—Pres and Dottie as they were known. Sensing Barbara's great discomfort, Dottie took matters into her own hands. "She put me in the back seat of a car and gave me a little bottle of castor oil to drink," Barbara recalled. "Then we drove up to New Haven with me moaning and groaning in the back seat as George and his mother said, 'Don't worry, everything's perfectly normal.'"

The party arrived at Grace-New Haven Hospital back in New Haven at half-past midnight, where Barbara gave birth in "a flash of light" just under seven hours later, at 7:26 a.m., on Saturday, July 6. The first and most fruitful year of the eighteen-year baby boom, 1946 saw the registered births of 3,288,672 American children. Among them was George Walker Bush.

Barbara and her newborn remained at the hospital for what felt like an interminable eight days after she was told by her doctors that she couldn't walk up and down the stairs of their one-bedroom apartment on the edge of the Yale campus. "Of course, George couldn't carry me up the stairs, I outweighed him," she recalled. "I was thinking I'd never get out again." Once again, Dottie Bush came to the rescue, inviting her daughter-in-law to join the family at their summer home in Kennebunkport, Maine, where she had hired a nurse to tend to Barbara and her newborn.

George Herbert Walker Bush gave his son three-quarters of his name—minus the Herbert—making it dissimilar enough that he and Barbara would quickly point out that their son was neither a "Junior" nor a George Bush II. It wouldn't much matter then, nor would it matter later; the *George* and the *Bush* were enough. The name came with a paternal yardstick by which he would forever be measured. In his youth, he became known as Georgie or Little George. Translation: Junior.

Wellborn and well-bred in upper-crust New England, the second of Pres and Dottie's five children—four sons and a daughter—George Herbert Walker Bush had his own namesake: his maternal grandfather, George Herbert Walker, from whom he also derived the nickname by which he was known from childhood to early adulthood. George Herbert Walker was lovingly called "Pop" by his four sons—his grandson and namesake became known as "Little Pop" or, more often, "Poppy."

The four years and one month that had passed between his high school graduation and the birth of George W. had been a virtual lifetime for Poppy Bush. The president of his class, he graduated from Phillips Academy in Andover, Massachusetts, on June 12, 1942, his eighteenth birthday, and began forging his own path the same day. On hand to deliver the school's convocation on graduation day was Franklin Roosevelt's secretary of war, Henry Stimson, a former Wall Street lawyer and aging emblem of public service who had taken up posts in the cabinets of William Howard Taft, Herbert Hoover, and

FDR and would go on to serve Harry Truman. As World War II played out in the European and Pacific theaters six months after the bombing of Pearl Harbor, the seventy-four-year-old Stimson, a graduate of Andover himself some fifty-eight years earlier, spoke of his hope that the outgoing class would temporarily forgo the war and instead go on to college. The war would be a long one, he maintained, and while America needed men on the front lines, there would be plenty of time for them to serve. What the country needed was leaders, and what they required was the knowledge that a university education would bring.

Pres Bush, on hand with his family for his son's graduation, listened to Stimson's sage words hoping that they would strike a chord with Poppy, who had earlier told him of his intention to enlist upon graduation. Soon after the bombing of Pearl Harbor, Poppy was one of many who thought, *We had better do something about this*, and decided that he would do his part at the earliest opportunity as a navy aviator. Pres had other ideas for his second son. His oldest son, Prescott Jr., had graduated Andover a year earlier and went on to Yale, Pres's alma mater, where he hoped Poppy, who had already been accepted, would soon follow.

The managing partner of the New York–based Brown Brothers Harriman, the nation's largest private bank, Prescott Bush was not a man to whom people said no. Poppy would remember him later as "an imposing presence, six feet four, with deep-set gray blue eyes and a resonant voice." Still, when Pres asked him in a crowded hallway after the ceremony if he had a change of heart upon hearing Stimson's counsel, Poppy replied, "No, sir, I'm going in." His father nodded his assent and shook his hand. Afterward, Poppy went to Boston where he enlisted in the U.S. Navy. Later the same day, as Pres drove with the rest of the family from Andover back to their home in Greenwich, he wept. It was the first time his daughter Nancy, two years Poppy's junior, had seen him cry.

Two months later, on August 6, Pres accompanied Poppy to New

York's Penn Station, where he put him aboard a train that would take him to basic training as a seaman second class in Chapel Hill, North Carolina. There was no certainty as to what lay ahead for Poppy as he left his father's side, "the youngest guy on the train," bound, ultimately, for war half a world away. Prescott Bush cried again that day.

The following June, three days short of his nineteenth birthday, Poppy became the youngest pilot in the U.S. Navy when he earned his wings in Corpus Christi, Texas. His father and mother journeyed to Texas for the ceremony, where Pres presented his son with a pair of gold cufflinks, which would become his most treasured possession, heirlooms he would give to his own son, forty-seven-year-old George W. Bush, over fifty years later before the latter's inauguration as governor of Texas. From there, Poppy went on to ten months of training to fly torpedo bombers at bases throughout the East Coast, then, in late March 1944, to the South Pacific aboard the USS *San Jacinto*, where he would see action against the Japanese in his first mission on May 23.

By the summer of the same year, now a lieutenant junior grade, Poppy had seen his share of war. On June 26, after flying his Grumman Avenger on numerous combat missions, he penned a letter to his parents declaring that the "glory of being a carrier pilot" had "worn off." Referencing his two younger brothers back home, he wrote, "I hope John and Buck and my own children never have to fight a war. Friends disappearing, lives being extinguished. It's just not right." But on the morning of September 2, as Poppy boarded his torpedo bomber with his crewmates radioman John "Del" Delaney and gunner Ted White, war's iniquities would become far clearer in his mind. And so would its fate.

The three men were charged with a mission to hit a Japanese radio tower on Chichi Jima, a barrier island off Japan's coast that held enormous strategic significance to the enemy. As their plane approached the tower, it was struck by a hail of antiaircraft fire sending it reeling as Poppy continued the plane's dive, dropping its payload on the target before retreating impotently back out over the Pacific. Shortly afterward, he shouted orders to Delaney and White to bail out before escaping himself. Minutes later, Poppy found himself alone in the water in a yellow one-man life raft, four miles northeast of Chichi Jima, bleeding, vomiting, stung by a Portuguese man-of-war, and searching in vain for Delaney and White. Using his hands, he feverishly paddled away from Chichi Jima, where the prevailing winds were blowing his raft, to evade certain enemy capture or death. As he did, he thought of "family and survival."

Throughout his life, luck had a way of finding George Bush. So it was in his most desperate hour. After three hours and thirteen minutes in the water, the USS *Finback*, a *Gato*-class U.S. submarine, miraculously peeked up from the water, as five crewmen scrambled to scoop Lieutenant Bush to safety on its deck before the craft dived back under the ocean. Luck was not with Delaney and White, who never rose from the depths of the Pacific.

Bush spent the next four weeks aboard the *Finback*, and they became the most reflective of his young life. Often, as he stood watch on deck from midnight to 4:00 a.m., when the submarine surfaced out on the Pacific, he felt a calm descend on him as he stared into the distance from the conning tower under a profusion of stars. Those moments of solitude left a deep impression on him that he would recall often in his later years. "As you get older, and try to retrace the steps that made you the person who you are, the signposts to look for are those special times of insight," he said nearly half a century later. "I remember my days and nights aboard the Finback—maybe the most important of them all. In my view, there's got to be some kind of destiny and I was being spared for something on earth."

His destiny would become clearer in time. But one of the things that would shape it came more consciously as Poppy contemplated his future; in the face of death he came to appreciate more fully his family and the values handed to him by his parents. And he realized how much he loved the girl whose name graced the side of his downed airplane, the girl—the woman—with whom he would make his future.

2

"BAR"

POPPY BUSH HAD MET BARBARA Pierce just before Christmas of
1941 at a dance in her hometown of Rye, New York, an affluent,
leafy New York City suburb in Westchester County on the Long
Island Sound just south of Greenwich. Dressed in white tie and tails,
Poppy knew he was no Fred Astaire but screwed up the courage to
ask the sixteen-year-old high school junior to dance after being introduced to her by Jack Wozencraft, a mutual friend. Barbara wore
a new brightly colored red-and-green dress in keeping with the season. Her suitor later recalled her as "the most beautiful girl you've
ever seen." As the band played strains of Glenn Miller, they danced,
talking idly before Bush suggested they sit out the next dance, a
waltz he didn't know. They talked further, fifteen minutes or so,
enough to lead to a meeting at a dance in Rye the following night
and on to courtship and love.

The two were a good social match, comparable shades of blue
blood. Barbara's mother, Pauline, quickly confirmed as much with
several phone calls after first hearing about the young George
Bush, much to her daughter's mortification. Social standing mattered
to Pauline Pierce, a woman not given to maternal warmth and to
whom Barbara was not particularly close. Her father, Marvin Pierce,
whom Barbara adored, was the president of McCall Corporation,
a magazine publisher of the popular women's titles *McCall's* and
Redbook, and could trace his family roots back to the fourteenth
president, Franklin Pierce. (The latter wasn't a point of pride for

Barbara, who was "humiliated" upon learning in elementary school that Pierce had achieved no great distinction in the White House.)

By the summer of 1943, while Bush was on a seventeen-day leave, the pair became "secretly engaged" during a holiday at his family's summer retreat on Walker's Point in Kennebunkport, Maine. The place had a special significance to Bush, holding as many family memories for him as there were jagged rocks on its craggy coastline. His maternal grandfather, George Herbert Walker, bought the plot of land in 1902 and later built a compound of homes including a bungalow that he gave Bush's mother as a wedding present. Later that summer, Barbara—now teasingly nicknamed "Bar" after Barsil, a family horse, by Bush's older brother, Pressy—accompanied her fiancé to Philadelphia where he slipped her an engagement ring before shipping off to the Pacific on board the USS *San Jacinto*, as she went on to Smith College in Northampton, Massachusetts. The young couple married at the First Presbyterian Church in Rye on January 6, 1945, shortly after Bush returned stateside, where he would spend the remainder of the war. The bride was twenty; the bridegroom, twenty-one.

Like thousands of American soldiers returning home after the war, Poppy came back older than his years. He was a man now, six feet two, fit and trim at 160 pounds, and no longer the judgmental, protected child of privilege who boarded the train for basic training three years earlier. "Although my childhood was very happy, my upbringing was also strict, indeed, puritanical," he wrote later. "As a result my world was very narrow. Like most young people, my horizon needed expanding." He had flown fifty-eight combat missions and barely eluded death. But as with the Distinguished Flying Cross, Air Medal, and two Gold Stars he earned and put away upon becoming a civilian, he quickly put the formative experience of war behind him and moved on—except for the memory of John Delaney and Ted White. "I think about those guys all the time," the former president said more than five decades after seeing them for the last time.

After being discharged, he and Barbara, who left Smith College to tend to her new husband, went to New Haven, where he enrolled at Yale in the fall of 1945 as part of the GI Bill. Bush had thought about skipping college altogether, much to his father's distress. It may just have been an ambitious young man's fancy to get out in the world as quickly as possible to make his mark. Eventually Bush came to his senses but fast-tracked college, graduating Phi Beta Kappa in two and a half years with an economics degree while, despite a heavy course load, actively pursuing a spate of extracurricular pursuits. Among other things, he played first base on Yale's varsity baseball team, becoming the team's captain in his senior year, served as chapter president of the Delta Kappa Epsilon fraternity, and raised funds for the United Negro College Fund. He was also tapped into the exclusive Skull and Bones secret society, becoming one of only fifteen members enrolled at the time, just as his father had nearly three decades earlier.

Part of moving on with life meant raising a family, which the newlyweds were eager to do. Just after their Little George's birth in the summer of 1946, they moved into a one-family dwelling that had been converted into apartments for veterans in answer to the postwar housing shortage. Thirty-seven Hillhouse Avenue became George W. Bush's first real home, and he had plenty of company. A dozen families crammed into the house, each with one child except for the one with twins. George and Barbara lived happily though sparingly among the throng on the nest egg they had squirreled away from Bush's navy pay. Though the home stood in a tony section of town—Yale president Charles Seymour lived next door—the modest accommodations belied the means for a bigger space Poppy could have secured through his father's bank account. But making his own way meant financial independence.

Like most things, fatherhood seemed to come easily to Poppy. "George loved that baby, he was a cute little fatso," recalled Bar. A pattern emerged then that would hold throughout George W.'s years under his parents' roof: As Bar played the role of hands-on mother, Poppy went freely about the business of building his future, unencumbered by the day-to-day, often mundane, routines of parenting that fell to his wife. Though their roles were largely traditional for their generation, Poppy's frenetic pace and far-reaching ambitions would often mean long days and absences from his family as Barbara zealously minded the Bush home. Her children and grandchildren would come to refer to her as "the enforcer."

Family and friends descended on New Haven to meet Little George, attending his christening and a lawn party in his honor. But Barbara's mother, Pauline, expressed a nagging reservation: "My mother hated to be in the same room with the baby," Bar said, "for if she took her eyes off him, [he] looked hurt."

The baby needn't have fussed. As the first son of the next generation of the Bush family, eyes would be on him for the foreseeable future.

3
GO WEST

AMONG THE SEVEN FRAMED PHOTOGRAPHS of family that stood on a table behind George W. Bush's Oval Office desk—and later behind the desk in his post-presidential office—was a well-known color shot of him with his parents and paternal grandparents, Pres and Dottie, smiling in the brilliant Houston sun in front of a corporate prop plane in 1949. If George W. wondered why his father left East Coast privilege to venture out on his own in Texas, the answer may have been right there. Pres looks the very picture of the starched, patrician East Coast investment banker and future U.S. senator that he was, dressed in a tan suit and solid blue tie, a handkerchief sprouting jauntily from his breast pocket and a fedora atop his head. His son, the young oilman, stands informally at the center of the photo in a checkered open-collared shirt and an unzipped tan jacket holding his toddler son, clad in shorts and cowboy boots. George H. W. Bush wanted something different, an alternative to the traditional and safe path from the Ivy Leagues to moneyed Manhattan. He found it in a state aggressive in its independence—its Capitol dome in Austin, at 302.6 feet, pointedly stands twenty feet taller than the nation's Capitol in Washington—a place where self-discovery and reinvention is as commonplace as dust storms on the West Texas plains, and where opportunity is as boundless as its sky.

Moreover, his success would be *his* success, far beyond the shadow cast by his father half a continent away. He went for one reason, as he said sixty-five years later: "To make it on my own—and I never looked back." Regardless of the experience of war, which had tested and expanded him, he believed later that he still would have found his way to Texas; he simply "didn't want to go to Wall Street" like his father and members of his family. Here, for Bush, there was filial symmetry between him and his immediate forebears. In finding his own way he was beginning to follow a pattern of tacit expectations in the Bush family: Chart your own course; establish yourself; make some money; take care of your family. Then do some good—serve others, serve your country. It's what Pres had done, and it's what Pres's father, Samuel P. Bush had done before him.

Samuel P. Bush, "S.P." as he was known, had migrated from the New York City area to Columbus, Ohio, where he rose to become the president of the Buckeye Steel Castings Company, director of the Federal Reserve Bank of Cleveland, and an adviser to Republican president Herbert Hoover as an appointee on his President's Organization on Unemployment Relief after the Great Depression struck. The latter association was apt; the spirit of Hoover's rugged individualism was in keeping with the tacit up-from-the-bootstraps tenet that marked the Bush family and its conservative ideological underpinnings.

Pres had found his own way, too. Despite growing up with every advantage, much in the manner of his own children, he refused to ride his father's coattails into the corporate world. After graduating from Yale, he broke free from his father financially, establishing himself as a hardware salesman, selling Keen Kutter tools, in Saint Louis, where he met and married Dorothy "Dottie" Walker, the daughter of George Herbert Walker, who had made his own fortune as a banker. Despite a distant and sometimes glacial relationship with S.P., Pres would work briefly with his father at his behest, helping in vain to turn around a Buckeye Steel subsidiary. Later he would go on to work with his father-in-law in establishing the prosperous Brown Brothers Harriman. But he never took a penny from either his father or father-in-law. Even after S.P. died, Pres made sure that his father's substantial estate was divided between his two sisters.

While Poppy may have wanted to prove himself to his father in making his own way, he also may have set his own course to honor him. Pres Bush was hardly a cuddly family man. His ramrod straight posture, which at six feet four made him all the more imposing, was matched by a rigid formality. Pres asked his grandchildren to call him "Senator." George W., his first grandson, recalled him as "a very dignified person. When we went to dinner at his house, we wore a tie. I never wore a tie, only to church, barely." But Poppy looked to his father's example even if he never felt he quite measured up; as Pres Bush's biographer, Mickey Herskowitz, wrote, "He'll say, 'My father was a great man.' Even after he became president, he never felt he topped what his father had."

As the second of his middle names suggested, George Herbert Walker Bush was as much a Walker as a Bush. If the Bush family was characterized by a WASP Midwestern modesty, frugality, and egalitarianism, the Walkers were marked by bravado, extravagance, and clannishness. The Walkers convened each summer at George Herbert Walker's compound in Kennebunkport, vying against each other in an endless series of competitions—tennis, golf, boating, fishing—while establishing familial traditions and rites of passage, like skinny-dipping in the icy Atlantic, that bound them to each other.

Warm and compassionate, Dottie Walker grew up among four brothers and a sister in a household where competition played out over everything big and small. George later called his mother "a real fighter and a great competitor," while Bar did him one better, labeling her "the most competitive woman I've ever known." Dottie encouraged the same spirit in the sprawling nine-bedroom Bush home—a Ping-Pong table stood in the front hall awaiting family matches—and urged her children to always play to win.

Winning was looked upon as a quiet virtue, but it came with implicit directives: In victory, be neither boastful nor aggrandizing; let humility be the hallmark of your triumph. Give credit to others. Extend your hand to the vanquished—make a friend, if you can. "Don't be a braggadocio," Poppy heard his mother say over and over, and it lodged in his psyche. Losing was to be handled with grace, and if not shameful, was looked upon as something to overcome. Poppy would personify his mother's spirit in youth and beyond, and along with her strong sense of family togetherness, would later instill it in his children. "He talked all the time about his mother, rarely about his father, and the principles he learned at her knee, and that he would cite all the time," recalled Brent Scowcroft, Bush's longtime friend and national security adviser during his presidency.

For Poppy Bush, finding his own success on his own terms was a by-product of the competitiveness that permeated the Walker-Bush household. Though he pointedly forged a career path outside of his father's, he had plenty of connections to take on the role of sponsor. Among them was Neil Mallon, a Yale classmate of Pres's and a fellow Skull and Bones man, whom Pres held in the highest regard. Mallon had been tapped by Pres to head up the oil conglomerate Dresser Industries, an interest in which Pres and his investors had taken a lead position. It was Mallon who counseled Bush on his career on a plane ride to Columbus, Ohio, for the funeral of S. P. Bush in 1948. Mallon, who according to Bush's brother Bucky, could "charm the fangs off a cobra," pointed Bush toward the prospect of riches in the oil patches of Texas, while holding out the promise that if he came to work at International Derrick and Equipment Company (IDECO), a Dresser subsidiary, and applied himself, he could someday run the company. (George and Bar's affection for Mallon became evident when they named their third son Neil Mallon.)

The decision hadn't come easy. "Right now I am bewildered to say the least," Poppy wrote in 1948, to Gerry Bemiss, a friend in

Richmond, Virginia, whose family also summered in Kennebunkport. "I want to do something of value and yet I have to and want to make money—after Georgie goes through three squares a day, one's wallet becomes thin and worn." He considered other professions, rejecting going to work for his mother's oldest brother, his uncle Herbie. Poppy didn't want to "capitalize completely on the benefits I received at birth—that is on the benefits of my social position." He also entertained becoming a teacher, but believed it would be "too confining and not challenging enough." Likewise, he thought about farming, even, along with Barbara, reading Louis Bromfield's *The Farm*, but abandoned the idea when he realized it wouldn't lead to riches. He told Bar the oil business was something he could "touch and feel," and that offered him the chance to prove himself.

With that prospect brimming in his mind in June 1948, he hopped in his brand-new red 1947 Studebaker, a graduation gift from his parents, and drove from Kennebunkport 2,200 miles— west on Route 80, then south—to West Texas, past Pearl Beer signs that glowed in the desert, where he would begin the next chapter of his life in Odessa. Bar and Little George followed him there a couple of weeks later by plane, enduring a twelve-hour ride from Maine. Bar "would have followed George Bush anywhere," even to hot, flat, hardscrabble West Texas. "I've always wanted to live in Odessa," she replied dryly after he first presented the notion. She later said of the move, "It was the first time in our lives that we lived in a place where nobody said, 'You're Marvin Pierce's daughter or Pres Bush's son.' It's pretty nice to be judged on your own."

Still, from the standpoint of two Brahmins in their early twenties, Odessa might as well have been Neptune. In answer to the housing shortage that came with the town's booming oil business—62,249,000 barrels of oil were extracted from county lands in 1948—they moved into a two-bedroom apartment on East Seventh Street, half of a duplex, the other half of which they would eventually occupy with a mother-daughter team of prostitutes who had gentleman callers coming and going at all hours. Occasionally, the Bushes would find themselves locked out of the Jack-and-Jill bathroom they shared with the working pair when one of their clients used the facilities and neglected to unlock the Bush's access door. George W. said later that his family's 1948 living arrangement should have dispelled the "myth" that his father was a spoiled rich kid, saying flatly, "Trust fund babies don't share duplexes with prostitutes." Bush's Ivy League pedigree only went so far with the Odessa natives. When asked by a coworker at IDECO if he had gone to college, he replied that he had graduated from Yale. "Never heard of it," the man replied. The town was "as different from Rye, New York, as any place imaginable," wrote Barbara later. "Nothing comes easy in West Texas."

And yet, despite the differences—maybe because of them—fitting in did come easy for George and Barbara. "It was the people," Bush said later. "They were accepting, friendly; didn't care about your background or your heritage." Fittingly perhaps, Bush became known in Texas by his given name, George, shedding the nickname Poppy by which he had been known to that point in his life and starting fresh.

Adapting and thriving in a new place were characteristic of the Bush family. Bush's grandfather had done it, pulling up stakes in the New York City area and finding success in Ohio, and his father had left the Midwest for the green pastures of the East Coast, settling prosperously in Greenwich. Texas proved to be a fit for George and Barbara, a good place to build their lives and George's career. A good place to raise a family, too. In a letter to Gerry Bemiss in Kennebunkport a few weeks after settling in Odessa, Bush wrote of his two-year-old boy, "You should see Georgie now, nothing like bragging about one's own kid. He really is cute, I feel. Whenever I come home he greets me and talks a blue streak, sentences disjointed of course but enthusiasm and spirit boundless. He is a real blond and pot-bellied. He tries to say everything and the results are often hilarious. How he would love to be there in K'port. The great thing is that he seems to be very happy wherever he is."

Texas would smile on George and Bar. But the move proved to be particularly fortuitous for Little George, who was New Haven born but West Texas bred and branded. Texas would shape him in a way that the East Coast couldn't, defining him in a way that would distinguish him from his namesake.

4

"THE SKY'S THE LIMIT"

GEORGE W. BUSH'S FIRST MEMORIES were borne not of Odessa but of Midland, twenty-three miles northwest of Odessa, where the Bushes settled in 1950. It was Midland where he spent his formative years, from age four to thirteen, and where he would later return to follow his father into the oil business. It was where he met and married Laura Welch, a Midland native, in 1977, and where their twin daughters Barbara and Jenna were raised in their early years. It was where he announced his bid for the presidency in 1999, and it was his first stop upon leaving the White House a decade later. "You're a product of who raised you and where you were raised," he said. "And I view Midland as where I was raised." The key to understanding George W. Bush is to appreciate the influences of George H. W. and Barbara Bush, and Midland, Texas. And Midland is also a key to understanding differences between the Bush father and son.

For George H. W. Bush, Midland was the last stop of an intensive, peripatetic training program Neil Mallon had set up to help his young protégé get a feel for the oil business. It had started "at the very bottom of the corporate ladder" in Odessa, where Bush did everything from sweeping floors and working in the factory to selling oil field supplies. Eight months later he was dispatched for a year to Southern California, where the family moved five times as George worked as an assemblyman then peddled drill bits for IDECO subsidiaries. Then it was on to Midland to work as a city salesman.

The Bushes returned to West Texas as a family of four. Pauline Robinson—"Robin"—was born in Compton, California, in December 1949, named after Barbara's mother, Pauline, who had died suddenly in a car crash in Rye two months earlier at age fifty-three. Though the Bushes had adapted to roughneck, blue-collar Odessa, Midland proved a better cultural fit for the family. Named for the speck on the map between El Paso and Fort Worth, the town was home to 21,713 souls in 1950, a population that had swelled five times in size since the 1920s and would grow three times more in the decade ahead as oil and gas production boomed in West Texas's Permian Basin, a 125-square-mile area of sediment and salt topped by rock and rife with opportunity. Odessa, it was said, was where you raised hell; Midland was where you raised a family. Both father and son Bushes would raise families in Midland, though the son would raise a little hell there, too.

Country-fried, God-fearing, and dry in climate and drink, the town was a magnet in the 1950s for Ivy Leaguers and aristocratic outliers from the Northeast who, like George, looked to the Permian Basin as a frontier to add to or seek their own fortunes. Scions of the Getty, Mellon, and Rockefeller clans had all carpetbagged Midland, which though still boasting a blacksmith shop on its Main Street catering to local ranches, featured other street names that transcended provinciality—Harvard Avenue, Princeton Avenue, Dartmouth Drive—as well as boasting clubs for Ivy League alumni. But Midland society, like its landscape, was flat. "I didn't feel any sense of being different," George W. said. "Midland is pretty egalitarian. My folks, like anybody else, were trying to get along in a postwar environment, in a society in which there were no elites."

The greatest gifts his father gave him, he said, were "a big name, no money, and a West Texas background." The Bush brand name would come into play often in George W.'s life as he built his career, but it meant little in Midland when George W. was in his formative years. And while he had no worries about where his next meal would come from or whether college would be paid for, money was never front and center in the Bush family home. When asked as he was beginning his political career how he differed from his father, he often remarked, only half-jokingly, "my dad went to Greenwich Country Day School in Connecticut and I went to San Jacinto Junior High in Midland." Never mind that he would go there for a single year before being enrolled in a private school in Houston, then following in his father's footsteps to Andover and Yale and going on to Harvard Business School, it was his years in Midland that explained him.

During those later academic experiences in the Northeast, he would come to know the children of privilege whose identities were tied up in their family's wealth and social positions, which engendered a sense of entitlement. His disdain for them—or anyone who was putting on airs, for that matter—came from the West Texas lens through which he saw the world. Among them he would often wield his West Texas persona defiantly. When he was four, his father wrote of him, "Georgie has grown to be a near-man, talks dirty once in a while and occasionally swears. He lives in cowboy clothes." As he got older, not much changed. At Harvard Business School in the mid-1970s, he would strut around Cambridge in his National Guard flight jacket, beat-up Levi's, and cowboy boots, a wad of chewing tobacco stuffed in his cheek that he would spit into the paper cup he clutched in his hand. Amid the ivy-strewn environs, his incongruous demeanor asserted, "I'm not one of you. I'm a Texan."

George W. was taken with Midland's openness. One sees it immediately in the landscape, which is dominated by sky, unlike the northeastern topography, where titanic trees spring heavenward, reigning in streets and veiling homes, and constraining one's perspective. "In Midland, the sky sits overhead like a flawless dome, bowing up from the earth at the edge of each horizon," wrote Laura Bush, the daughter of a Midland home builder, who grew up Laura Welch in numerous houses constructed by her father, one as few as ten blocks away from her future husband, whom she wouldn't come to know until 1977. When George W. was coming up in the oil business in the mid-1970s, the city's slogan was "The Sky's the Limit." It spoke not only to the entrepreneurial spirit, he believed, but also to the mind-set of its inhabitants. "There are no trees," he said of the town that bore him. "You can see a long way. It makes you accept people for who they are."

While his father had grown up with three maids and a chauffeur who stood ready to tend to the family's needs, George W. grew up comfortably though decidedly middle class. His friends, many of whom would remain in his life well beyond his Midland years, were the sons of plumbers, house painters, and accountants who lived no differently than he did; they rode bikes, played in Little League, and attended Cub Scout meetings, Friday night high school football games, and Sunday school classes. It was a halcyon small-town boyhood for George W., one that he later believed embodied the American Dream, though one distinctly homogenous. More southern than western in matters of race, Midland was segregated, and George W. grew up not knowing any of the few African Americans who lived within its city limits.

The Bush family first settled in a housing development on East Maple Avenue called "Easter Egg Row," where the small uniform homes stood together like pieces on a Monopoly board, distinguished only by their pastel colors, theirs a light blue, before moving to a slightly larger home on West Ohio Avenue that the Bushes purchased in 1951 for $9,000. Even with the housing upgraded, the three-bedroom, one-bathroom ranch-style home—1,547 square feet—contained no shower or air-conditioning but would house as many as five as the family burgeoned; John Ellis—called "Jeb," around his initials J-E-B—was born in 1953, followed by Neil Mallon in 1955, the same year the family graduated to a ranch-style home twice as big on Sentinel, which would become the first home of Marvin Pierce, the last of the Bush brothers, born in 1956.

The "gift" of a West Texas background was largely given to George W. alone. None of his siblings lived in West Texas for more than their first six years—the family moved to the Houston suburbs in 1959, where George W. lived for just a year before being sent off to Phillips Academy Andover. Midland wasn't at their core as it was for their oldest brother; it was for Jeb what New Haven was for George W.—the place where he was born and not much more. But one could hear Midland in the twang of George W.'s voice and in his tone, which unlike that of his genteel, northeastern-bred father, could be curt or rude. As Laura Bush explained in her memoir, "There is a plainness to the way West Texas looks that translates into how people act and what they value. Those who live there are direct and blunt to the point of hurt sometimes. There is no time for artifice; it looks and sounds ridiculous amid the barren landscape." One could also see West Texas in the swagger of George W.'s step and, later, sense it in his unabashed embrace of faith. His father saw it in his free-spiritedness. It made George W. different from him—and from the rest of his progeny. Indeed, George W.'s inherent Texas character was an incalculable gift to a future politician with designs on the governor's mansion. During his initial bid for the Texas governorship, the *New York Times* described George W. as "a square-jawed, curly-headed man whom John Ford may have cast as a cowboy if he were a little taller."

Unlike his brothers, father, and paternal grandfather and great-grandfather, George W. felt no need to leave his home state to find his own identity. Texas *was* his identity. The only times he moved away were to venture east to attend school, and to go to Washington to help run his father's campaign for the presidency in 1988, and after he took the presidency himself in 2001. Even then, as he said himself, though he had left Texas, it never left him.

5

"MORE THAN TONGUE CAN TELL"

THE FIRST MEMORIES OF GEORGE W. Bush all revolve around his father, whom he saw as "a heroic figure." His impressions, he surmised, had much to do with his mother's devotion to her husband. "Mother adored him, and as a result, she imparted that adoration to me," he recalled. It helped that George H. W. Bush had all the qualities of a midcentury American hero in the Jack Armstrong vein; tall, handsome, athletic, humble, and squeaky-clean, he looked and acted the part. Little George would pore over scrapbooks his mother had carefully assembled chronicling her husband's life, everything from photos and the box scores she kept meticulously during his days as Yale's first baseman and team captain to a piece of the yellow rubber life raft he desperately clung to in the Pacific after getting shot down in the war.

George—also focused and scrupulously ambitious—was the very picture of a young man in a hurry. Busyness defined him. "George was a great father," Barbara bragged after their children were well into middle age, to which George responded wistfully, "I was busy, though." It was always a special occasion for George W. to steal childhood moments with his dad. "Keep in mind when analyzing this guy's life that he was really busy," he said. "Therefore, when he made time for you, it meant a lot."

Two of George W.'s early memories involving his father portended his professional future in the oil business and later in Major League Baseball. He recalled the "so-called camping trip," when his dad took him on a drive across the desolate plains of West Texas to visit pumping units, stopping at a diner on the way. Then there were his recollections of playing baseball with his dad and neighborhood friends in his tiny backyard on West Ohio Avenue—though not so small to him then—as his father, who also served as their Little League coach, threw to them and caught pop-ups behind his back, a trick he mastered at Yale. Baseball held particular significance for father and son. Though football dominated West Texas sports culture, it was baseball that captured young George's imagination. He remembers his pride swelling when his father, after the two played catch, said, "Son, you've arrived. I can throw it to you as hard as I want."

His starkest memory, though, is far darker: the day in second grade, when his parents pulled up to his school unexpectedly in their pea-green Oldsmobile and he climbed in as his mother, holding him tight, told him that his sister, Robin—just two months shy of her fourth birthday—had died. Georgie had been lugging a bulky wooden Victrola down a covered walkway on an errand to the principal's office when he saw the car coming up Sam Houston Elementary School's gravel driveway. Racing toward it, he had thought he saw Robin's blond curls in the back seat. As the three of them drove to their home on West Ohio Avenue less than a mile away, he watched helplessly as his parents wept.

Just seven months earlier, Robin had been diagnosed with advanced leukemia. Barbara suspected something might be wrong with her after Robin woke up enervated on a late-winter morning and said, "I don't know what to do this morning. I might go out and lie on the grass and watch the cars go by, or I might just stay in bed." Tests quickly revealed her condition, which in 1953 was largely untreatable and meant almost certain death. George and Barbara made the decision to take her back east to the hospital that would later be known as Memorial Sloan Kettering Cancer Center in New

York City, where George's uncle John Walker worked as a doctor. Walker thought every effort should be made to extend her life if a research development might offer hope, and the Bushes agreed.

The next seven months would find Barbara and Robin in New York for weeks on end, as George went back and forth on weekends while contending with a heavy travel schedule as he built his professional future. In 1950, he had made the decision to leave IDECO and the tutelage of Neil Mallon to form an oil development company with a neighbor, John Overbey, raising more than $350,000 in investment capital by tapping into connections on the East Coast, including his father and uncle Herbert Walker. His talent for making friends led him and Overbey to Hugh and Bill Liedtke, a pair of brothers who ran a small oil-drilling business. Just before Robin's diagnosis, he and Overbey merged their company with the Liedtkes', forming Zapata Petroleum (named after the 1952 movie *Viva Zapata!* starring Marlon Brando as Mexican revolutionary Emiliano Zapata, which was playing in Midland's local cinema) with the prospect of striking it rich as wildcatters.

With their parents and sister gone for extended periods, Georgie and Jeb spent much of 1953 in the care of neighborhood friends or a family nurse dispatched to Midland by Dottie Walker, who had once again come to the aid of George and Barbara. The only time Georgie would see his sister for any length of time was when the Walker and Bush families made their annual summer pilgrimage to Kennebunkport. Even then, he was ordered by his parents not to play with her due to their concern that rough play would lead to bruising. She passed away at Sloan Kettering on October 11, 1953, with her parents at her bedside, a picture of her brothers Georgie and baby Jeb taped above the headboard of her hospital bed. George and Barbara left the grim task of burying their daughter to Dottie Bush and George's Yale classmate and friend, Lud Ashley, and returned to Midland to break the news to Little George.

It came as a shock. Prior to her death, the Bushes chose to let Georgie know only that Robin was sick but that everything would be okay. He heard it repeatedly when he asked his father where his mother and Robin were. "He didn't forgive us for quite a while," Barbara recalled. "But how do you tell a six-year-old, 'Your sister is going to die'? It just doesn't work that way." Their posture reflected their own hope, however distant, that Robin would pull through; George initially refused to accept the doctors' prognosis and visited Midland's First Presbyterian Church to pray every morning at six thirty when he wasn't on the road.

Neither George nor Barbara had yet reached thirty when their daughter died, and it took its toll. Years later, George W. would joke that it was his reckless youth that resulted in turning his mother's hair its signature shade of white. In truth, her hair transformed from light brown to shock white throughout the torturous days of 1953. While she had been a rock before Robin's death, she "fell apart" upon her return to Midland. One of the ways she dealt with the agony was to throw herself into Georgie's life. "Mother's reaction was to envelop herself totally around me. She kind of smothered me, then realized it was the wrong thing to do," George W. said. It occurred to her when she overheard Georgie decline to play with a friend because his mother needed him. That did it. "I was thinking I was there for him," Barbara said. "But the truth was he was being there for me."

George struggled, too. *Dear God*, he thought, *why does this child have to die? She was the epitome of innocence to us, and there was no explanation.* At the same time, he believed "these things contribute to your life, your character, what you stand for." He spoke little about Robin, keeping it inside and moving on stoically much as he had after the war. Later he saw it as a weakness. "For forty years I wasn't able to talk about it," he said. "I was too weak."

While contending with nightmares about Robin's death, Little

George dealt with the loss by peppering his parents with questions about her—*Where is she? If the earth is rotating, where is she buried? Is she standing on her head?*—bringing the issue out in the open and making it a little easier for them to bear over time. During a Friday night high school football game with his father and some of his father's friends, he commented that he wished he were with Robin. When George asked why, he replied "I'll bet she can see the game better up there than we can from down here." His parents believed the loss changed him; it revealed his compassion and sensitivity. Known eventually as the "family clown," Little George tried to ease his parents' minds by making them laugh. As Barbara recalled much later, "After Robin died, George was wonderful to us. Sort of comforting for both of us. But he's got more heart than people give him credit for." The ability to buoy the spirits of those around him would come to bear throughout his life—leading cheers for the Phillips Andover football team in high school; lifting campaign staffers after his father lost the all-important Iowa caucus in his 1988 presidential bid; ascending a pile of wreckage at Ground Zero in lower Manhattan and rallying the nation through a bullhorn after 9/11. It became part of his best self.

Gradually in the wake of his sister's death, George W. would move back into the normal patterns of boyhood, though one less innocent, less pliant to his will. But Robin's absence made him feel like an only child—Jeb, only eight months old when she died, was almost seven years younger. In many ways, with Robin gone, George W.'s childhood was his alone, devoid of shared memories with his siblings. The years in Midland, with its chapter of loss and heartache, bound him uniquely to his parents. It also made him different from his siblings, more aware of life's capriciousness, imbued with palpable energy poured into living life in the moment.

Shortly after George's mother, Dottie, died in 1992, George's brother Jonathan found a letter to her written in 1959 by George, then a father of four boys, which read in part: "There is about our house a need. We need some starched crisp frocks to go with all our torn-kneed blue jeans and helmets. We need some soft blond hair to offset those crew cuts. We need a dollhouse to stand firm against our forts and rackets and thousand baseball cards. We need someone to cry when I get mad—not argue. We need a little one who can kiss without leaving egg or jam or gum. We need a girl."

The need was eventually met. The last of the Bushes' children, Dorothy Walker or "Doro," was born in August 1959. By that time, the family had moved from Midland to Houston, the hub of the oil business, as the family's fortunes climbed with the success of Zapata; every one of the seventy-one wells the company had drilled in the Permian Basin gushed with oil. The Bushes, now a family of seven, settled into a four bedroom, 4,300-square-foot home at 5525 Briar Drive on an oversized lot in Houston's posh Tanglewood neighborhood.

Still, Robin remained a presence in the Bush family. A large portrait of her would hang prominently in the Bush family living room throughout the balance of their lives. George said her death made "every child more valuable." All the Bush children, even Neil, Marvin, and Doro, who weren't born when she died, would later say that she was a factor in their lives. Protective of their parents, they grew up conscious of how harrowing the loss of Robin had been. Though her death became a hole in their lives that could never be filled, she became a symbol of family love and loyalty.

"I love you more than tongue can tell," Robin whispered to her father shortly before she died. George could hear her voice as though she were right there, even well after she was gone. Years later, when all his children were grown, he was able to talk more easily about her loss as the Depression-era stoicism and Yankee emotional repression that marked the world he grew up in had given way to the openness of his children's generation. By then, he was far more expressive and demonstrative with his children. He hugged them more readily and cried easily, allowing himself permission to let the tears flow; "I stopped being a sissy about it," he said. In regular "check-in" telephone conversations during his son's presidency, George W. would invariably end their calls by saying, "I love you," to which his father, invoking the family's most hallowed memory, would often respond, "I love you more than tongue can tell."

6

AGGRAVATION AND PRIDE

IN HIS LATE EIGHTIES, LONG after his memory for specifics had begun regularly failing him, George vividly recalled an episode from George W.'s childhood: his obdurate young son enflamed in a fit of temper aimed at him while he and Barbara walked with him on a street in Midland. As Georgie's arms flailed away attempting to land a blow, George impassively held him at bay by placing his hand on Georgie's flushed forehead. As his father laughed, Georgie became angrier, his arms swirling like a windmill. Eventually, tuckered out, Georgie's head cooled, and the two walked on with Barbara as though nothing had happened.

It was George's expression of disappointment, not anger, that was his strongest and most effective weapon as a parent. Used sparingly, it could send any of his children reeling with shame. "He was hands-off except to the extent that he set boundaries—humility, honesty, hard work, share credit," George W. observed, adding, "I wasn't always good at 'share credit.'" They were boundaries inherited from George's own background, mainly from his mother, and overstepping them was tantamount to letting him down. "I think all of us would tell you the biggest words we didn't want to hear were 'You've disappointed me,' which speaks to an interesting way of being a father, doesn't it?" observed George W., while Jeb recalled, "I can remember screwing up and my dad telling me he was disappointed. I can still feel the emotion well up. Just his disappointment was enough to wreck one for weeks."

That's not to suggest that George W. didn't ever get to his father. "George [W.] aggravates the hell out of me at times. I am sure I do the same for him," he wrote of his nine-year-old son to his father-in-law, whom he referred to as "Mr. Pierce," even after his marriage to Barbara. "But then at times I am so proud of him I could die." From an early age, George W. exhibited a recalcitrance that was outside the Bush mold. A schoolmate of George W.'s, who would go on to work for George, said, "It's not that W. rebelled; he just was wilder than the old man expected—it was a rowdiness. Not doing well in school when you could, being class funnyman—those were detours from the code." George recalled his son as "a rambunctious lad, full of piss and vinegar," a nod to his brashness and penchant for mischief making, while Barbara called him "a wonderful, incorrigible child who spent many afternoons sitting in his room waiting for his father to come home to speak with him about his latest transgression."

As a preteen, he pilfered his mother's cigarettes, puffing away in public before occasionally getting caught, as was the case in an alleyway by the middle school football coach and in the dining room of the Nonantum Hotel in Kennebunkport by his paternal grandparents. During an outing at church, rather than greet Mrs. Witherspoon, a parishioner, with an appropriate salutation, it was "Hiya, little lady, lookin' sexy!" He was sentenced to three whacks of the principal's "board of education" when, egged on by the laughter of his classmates, he penned a beard, mustache, and long sideburns on his face. While Barbara at first objected to the principal's use of corporal punishment, she came around to his side when he explained that George W. had "swaggered in as though he had done the most wonderful thing in the world."

If George expressed disappointment around his son's wrongdoing, it fell to Barbara to rein him in. "I don't remember [my dad] ever striking my brothers or sister," Jeb recalled. "I don't ever remember him getting angry. My mom got angry regularly." Indeed, George W.

was often the target of his mother's wrath including the time he uri-
nated in the bushes outside his home and his mother paradoxically
washed his mouth out with soap. They clashed often, their swords
crossing in part because they were so much alike: tart-tongued,
headstrong, quick-witted. "I have my father's eyes and my mother's
mouth," George W. would later say. His mother, he reflected, was
"easier to vent with" than his father because she would "fire right
back."

Mild antagonism would suffuse their relationship even well after
George W. left the nest. Andy Card, who served in both father and
son Bush administrations, quipped later, "George W. Bush is the
most disciplined person I've ever met, but when I met him most
of the discipline came from his mother." He would long remember first meeting George W., just as he knew George W. had long
forgotten it. An early supporter of George's 1980 presidential campaign, Card drove the elder Bush from Card's native Massachusetts
to the Bush summer home in Kennebunkport after a long day of
campaigning. "He was all excited because his son was going to be
there," Card recalled, painting a vivid picture of what followed.
"And he gets out of the car and bounds up the stairs to go into
the house . . . and there was George W. Bush having an argument
with his mother and the language was pretty crass. And he's got a
Styrofoam cup in one hand and he's wearing a flannel shirt that I
remember being misbuttoned, and jeans with a tear in the left knee,
and brown drool was creeping out of the side of his mouth and he
was getting ready to spit in a cup, and his mother was yelling at
him, 'You're not going to do that in this house!' His mother was
giving him the 'what for.'"

But as exasperated as his parents may have been at times, their
pride in their son, as George suggested to his father-in-law, abounded.
George W., like his father, was loyal, good-natured, and quick to
make a friend. "He does these amazingly loving things when some-

one is in need that nobody knows about, and it's sort of always a surprise," Barbara observed. Also like his father, he was a natural leader as he showed in his last year of school in Midland, when he was elected class president of his seventh-grade class. His father's exasperation would wane in time; the pride was constant.

7

"A CERTAIN SORT OF EXPECTATION"

THOUGH GEORGE W. WASN'T ALWAYS a chip off the old block in comportment, he would become one in pedigree. After he attended ninth grade at the Kinkaid School, Houston's oldest coed private school, his parents packed him up in the fall of 1961 and sent him off to Phillips Academy Andover, in Massachusetts, where his father and grandfather had gone before him. To George W., who had no desire to leave Texas for the confines of an elitist all-boys northeastern prep school, it was "an obligation," a three-year sentence imposed by his parents and a high school experience that George W. found "cold and distant and difficult."

Outside of summer sojourns in Kennebunkport, those years at Andover marked the first time he was confronted by his father's long shadow. Texas was a blank slate. No Bush had blazed trails there before his father arrived in Odessa in 1949. But at Andover, where George had attended school just nineteen years before George W.'s matriculation, George's tracks at the same age were memorable to the point of lore. George Herbert Walker Bush had been the Andover ideal, epitomizing one of Andover's two mottoes, *Non sibi*, Latin for "Not for self," which could well have been Dorothy Walker Bush's maxim for the Bush family. He was recalled fondly by teachers and coaches, who would readily call to mind the prodigious list of achievements left in his wake: president of the senior class, captain of the baseball and soccer teams, treasurer then secretary of the student council, and on and on. He was the natural choice when he was elected the school's Best All-Around Fellow as a senior.

"For George [W.], everywhere he went his father was there," said his cousin John Ellis, six years his junior. "[His father] was just really high up there on a pedestal. And, of course, every year the pedestal seemed to get bigger and bigger because he continued to achieve and achieve. How do you compete with that?" One of the walls of Benner House, where students smoked and snacked between classes, featured a photograph of young Poppy Bush exuding spring glory in his Andover baseball uniform. Tex Robertson, one of George W.'s classmates, recalled it as an icon. "Everybody knew that this was the guy who'd been captain of the baseball team and who had gone up to Secretary of War Stimson at graduation and said, 'I don't wanna go to Yale right now, I wanna fight for my country.' I mean, everybody knew about it."

But if Andover's young charges knew about George Herbert Walker Bush, it didn't come from George Walker Bush. "There was very little reference to [his dad]—as a fifteen-year-old, as an eighteen-year-old, as a twenty-seven-year-old. He never brought up his dad's name," said Clay Johnson, a fellow Texan and Andover classmate of George W.'s who would go on to be his roommate at Yale and a lifelong friend. Nor did George W. bring up his grandfather Prescott, who had served as U.S. senator from neighboring Connecticut since 1952. Maybe it was normal teenage self-absorption, Johnson surmised, but "there was no reference to the Bush family or his dad, or his grandfather. There was never an attempt to cloak himself as the son of this very prominent person, even as [his father] got *really* prominent."

Moreover, if there was hope that George W. would ascend the heights George had once climbed, they didn't come from George, who was content to let his son find his own way at Andover—and, ultimately, beyond. Foisting his own ambitions on his son would have been uncharacteristic. "If you're humble and successful and you don't talk about yourself, you don't set unreasonable expectations for your children," George W. said, adding that his parents "didn't steer the direction of their children but tried to inculcate character."

George H. W., in other words, let George W. *be* George W. "The greatest inheritance I got from my father was unconditional love, which is the greatest gift a father can give a child," the son stated. "The gift of unconditional love mitigates failure, in a way. Fear of failure can prevent you from living life to the fullest, and so ultimately I never feared failure—otherwise I wouldn't have done some of the things I did." That, he believed, was a key to understanding his father's influence on him from the start.

If there was self-imposed pressure for George W. to reach the heights of his namesake, at Andover and later, it wasn't evident to those around him. "I didn't see any desire [on his part] to emulate [his father], to measure up to him, or rebel against him, which as I think about it, is remarkable," Clay Johnson said. Another Andover classmate, Jeff Stipling, recalled, "He was already who he was," but observed, "It's very hard to get below the surface of George." Under the surface, though, there was something more, an edginess that occasionally came out in his nature. "He was a flip, in-your-face kind of kid," recalled a close family friend. "And he kind of had a chip on his shoulder."

Still, George W. made his own mark at Andover—albeit less auspiciously. Though he lacked his father's consistent presence on the honor roll and his athletic grace—a key attribute at an all-boys prep school—he made himself known on campus by sheer force of personality. Mildly rebellious with an easy confidence, he was, according to one of his classmates, "one of the cool guys. He had a way about him, and he fit easily in." He knocked the school's mandatory coat-and-tie dress uniform, sporting rumpled shirts, skewed ties, and sneakers without socks, and donning a beat-up army jacket to combat New England winters. To much fanfare, he created an intramural stickball league in the spring, to which he named himself high commissioner, relieving the tensions that hung over campus as the school year wound down. His appointment as Andover's head cheerleader—something, he conceded later, he didn't exactly brag about when he went back to Texas—owed to his jocularity and buoyancy.

Some of George W.'s deviations from his father's example can be explained by the changing times. While the cultural differences between the "greatest generation" and the "baby boomers" would become increasingly pronounced throughout the 1960s, widening the gap between them, the intrinsic conformity of the former generation was slowly giving way to the burgeoning rebelliousness of the latter, the rumblings of which could be felt even on Andover's insulated campus. Nineteen sixty-seven's *Cool Hand Luke*, the story of a recalcitrant and incorrigible chain-gang prisoner, would later become George W.'s favorite movie, a reflection of his generation's bent toward the antihero. George W.'s wisecracking brashness and gentle subversion burnished his popularity among his peers, just as his father's selflessness and feats on the baseball diamond once had with his own set. The means of being popular, of fitting in—a common trait in the Bush tribe—had changed with the zeitgeist.

To be sure, George W. also saw "the value in lowering expectations," according to Elsie Walker Kilbourne, a Bush relative. "He became a master at it. The whole family clown thing became as much about lowering expectations so he wouldn't disappoint as anything." It was a formula that would serve him throughout his life: Manage expectations downward causing those around you to underestimate your abilities, then—*Surprise!*—prove them wrong. In his first year at Andover, he showed the stubborn discipline and penchant toward self-improvement that was at the root of the formula's success. After his grades weren't up to snuff as he struggled to keep up with Andover's academic rigor, he studied late into the night after a mandatory lights-out at 10:00 p.m. by drawing on the hall lights that streamed in beneath his dorm room door. He was motivated, at least in part, by self-imposed expectations of what it was to be a Bush. Clay Johnson recalled that George W. "had this fear that generation after generation had gone to Andover and he would fail after three weeks." His grades went up, though he wouldn't be among the 110 in his class of 264 who would make the honor roll at least once.

George W., like his father, bucked the trend of hazing fellow students prevalent among the most popular ranks of all-boys prep schools. When George H. W. Bush was at Andover, he made a life-long friend when he came to the aid of a schoolmate who was being bullied. George W. was the good guy, too. Lanny Davis, who would eventually serve in Bill Clinton's administration, remembered Bush later at Yale telling a fellow student to knock it off after he disparaged another student, believed to be gay. "Why don't you try walking in his shoes and seeing how it feels?" he said. It helps to explain him polling second in his senior year as Andover's "big man on campus," a designation almost exclusively reserved for the school's star athletes, which despite a deep interest in sports and positions on the varsity baseball and basketball squads, George W. would never be.

While attending Andover was a paternal dictate of which George W. made the best, his choice of where to attend college was his alone. When considering his options, he entertained only two: Yale, the alma mater of his father and paternal grandfather, and the University of Texas at Austin, the crown jewel of Texas's state university system. Getting into Yale was by no means a certainty. He was encouraged to apply to UT as a "safety school" by Andover's dean, who noted George W.'s subpar grades and board scores. "I was fully prepared to go to Texas to the point that Dad and I went over to take the campus tour, which was kind of a disaster," recalled Bush. The two of them surveyed the campus's famed "forty acres" without setting up meetings or a formal visit. They were, to George W.'s memory, "just two guys wandering around this giant place unescorted," including impromptu visits to dormitories at which father and son asked strangers to see their rooms.

In the end, George W. had his pick of either. Yale accepted him even though his board scores—566 for verbal and 640 for math—were well below the median scores for his class: 668 and 718 respectively. Doubtless, his acceptance owed much to his bloodline; George and Pres had sailed through Yale with stratospheric distinction, and the latter sat on the university's board of trustees as a former U.S. senator from Connecticut. George W. chose Yale but claims it wasn't a certainty. Though he had followed Yale football as a young boy through newsletters a friend of his father's sent to him in Midland, he "wasn't dying to go." But he conceded that "there had to be something there," because when Clay Johnson proposed the idea of attending Yale together as roommates, he quickly agreed, heading off to New Haven and implicitly embracing the northeastern Ivy League elitism that had seemed anathema to him.

Though his father had been there before him, as at Andover, he didn't appear to friends to occupy an active place in George W.'s consciousness. Just two months into his freshman year, on October 31, 1964, a Saturday, George W. packed a suitcase and announced to Johnson that he was flying home to Houston the next day for a four-day visit.

"What are you going to Houston for?" Johnson asked.

"The election," George W. replied. "Dad's running for Senate."

Johnson admitted to somehow missing the fact that his roommate's father was on the ballot in his home state, the Republican candidate out to unseat the incumbent Democratic senator, Ralph Yarborough, but he had every reason to be taken aback; it was the first time George W. had mentioned it.

Though he didn't wear the influence and example of his father on his sleeve, it was never far from his mind. The year before, when he returned to Andover from a trip home to Houston in his senior year, his roommate, fellow proctor John Kidde, noticed Bush carrying a book that didn't appear in any of the Andover syllabuses: *The Conscience of a Conservative* by Barry Goldwater. "What the hell is this?" Kidde asked him. Bush hadn't shown any outward interest in politics and their heavy course demands didn't allow for much extracurricular reading. Bush explained that it was something his parents had asked him to read, and according to Kidde, something he "seemed honestly interested in."

The young man's interest in politics and conservative leanings would evolve in time, but in his formative years, his father's influence on his conscience would be far greater.

PART II

"A Debt to Pay"

8

"TO DO SOMETHING OF SERVICE"

TO THOSE WHO KNEW GEORGE H. W. Bush, it was no surprise that he would get involved in politics. Years earlier, shortly after landing in Texas in 1948 as he started in the oil business, he wrote his friend Gerry Bemiss, "I have in the back of my mind a desire to be in politics, or at least the desire to do something of service to this country." Yes, he wanted to make something of himself, to prove himself, to make money, but he was also anxious to give back. It was—and would remain—the Bush way. Like his father, he looked upon public service as a higher calling. "I knew what motivated him," he wrote of his father. "He'd made his mark in the business world. Now he had a debt to pay."

A progressive on social issues and a conservative on fiscal matters, Prescott Bush was an archetype of the establishment northeastern Republican—a moderate before the word was co-opted by conservative ideologues as an epithet—and like his son after him, a model of civility, tolerance, and decency. While serving as Connecticut's Republican Party chairman from 1947 to 1950, he also chaired his state's branch of the United Negro College Fund and acted as treasurer for a national Planned Parenthood fundraising. He brought the same activist instincts to the Senate, supporting the Peace Corps and the Civil Rights Acts of 1957 and 1960, largely toothless but for their symbolism as the first civil rights legislation in nearly a century. In the mid-1950s, when the anti-communist crusade of Republican senator Joseph McCarthy had overreached to the point of witch-hunt extremism, Pres condemned his fellow Republican for causing "dangerous divisions among the American people because of his attitude and the attitude he has encouraged among the American people." After providing a voice of reason, his vote was among the sixty-seven censuring McCarthy, versus the twenty-two votes, all cast by Republicans, who opposed the measure.

During his tenure in the Senate, he earned the respect of his colleagues while becoming a close associate of Dwight Eisenhower, who occupied the White House throughout the first eight of Bush's ten years in the Senate. Eisenhower, a golfing buddy, placed Bush on a handwritten short list of those Republicans he deemed worthy of succeeding him as president.

As Pres left elective politics, opting not to run for reelection in 1962, George H. W. Bush hurled himself into the realm the same year, vying for election in Houston as the chairman of the Harris County Republican Party. His charge was to keep another kind of extremist threat to the party at bay: the John Birch Society, a burgeoning right-wing advocacy group founded around a doctrine of limited government and anti-Communism. "I'm not going to vote for 'nother country club asshole," one Bircher said of Bush, in keeping with the sentiments of other Birchers. Bush won anyway, preventing the Birchers from wresting control of the party in Houston. But in typical fashion, Bush reached out to them, making them stakeholders by offering them positions within the party. "George Bush's instinct politically is to bring people together, to be a uniter," explained his friend Victor Gold. "And so he didn't come in in a confrontational style, slam the door and throw all the Birchers out. His idea was, 'Let's get the Birchers and have some common meeting ground with them, because if we want to beat Democrats, we need those people.'"

The post suited Bush's superlative ability to foster friendships and alliances that would serve him well, as he raised $90,000 for the party's war chest along with his visibility across the state. It set him up nicely for his next political quest: a seat in the U.S. Senate, which he set his sights on in 1964. George W. surmised that his ambition to achieve a place in Congress's upper chamber had much to do with the fact that his own father had done so twelve years earlier, relinquishing his seat in 1962 after holding it for ten years. "He admired his dad so much that the idea of serving in the Senate appealed to him," George W. believed.

After winning the GOP primary in a runoff against Jack Cox, a former Texas gubernatorial candidate, Bush went on to face the incumbent senator Ralph Yarborough. A New Deal liberal, Yarborough had earlier in the year voted for the landmark Civil Rights Act of 1964, known as the Public Accommodations Act, which put a legal end to Jim Crow segregation that pervaded the Deep South. It was not a vote that was looked upon kindly by the Texas electorate, creating a political soft spot that Bush could seize at Yarborough's expense. But it wasn't that easy, as Bush's conscience, as would often happen in his political career, was put squarely at odds with his instinct to win.

Like the Republican presidential nominee, Barry Goldwater, Bush opposed the Civil Rights Act on constitutional grounds, contending that desegregation was not a legal issue to be determined at a federal level but by individual states. As he later explained, "my reasons for not supporting the bill were very different from those who hated the bill for racist reasons." He wrote of his dilemma to a Jewish community leader in Houston: "I want to win but not at the expense of justice, not at the expense of the dignity of any man— not at the expense of hurting a friend, not teaching my children a prejudice which I do not feel . . ." Certainly, Bush didn't oppose the Civil Rights Act on racist grounds, but he also knew that his own opposition to the law, regardless of his reasons, would resonate with those who did. On the campaign trail, Bush contended that the law was "politically inspired and is bad legislation in that it transcends the Constitution," while privately conceding later, "I took some far right positions to get elected. I hope not to do it again. I regret it."

His conservative positions didn't much matter. Painted by Yarborough as an effete Yankee carpetbagger, Bush faced a deeply entrenched Democratic electorate in Texas during a year when President Lyndon Baines Johnson, a year after the assassination of John F. Kennedy, was riding high. LBJ won the state by 63 percent against Goldwater, as Yarborough, riding the Democratic tide, outpolled Bush by over 300,000 votes, taking 56 percent of the vote. Little consolation came from the fact that Bush had yielded over a million votes, the most any Republican had ever won in a statewide race to that point.

Bush's loss came at a time when the GOP was at a crossroads. In the summer of 1964, he had been among the party faithful at San Francisco's Cow Palace for the Republican National Convention, where the body divided viciously over two potential standard-bearers: the moderate establishment candidate, New York governor Nelson Rockefeller, and Goldwater, the senator from Arizona and conservative firebrand. Goldwater had charged up an increasingly large and vocal conservative wing of the party with his declaration, "Extremism in the defense of liberty is no vice . . . moderation in the face of justice is no virtue." After a hostile convention floor battle, Goldwater emerged as the party's nominee and marched onto the campaign trail against the incumbent Johnson with his slogan "In your heart you know he's right."

The vast majority of Americans felt otherwise, instead going "All the way with LBJ." Johnson pulled 61 percent of the popular vote, racking up, to that point, the largest electoral mandate in American history. Goldwater, seen as a trigger-happy extremist at a time when Cold War tensions with the Soviet Union predicated foreign-policy doctrine, proved a disastrous candidate. The GOP, however, learned its lesson. For the next half a century, the party would not go passionately with its heart, but cautiously with its head, prudently choosing viable candidates for a national electorate: incumbent presidents, runners-up in the previous presidential primary cycle, or an otherwise establishment Republican brand name. It was a trend that would well suit George H. W. Bush and George W. Bush, who between them, would grace six national tickets in the course of less than a half a century, but would stop dead with Jeb Bush's aborted campaign for president in 2016, which was crushed by the populist rise of Donald J. Trump.

The political education of George W. Bush began alongside his father during the latter's failed Senate bid. The summer before entering Yale, George W., eighteen and eager, pitched in by doing odd jobs, including research at the campaign headquarters in Houston and traveling with his dad on the hustings back in West Texas. George had expressed to friends that he wanted his oldest boy with him during the race, providing a bonding opportunity for the two of them, comrades in arms, as the remainder of the Bush children stayed at home, too young—the oldest of them was Jeb at age eleven—to make a meaningful contribution. "Watching him on the campaign trail intrigued me and obviously sparked an interest in the political process," said George W., adding, "if the political process corrupted his ability to be a good father, I think it would have turned us off to politics." Moreover, he related, his father gave him a guiding principle that he would apply to his own life: "If life is to be led to the fullest, you take risks. Winning isn't everything in life. In fact, you can gain from losing at times." In this way, the Bushes were a living embodiment of Theodore Roosevelt's manifesto celebrating the man "in the arena," who "if he fails, fails while daring greatly, so that his place shall never be with those cold and timid souls who neither know victory nor defeat."

But defeat never defined George H. W. Bush. Two years later, he would take another risk, venturing back into the arena, where he would gain by winning.

9

YOUNG AND FOOLISH

GEORGE H. W. BUSH DIDN'T send his son off to Yale with words of advice or expectation. There wasn't much ceremony in it at all. George W. had, after all, been at Andover for three years and felt as if he had been on his own since he was fifteen, checking into the family home in Houston and, later, Washington, D.C., in between semesters for a few weeks at a time, rarely more. "I was a pretty independent guy," he recalled. "I'd come home for vacations, but I was pretty well gone." Heading off to Yale meant just packing another bag and setting up a life on another campus.

Regardless, George W. didn't need his father's words to guide him at Yale; his father's example was always there, quietly looming large. Though insisting in a 2001 interview that he didn't feel the "weight" of his father's prototype at Yale, he followed it—minus the accelerated pace and Phi Beta Kappa grades—almost to the letter.

After returning to Yale after his father's 1964 Senate loss, George W., at George's suggestion, went to see the school's chaplain, William Sloane Coffin. George and Coffin had grown up in the same rarefied world, attending Andover together, then going on to Yale, where George had seen to it that Coffin was tapped for Skull and Bones, the university's prestigious secret society. It was natural for George to encourage his son to see an old friend, one who might be helpful to him, and he did it often. But during their visit when the subject of his father's recent Senate defeat came up, Coffin's response hit George W. between the eyes; "I knew your father," Coffin said, "and your father lost to a better man." Perhaps it was just the preferred liberal ideology of George Bush's Democratic opponent, Ralph Yarborough, that was at the root of Coffin's denouncement of his old friend. Though the elder Bush and Coffin had traveled the same roads earlier in their lives, Coffin had gone back to Yale to attend divinity school, then rallied around the cause of the antiwar movement. Still, Coffin's comment made George W. seethe, turning his back on Coffin for the balance of his tenure at Yale. He wasn't used to people talking about his father like that. And maybe it was Coffin's phrasing, too: "a better man." The ultimate compliment the Bushes extended was designating a worthy subject "a good man"; it was part of the Bush lexicon, and the family looked upon their patriarch, George Herbert Walker Bush, as the paragon of a good man.

The incident foretold cracks in the culture that made their way to Yale during the course of George W.'s four years on campus. The university would bifurcate in much the same way as George H. W. Bush and William Sloane Coffin had evolved. Some students would cling to the carefree traditions of an earlier era—boozy frat parties, gauzy sorority dances, football games, and glee clubs—while others would get swept up in the 1960s counterculture—protest marches, sit-ins, and consciousness raising often in a haze of marijuana smoke. The latter exploded in his last semester in 1968, an annus horribilis that shook the nation to its core as the Vietnam War raged hopelessly, Martin Luther King and Robert Kennedy were gunned down, and cities were set ablaze in protest.

George W. took the traditional path. On campus, he once again showed he had inherited his dad's egalitarianism and his talent for remembering names and making friends, many of whom would come into play later in his life. One of them, Roland Betts, recalled, "George probably knew 1,000 of the 4,000 undergraduates at Yale. There was probably no one else who knew 200." If Yale was a birthright for him, he didn't let it show. "He was third generation Yale. I was first generation Yale—first generation college, for that matter," recalled Calvin Hill, Yale's star running back, who would later play for the Dallas Cowboys. "Yet nothing about him suggested he thought he was better than other people. I guess you'd say his mother and father raised him right."

He gravitated to his father's fraternity, Delta Kappa Epsilon (DKE), pledging in his sophomore year along with his four roommates, Robert Dieter, Ted Livingston, and brothers Clay and Terry Johnson. Much of his college life revolved around DKE, which he would lead as the house's president in his senior year, upholding its traditions. When the *New York Times* ran a 1967 article about the "sadistic" fraternity ritual of branding its pledges with a coat hanger, George W. defended the practice, describing the resulting wound as "only a cigarette burn."

Another tradition was drinking. DKE had a reputation for being the hardest-drinking frat on campus, and George W. did his part to burnish its reputation, consuming alcohol copiously, like many college students, for the first time. He got into a little trouble, perpetrating booze-fueled transgressions that never rose above the level of sophomoric hijinks and made for good fraternity lore. After a Yale football win at Princeton, he was among a rowdy bunch that pulled down the goalposts and who were then taken to the campus police station where, Wild West style, they were given ten minutes to get out of town. Later, he evaded a disorderly conduct conviction after being arrested for "borrowing" a Christmas wreath from a New Haven store, and charges were dropped.

"I liked to drink. I wasn't a knee-walking drunk, but I liked to drink and have a good time," George W. admitted later. Jeb remembered his brother as not being "particularly rebellious" during his college and later years. "He was a little wild," he said, "but so was everybody else. He was inside the lines of what would be considered acceptable behavior, I think. He wasn't a danger to society."

If alcohol was getting the better of him, his father, characteristically, never felt compelled to rein him in. "I don't ever remember my father saying, 'Hey, don't you think you need to dial it back,'" he said. Then again, for the most part, George W. revealed his best self to his father. It was less an Eddie Haskell–like contrivance than an earnest attempt to win an approving nod from the man he worshipped. Cousin Elsie Walker observed that W. "sometimes was sort of sensitive to the fact that he was too coarse, or, you know, rough-edged for his father."

The rough edges, though, were fair game for his mother, who saw a more unrestrained, uncouth version of her son during his college years. Family bonding time was often found in sports, where competitive juices could flow and members could manifest athletic prowess and character. During a golf outing with George W. at Houston Country Club, however, after repeated reprimands for using foul language, Barbara demanded, "Take your clubs and just walk home!" leaving her "humiliated" as she finished the round alone with their caddy. Another incident at the club occurred during a doubles match with George W. and two of her friends. After bearing witness to her son's repeated offensive utterances, she called him to the net and threatened to send him home if he didn't curb his behavior. She then "turned around to pick up and serve, and he had just walked off," she recalled.

In affairs of the heart, though, young George W. seemed to follow the course set by his father. During Christmas break at home in Houston in his junior year, he popped the question to his girlfriend, Cathryn Wolfman, a junior at Rice University who had transferred from Smith College, where Barbara had attended her freshman year before dropping out to marry George. Cathy, from a well-heeled Jewish family in Houston, held a beauty, sophistication, and intelligence that made her a keeper. Doug Hannah, a Houston friend of George W.'s, described her as "the pick of the litter . . . If George W. wanted a trophy and that was his goal, that might have been what he was going for."

What he appeared to be going for was Dad's template: propose to a good social match at age twenty, marry, and move your newlywed wife to New Haven where you'll set up house as you finish up at Yale. That was the plan for George W. and his fiancée, and the filial symmetry was not lost on her. Friends speculated that she resisted rushing down the aisle so as not to play into George W.'s instinct— perhaps subconscious—to emulate his dad. "The whole process of getting engaged as a junior, which was just duplicating what his father had done—I think she realized at some point that George was just trying to do exactly what he did," surmised Hannah. "She didn't want to be the next-generation duplicate, she said, 'Let's wait.'"

They did, for a year. Delays in the wedding planning led to the dissolution of the relationship altogether after George W. had left Yale and was enlisted in the Texas Air National Guard. While the parting was amicable, it seems to have left George W. shaken. Hannah recalled the breakup coming as close to "undoing" George W. "as anything that had ever happened to him." Barbara thought the young couple "just sort of panicked," while George W. said later that he "wasn't ready" for married life. It was not something he would contemplate again for another decade, when he was thirty-one, well past the age when most of his Texas friends had tied the knot.

While George W. could have been one of the many youth at Yale who grew up in conservative privilege only to gravitate to liberal causes, his college years only seemed to reaffirm the values and positions espoused by his father. As protests against America's involvement in Vietnam escalated throughout his college tenure, many on the Yale campus, Bush supported his father's congressional support of the war. "He believed that his father's position was correct," Robert J. Dieter, one of his roommates of four years, recalled. "We're involved, so we should support the national effort rather than protest it." George knew his son could have been drawn to the youthful vices of the times—drugs, subversion, cynicism—but wasn't. Instead, he had stuck to themes that could have been plucked from George's own youth: friendship, loyalty, and faith in government and traditional institutions that had helped to build a great country. His father saw it as a sign of his boy's character. In an undated letter written sometime between November 1968 and November 1969, George wrote Allie Page Matthews, a member of his congressional staff, making reference to George W., who was, by then, enlisted in the Texas Air National Guard:

> Last weekend our son came home . . . and had lots of college friends there, for a wedding weekend. Some had the long hair and the bell bottoms, *all* seemed close and concerned.
>
> Yesterday we got a letter from George and it thanked us for having his friends—and then it went on to say "Someday Jeb, and Neil, Marvin, and Doro will know how lucky they are— someday they'll know what it is to be surrounded by love"—and we thought back to Yale and turmoil as we watched George try to figure which way to go—how far to reach—and we wiped back our tears and we said, "We are very lucky to have a boy today who understands."

If George W. rebelled, it wasn't around the causes taken up by many of his classmates, but against the classmates themselves. More "y'all" than Yale, he bristled at their "intellectual arrogance" and the judgments rendered around his beloved home state, the "You're from Texas, therefore . . ." attitudes. "I still believe that just because somebody's got an Ivy League title by their name, it doesn't make them smarter than anybody else," he asserted in 2001.

He also resented it when his fellow students "wrote him off as a legacy kid," observed classmate Robert Birge, at a time when Yale's diminishing favoritism toward its alumni was being called into question. This served as another reflection of the changing times as the grand sweep of Lyndon Johnson's Great Society programs aimed at leveling the socioeconomic playing field. The year George W. was enrolled in Yale, the acceptance rate among legacy applicants was 51 percent, a number that dropped precipitously to 12 percent three years later. Just as George W. didn't feel a sense of entitlement around growing up a son of privilege, neither did he feel the need to apologize for it. He felt nothing but pride in what his family—his father—had attained through talent, character, and just plain hard work, and looked contemptuously at those of similar backgrounds who "felt guilty about their lot in life because others were suffering, or people who felt guilty because they hadn't done a dang thing to deserve it."

Despite the backlash that came as part of the cultural shift, being a legacy gave George W. a boost with Skull and Bones, which reserved one of its fifteen annual recruiting spots for an undergraduate whose father had been summoned to their ranks before him. The underground secret society Skull and Bones, perhaps Yale's most august, was a deep tradition in the Bush-Walker clan; in addition to his father, grandfathers Prescott Bush and George Herbert Walker and Uncle Jonathan were Bonesmen. So were cherished Bush family friends Neil Mallon and Lud Ashley. As "Tap Night" approached in late April deep into the spring semester of his junior year, George W. anticipated being invited to be one of Skull and Bones' chosen few. But he flirted with bucking family tradition, entertaining a possible bid from a lighter, "above ground" alternative called Gin and Tonic. "This is what's going to work for me," he told his DKE fraternity brother Robert Reisner.

Rumor had it that late in the evening on Tap Night, George H. W. Bush, sensing his son's uncertainty, showed up at his door to implore him to accept the Skull and Bones bid and, implicitly, to uphold the family legacy. While no one could confirm it, the hearsay had its own significance. "To me, that was sort of symbolic—of the tradition and the sense of Walker's Point and the history that he had to live up to in his life, the idea that your father was that kind of presence in your life, as a mentor," Reisner said.

In fact, the rumor was just that. George H. W. Bush, now a rising first-term congressman, had not journeyed from Washington to New Haven to nudge his son, who accepted the Skull and Bones bid of his own accord. As George W. later confirmed, "It's not one of those things that they ever tell you, 'Stand by.' Nor did he ever show up on campus, nor was he ever there when I was there. We never were there together."

He could be forgiven if he forgot his father's two-hour appearance at his June 1968 graduation, most of it around the ceremony itself. "He hung out with my family for most of the two days," recalled Clay Johnson. "I remember as his dad left, he made some comment about [wishing his] dad didn't have these other obligations. 'It would have been great if my dad could have been here during the whole time,'" Johnson recalled George W. lamenting, observing of his roommate, "It wasn't said in passing. He [was] very aware of the toll that public service takes on family members."

Aware, yes, but as his own future would make clear, not deterred.

10

"A WHOLE DIFFERENT LIFE"

THERE WAS LITTLE DOUBT THAT George H. W. Bush would re-enter the political ring despite his failed Senate run in 1964. He was determined to live a purposeful life, and public service was the best route. In 1966, after concluding that he couldn't run for office and "do justice" to Zapata's shareholders, he made the "agonizing decision" to sell the company, garnering $1.1 million and going all in on the pursuit of a seat in Congress representing Texas's Seventh District against conservative Democratic challenger Frank Briscoe. Like his father, who had lost his first race for the Senate in 1950, he had shrugged off the loss, learned from it, and, undaunted, ventured back into the fray. Also like his father, he prevailed in the next contest. It was a pattern that would be repeated later in the political careers of George W. and Jeb, almost as a Bush family ritual, manifesting the grit that the Bushes quietly bore. Baptism by defeat.

Bush's win presaged the slow turning of the tide for Texas from blue state to red, as he became the first Republican to win the Seventh District's congressional seat. There was great satisfaction in it for Bush, who had come to Texas evangelizing the Republican cause to mostly deaf ears. Upon his arrival in Washington, the forty-three-year-old congressional legacy was looked upon as a comer; he quickly won the favor of influential colleagues who would help him advance the ranks aided by connections his father made in his decade as a senator before stepping down in 1963. Among them was Wilbur Mills, the Arkansas Democrat and longtime chairman of the powerful Ways and Means Committee, of which Mills made George a member, a prize for any congressman let alone a freshman. Like his father before him, he was an advocate of birth control and family planning. So much so that Mills tagged him with the nickname "Rubbers."

Atoning for his opposition to the Civil Rights Act of 1964, Bush supported an all-black girls' softball team in Houston—the Bush All-Stars—ensuring their inclusion in the city's tournaments, and over time earned the support of blacks and Hispanics throughout the district. As Yarborough had in supporting the Civil Rights Act of 1964, Bush alienated many white constituents when he voted for the Civil Rights Act of 1968, known as the Fair Housing Act. For Bush, it was a matter of conscience. Paying his own way, he had visited Vietnam for two weeks in 1967 to get a firsthand view of the war, where he saw "young black soldiers fighting for love of their country while affluent white kids ran away or got deferred." While he had the same constitutional concerns about the Civil Rights Act of 1968 as he did the '64 Act, he believed that African Americans should have the same rights to housing as whites. On April 4, the day of Martin Luther King's murder in Memphis, he explained his position and the political fallout in a handwritten note to a friend and former volunteer, Chase Untermeyer:

I'll vote for the bill on final passage—Have misgivings—giant political misgivings—also constitutional—also I know it won't solve much . . . but I'm for much of the bill and in my heart I know you're right on the symbolism of open housing.

The mail is more on this than any subject since I've been in Congress—all against except 2 letters—500 to 2 I'd guess.

But this will be my character builder.

Character intact, Bush made an impassioned defense of his position in a meeting with angry constituents at a Houston high school. "With one of the most conservative party voting records in the House, I am now accused of killing the Republican Party with this one vote," he said. Bush won the respect if not the accord of his audience, which gave him a standing ovation. For George W. it was "by far the most meaningful public event" of his father's early political career.

The episode portended the challenges Bush would have in convincing right-wingers that he was a bona fide conservative, not a moderate in conservative's clothing. Despite the fair housing vote, Bush easily won reelection in 1968, unopposed by a Democratic contender.

As with all other chapters in his life, Bush had a willing political partner in Barbara who gamely went to Washington prepared for "a whole different life for us, exciting, overwhelming, intimidating, interesting, exhausting." George W. credited her as the one who "kept the family together" through all their many moves. "She handled all the moves every time and, you know, I never saw them fight," he said, adding, "Now, they may have been behind closed doors chewing each other out—[but] I doubt it." The Bushes kept their home in Houston and bought a two-story brick town house in the moneyed Spring Valley section of Washington, where they settled with Marvin, Neil, and Doro at the start of 1967, as Jeb, living with Bush family friends, finished his freshman year at Kinkaid in Houston before beating a well-worn Bush path to Andover, and George W. continued on in his junior year at Yale.

For George H. W. Bush, throwing himself into a new, meaningful endeavor fit his life's ongoing quest of "broadening horizons," a term he and his oldest son would often invoke. His formative years were defined by experiences that enriched his perspective on the world, often opening new doors of opportunity. Most of them he consciously sought out himself. Going to war gave him exposure to men of all walks of life tangled up in a cause no less than democracy's salvation. He emerged less prone to judgment, more understanding of those of different backgrounds and life experiences. Striking out on his own in West Texas gave him the confidence of knowing that he could make his own way in the business world far out of range of his father's long shadow, and led to his entrepreneurial success with Zapata. By establishing himself outside the confines of the privileged orbit he enjoyed as an accident of birth, he stretched himself further and made himself better—and he wanted the same for his children.

George gently guided George W. toward opportunities during his teens and early twenties to broaden *his* horizons. When he told George W. after his first year at Andover, "I think you should work the summers," his son embraced the concept, dutifully working jobs designed to expand his range and depth of experience, each arranged by his dad. In various summers, he was a messenger for Baker, Botts, Shepherd & Coates, the august Houston law firm established during the railroad boom of the 1840s and later run by Jim Baker, the close Bush family friend who would eventually become George H. W. Bush's secretary of state; a salesman at Sears, Roebuck and Co., hawking everything from Ping-Pong balls to fishing lures; a desk jockey at Rauscher Pierce; a hand on a cattle ranch in Arizona; and a West Texas roughneck at the Circle Drilling Company in Lake Charles. All told, those experiences gave him glimpses into the world far beyond the rarefied worlds of Andover, Yale, and manicured Houston suburbs.

Two lessons he gleaned from his father during that period stood out. The first was a parenting lesson he picked up from his dad in the summer of his junior year at Andover, during the family's annual summer pilgrimage to Kennebunkport. When Barbara told her seventeen-year-old son that she and his father wanted to take him to dinner at the Old Grist Mill, a Kennebunkport staple, George W. thought, *What's going on? This doesn't happen much; something's up.*

He soon found out. Before the menus arrived, Barbara snapped, "You *smoke*! I discovered an ashtray under the bed upstairs." George W. confessed to the transgression, but the subject was quickly defused when his father quietly said, "So do you, Bar." As George W. recalled, "That was it, the end of the conversation. It speaks a lot about how he raised us. The guy basically said, 'Until you stop smoking, you can't tell anyone else not to smoke.'"

The second was his dad's reaction when he learned that his son had left his stint as a roughneck a week earlier than he had promised his foreman so that he could go to see his girlfriend. "He called me down to his office and he said, 'This is really disappointing, son. You made a verbal contract and you didn't honor it,'" George W. recalled. "I didn't honor the contract, the verbal contract, the rules. I felt terrible. He exposed my weakness in a gentle way— and the weakness was that I didn't honor my word. I didn't think it was a big deal at the time, but then when he talked to me about it, I realized it was a selfish move on my part." But confident that George W. had taken the message to heart, George didn't let his disappointment linger; later the same day he called his son to ask him and his girlfriend, Cathy, to join him at an Astros game. The gesture of faith would leave as deep an impression on George W. as his father's admonishment. Indeed, faith would mark George's attitude toward his son even when the latter's path was not clear.

11

UP IN THE AIR

THE VIETNAM WAR PERVADED 1968, lodging itself virulently in the American consciousness. It filtered through TV screens in dour evening newscasts and brayed across newspaper headlines day after day, as protest rallies clamored across college campuses amid growing opposition. It had worn down its chief protagonist, the titanic Lyndon Johnson, who, to the shock of the nation on the last day of March, opted to forgo a reelection bid in an anguished, elusive search for an honorable peace. And it divided the country into camps defined categorically by one's stance on the war: hawks or doves. But it was particularly pronounced for young men whose lives were subject to the whims of fate predicated by draft numbers, financial means, college acceptance and enrollment, and social and political connections.

While enrolled at Yale, George W. had registered in Houston with Texas Local Board No. 62, but like other college students, was deferred due to "activity in study." Now with his sheepskin in hand, the question of military service was front and center, just as it was for most male college graduates in 1968. As he wrote in his 1999 autobiography, *A Charge to Keep*, "We didn't have the luxury of looking for a job or taking time to consider what to do next. It was hard to get a job until your military status was resolved." But there was no question that he was going to serve in some capacity. Roland Betts, a friend and DKE fraternity brother, observed that George W. "felt that in order not to derail his father's political career he had to be in military service of some kind."

But the war itself was not a cause George W. felt passionate about, not like his father had about serving in World War II when he rejected his own father's pleas and enlisted in the navy on his eighteenth birthday. In fact, like many, George W. would become disillusioned with war in Vietnam as he realized, increasingly, "we could not explain the mission, had no exit strategy, and did not seem to be fighting to win."

An elegant alternative to serving in the war was to enlist in the Texas Air National Guard, on call principally to protect Texas's 367-mile Gulf Coast. His service would allow him to honorably fulfill his military service with little chance of getting shipped off to the war, where in 1968 American troops were dying at a rate of 350 a week. During Christmas break of his senior year, George W. discovered through friends that there were openings in the Guard, learning in a phone call to Colonel Walter "Buck" Staudt, the cigar-chomping commander of the 147th Fighter Group, what he needed to do to apply. When Staudt asked why he wanted to join the Guard, Bush replied, "I want to be a fighter pilot because my father was."

Dubbed the "Champagne Unit," the 147th's ranks during the latter years of the Vietnam War would be rife with the progeny of powerful Texans, including Lloyd Bentsen III, son of a U.S. senator from Texas; John Connally II, the son of the former governor and incumbent treasury secretary for Richard Nixon; and Al Hill Jr., grandson of oil magnate H. L. Hunt. It also included exalted Texans of a different sort: several members of the Dallas Cowboys. Spots in the "Champagne Unit," as its nickname suggested, were highly coveted and hard to come by. Wait lists were long, and for most, so were the odds of getting admitted. Nonetheless, on May 27, 1968, a dozen days before losing his draft deferment and two weeks before his graduation from Yale, George W. Bush went home to Houston and appeared at the Texas Air National Guard at Ellington Field with the hopes of joining its ranks. Despite scoring a 25 percent on the pilot aptitude test, the minimum acceptable grade, he was sworn in as a second lieutenant the same day. How he got there would become a question of some dispute.

Ben Barnes, the dynamic twenty-nine-year-old Democratic speaker of the Texas House of Representatives at the time, claimed that it was his call to the Guard's Brigadier General James Rose recommending young George W. Bush for the unit that likely resulted in his ready acceptance. "There is absolutely no way Bush could have gotten into the Texas Air National Guard so quickly unless he had special help," Barnes said later, estimating that there were as many as two thousand names ahead of Bush's on the list. His endorsement of George W. came at the request of Sidney Adger, a Houston oilman and well-connected Bush family friend, who asked Barnes if he could pull strings as a favor. As Barnes stated later, "Sid Adger was an oilman who liked being around important people. I assumed that I wasn't just doing a favor for Sid Adger but for the Bush family." A protégé of LBJ, Barnes knew the political utility in doing favors that might someday "pay back a dividend or two." He gladly placed the call, gave his backing, and collected a chit, just as he had for "dozens" of other influential Texans that included at least one other congressman: Texas congressman Frank Ikard, a Democrat from Texas's Thirteenth District, whose son Bill was admitted to the National Guard after Barnes intervened on his behalf.

In 2004, as allegations that George H. W. Bush used his influence to get his son into the Guard swirled around his son's presidential reelection campaign, the former president responded by calling them "a total lie." "Nobody's come up with any evidence but it comes up all the time," he claimed. True enough. Barnes clarified in his 2006 memoir that Bush "did not contact me personally to ask for the favor," which came solely through Adger. Six years earlier, in 1999, he said as much to Don Evans, George W.'s longtime friend and adviser during his governorship, who had been asked to look into the matter. Afterward, the governor sent a letter to Barnes that read, "Don Evans reported your conversation. Thank you for your candor and for killing the rumor about you and Dad ever discussing my status. Like you, he never remembered any conversation. I appreciate your help."

After spending the summer and early fall in basic training at Lackland Air Force Base in San Antonio, Bush took the first of several unusual leaves of absence that would mark his tenure in the Guard. At his parents' urging, he spent the bulk of the eight-week leave doing low-level work as a travel aide in the U.S. Senate campaign—ultimately successful—of conservative Florida Republican Edward Gurney. Afterward, Second Lieutenant Bush took off in his blue Triumph convertible for thirteen months of flight school training at Moody Air Force Base in Valdosta, Georgia, where Bush learned to fly T-41, T-37, and T-38 aircraft, and finally the F-102 Delta Dagger. He proved his natural skill as a pilot, flying single engine aircraft and jets on a simulator, while jovially fitting in with the other troops, much as his father had during World War II, where crewmates on the USS *Finback* called him "GeorgeHerbertWalker-Bush," gently mocking his rarefied pedigree. George W. reverted to form, buddying up with many of his fellow trainees, most ultimately bound for Vietnam, chugging down beers with them at the officers' club and tagging them with nicknames like "Fly" and "Chubby." Bush's flight commander, Major Norm Connant, recalled him as "a lively individual. He had a great personality and was eager to learn." But aside from the fact that Bush was the only one being trained for the Guard of the seventy men, there were a stream of reminders that George W. Bush wasn't just another rank-and-file military pilot.

The most memorable came in early 1969 when his father, who was throwing a party at Washington's Alibi Club for Apollo 8's commander, Frank Borman, called his son with a proposition. "How would you like to fly up to Washington for a dinner with an astronaut?" he asked in a phone call, adding that he had also invited first daughter Tricia Nixon. "I thought it might be fun if you took her to the party," he said. Taken aback and noncommittal at first, George W. accepted the blind date only after betting his doubtful flight school buddies that the invitation was real. He found himself fifty dollars richer when a military plane later arrived at Moody to whisk him off to Washington for the occasion.

In spite of his father's best matchmaking intentions, the date ended up a bust. What started less than auspiciously, with George W. entering through the gates of the White House for the first time in his life—behind the wheel of a purple AMC Gremlin, complete with Levi's denim seats, that he had borrowed from his parents—ended up with the president's eldest daughter asking to return home just after dinner. "Being a swashbuckling pilot, I had taken to drink," George W. recalled of the date much later. "I reached for some butter, knocked over a glass and watched in horror as a stain of red wine crept across the table. Then I fired up a cigarette prompting a polite suggestion from Tricia that I not smoke." When he returned to the dinner after escorting Ms. Nixon home, a friend of his father discreetly asked, "Get any?" "Not even close," he replied.

Congressman Bush presided over his son's graduation from flight school in December of the same year, serving as its keynote speaker and pinning his son's wings on the lapel of his National Guard dress uniform. A photo of the elder and younger Bushes grinning ear to ear on the occasion shows clearly a father's pride in his son. A few months later, the twenty-four-year-old congressman's son, back in Texas for five months of additional training on the F-102 fighter interceptor, was the subject in a National Guard antidrug public relations press release, which read in part:

> George Walker Bush is one of the members of the younger generation who doesn't get his kicks from pot or hashish or speed. Oh, he gets high alright, but not from narcotics. Bush is a second lieutenant attached to the 11th Combat Crew Training Squadron, 147th Commander Crew Training Group, Texas

Air National Guard at Houston . . . After a solo, a milestone in the career of any fighter pilot, Lt. Bush couldn't find enough words to adequately express the feeling of solo flights . . . Lt. Bush is the son of U.S. Representative George Bush, who is a candidate for the U.S. Senate seat of Senator Ralph Yarborough. The elder Bush was a Navy fighter pilot. Lt. Bush said his father was just as excited and enthusiastic about his solo flight as he was . . . As far as kicks are concerned, Lt. Bush gets his from the roaring afterburner of the F-102.

As the statement indicated, the aspiring George Herbert Walker Bush, after gaining reelection to the House of Representatives in 1968, had once again taken aim at a seat in the U.S. Senate, the prize that had eluded him in his race with Yarborough four years earlier. Bush was considered a rising star in Washington, gaining the notice of Richard Nixon, who, while holding Bush in judgment for his Ivy League pedigree, had nonetheless considered him as his running mate in 1968 before settling on Maryland's law-and-order governor Spiro Agnew. Nixon was in favor of Bush making a run for the Senate, despite putting at risk his seat in the lower house of Congress. A trip to the LBJ Ranch in Texas Hill Country to solicit the counsel of former president Lyndon Johnson confirmed what Bush likely already knew; "The difference between the Senate and the House is the difference between chicken salad and chicken shit," the thirty-sixth president put it bluntly.

It came with a risk. Bush would be forfeiting another term as a representative in what had become a safe congressional seat. There was no such security in a Senate run, but he was in. At the time, he expected a rematch with Ralph Yarborough, whose progressivism was becoming increasingly out of step with Texas's growing conservative bent. Instead he got Lloyd Bentsen, a successful forty-nine-year-old insurance and cattle magnate and former three-term U.S. representative whose affiliation with the Democratic Party belied conservatism more in keeping with the Texas electorate.

As with his congressional campaigns, George Bush's Senate race became a family affair, with members of the Bush brood doing what they could to pitch in. As always, politics was a bonding experience. "Being in politics either brings you together or sends you apart, and in our case, it brought us all together," Barbara said. "When the world was against my husband, it drew us closer together. We had something very big in common, which was we all were for George H. W. Bush." When not obligated by National Guard duty, George W. was often on the campaign trail with his father, just as he had in his father's first Senate foray, this time barnstorming across the state in a six-seat King Air. Six years older than he was during his father's first Senate run, George W. occasionally took a turn behind the microphone. Wearing the National Guard jacket that became a stock part of his wardrobe, he demonstrated his facility with campaigning in what amounted to a discernible imitation of the man he was promoting. Doug Hannah, a young campaign aide and Bush family friend, observed, "The funniest part of the whole thing was watching George try to be his father—and talk like him, and picking up the mannerisms. He ultimately did it perfectly. I don't know if it was conscious or not."

Despite an all-out effort, Bush's adopted state had once again rejected him. He found himself outpolled by Bentsen by 53.5 percent to his 46.6 percent. A high voter turnout, infused by Democrats in rural areas showing up at the polls to weigh in on a "liquor by the drink" proposition, was a deciding factor. "The loss hurt a lot more than 1964. That one could be explained by the Johnson landslide," George W. wrote later. "This one seemed like the death knell for George Bush's political career." The question arose as to what Bush's future would hold. The *Dallas Morning News* made it a headline: "What Will George Bush Do Next?" "The future—I don't know," he wrote a supporter, revealing to another, "We're torn between staying in politics in some way, or moving back to Houston and getting fairly immersed in business."

If the forty-six-year-old Bush's future was up in the air, so was the future of twenty-four-year-old George W., who had taken his father's defeat hard. "He was lost," Doug Hannah observed of his friend after the campaign. "He was getting fairly aloof, and part of it was the embarrassment of his father losing in the Senate race. It was the first time that he had kind of lost his anchor. He wasn't doing anything."

Always resilient, optimistic, and blessed with more than a little luck, George H. W. Bush would soon find his way back toward success. It would take his son a little longer to find his own way.

12

"LISTEN TO YOUR CONSCIENCE"

GEORGE H. W. BUSH DIDN'T think Richard Nixon owed him a thing. Though the president had implied that he would want Bush to remain in government regardless of the outcome of his Senate run, Bush wasn't counting on it. A sense of entitlement was anathema to him, just as it would be to his children. Besides, Nixon was ambivalent about him and he knew it. Though Bush's mind and loyalty impressed the president, he had reservations; Nixon, who put a premium on toughness, equated Bush's privileged upbringing and Ivy League diploma with softness and a lack of grit. But in an Oval Office meeting on December 9, less than a month after Bush's Senate loss, it was Nixon who would offer Bush a way forward by dangling before him several job possibilities: chairman of the Republican National Committee, a high-level White House staff job as a special assistant to the president, or U.S. ambassador to the United Nations. All of them offered Bush a political lifeline.

Bush ruled out the first job; he had no desire to head up the GOP. And while he found the notion of a top staff job in the Nixon White House intriguing, it was the UN post that most interested him. Nixon gave Bush the nod, nominating him for the position, which the Senate confirmed despite concern among several of its members over Bush's lack of foreign-policy experience. Once again, George and Barbara packed their bags, this time heading for Manhattan, where they set up house at the ambassador's resplendent official residence, apartment 42A in the Waldorf Astoria Towers—once the home of the imperious Douglas MacArthur—at Park Avenue and Fiftieth Street.

After settling into his new responsibilities, Bush judged the position as "broad, interesting, fantastic but . . . not as 'important' as some think because we have less policy input at this point than I would like to see it have." Part of it had to do with his status within the Nixon White House, where zealous insiders minimized Bush's influence. Bush found himself on the outside in 1971, when pursuing the most contested issue of his UN tenure, the body's recognition of Taiwan over the People's Republic of China as the legitimate representation of China. The matter prompted sharp debate and a vote in the General Assembly to determine if Taiwan should be expelled in order to offer the PRC a permanent seat. Bush pushed for preserving the status quo, in keeping with the U.S. policy of refraining from recognizing the communist PRC in favor of nationalist Taiwan. He vigorously lobbied other nations to do the same, little realizing that Nixon had dispatched his national security adviser, Henry Kissinger, on a secret diplomatic mission to Beijing, China's capital. Kissinger was to meet with Chinese leadership as a first step toward opening relations with China and disrupting the geopolitical balance with the Soviet Union. Bush wound up on the losing end of the vote, which came to 76 nations in favor versus 35 opposed.

Nonetheless, the post enhanced Bush's experience in the foreign-policy arena and expanded an already swollen Rolodex, as he and Barbara won friends by plying a distinct brand of personal diplomacy that would become a signature of the Bush clan. The ambassador would entertain often in the Waldorf apartment, or host his counterparts on outings that ranged from catching a Mets game at Shea Stadium to going for a round of golf or an afternoon of ice-skating. They took a Chinese delegation for lunch to the Bush home in Greenwich, where Dottie Bush wore black to make her plainly dressed guests feel at ease but accidently burned the Minute Rice she had prepared. When journalist Dick Schaap wrote a *New York*

Magazine article in which Ambassador Bush was among "The Ten Most Overrated New Yorkers," Bush embraced the dubious distinction by throwing a party for the additional nine on the list along with Schaap, who, he surmised, was "too humble to include himself." He also invited some UN counterparts as another chance to "put a human face on diplomacy."

Being back in New York, though, reaffirmed why he had left the area nearly a quarter century earlier to put a stake in the ground in Texas. As he wrote in his diary in 1971, "I am continually amazed at the arrogance of the intellectual elite in New York. They are so darn sure they are right on everything. It's unbelievable. Having lived in Texas for 23 years I have forgotten how concentrated this problem is, but it's sure there." But the proximity to his parents' home in Greenwich, Connecticut, forty miles or so north of the city, proved a blessing in September 1972, when Prescott Bush fell ill with lung cancer and complications around a prostate operation. After being hospitalized for a chronic cough at Sloan Kettering in New York, the same hospital where Robin Bush had died in 1953, the seventy-seven-year-old Bush passed on to the ages on October 8.

The Bush family gathered together at Greenwich's Christ Church for the funeral ceremony, where George W., Jeb, Neil, and Marvin served as pallbearers. As Barbara wrote, "The Senator dreaded the thought of his friends huffing and puffing under the burden of the coffin." Prescott Sheldon Bush was laid to rest in Putnam Cemetery, next to Robin, who had been buried there nearly two decades earlier. Among the tributes that poured in for the venerable former senator was one from Nixon. Bush responded with a letter thanking the president for the kind words he had offered. "My Dad was the real inspiration in my life," Bush wrote, "he was strong and strict, full of decency and integrity, but he was also kind, understanding and full of humor."

It was Bush's own reputation for integrity—and his loyalty to Nixon—that led to his next assignment. A month after his father's death, Bush was summoned to Camp David, where Nixon asked him to give up his post at the UN to become the chairman of the Republican National Committee. The president wanted a "good Nixon man—first." Bob Dole, the incumbent chairman apparently didn't fit the bill; Bush did. Not that Dole knew. When he visited Bush in New York later to inquire about his hypothetical interest in the position, it fell to Bush to gently tell Dole that he no longer had the chairmanship himself. Such was the dysfunction of the Nixon administration. Bush understood that the job as GOP head was a lateral move at best, but the president was asking . . . He accepted.

Integrity was not a hallmark of Nixon's White House as the strands of the Watergate scandal slowly, irrevocably unraveled. Five months earlier, on June 17, 1972, Washington police reported a simple bungled robbery of the Democratic National Committee offices in Washington's Watergate office building. The burglary resulted in the arrests of seven men, including one former FBI agent and two former White House aides who were working for the Committee to Re-elect the President, given the unfortunate acronym CREEP. Coverage of the story continued throughout the fall as the drip, drip of revelations implicated Nixon White House aides. But they stopped short of Nixon, who went on strong to win reelection in a rout against his Democratic challenger, South Dakota senator George McGovern, by garnering 61 percent of the popular vote.

Throughout Bush's thankless nineteen months in the job, he diligently traveled the country preaching the Republican gospel at key meetings and fundraisers, defending Nixon, whose credibility continued to erode and with it financial contributions to the party. The president, after all, had told Bush—and the country—that he was innocent of any wrongdoing in Watergate. Bush took him at his word.

In the fall of 1973, the White House and the Republican Party took another hit when Nixon's vice president, Spiro Agnew, was implicated in a bribery scandal dating back to his days as governor of Maryland. Agnew resigned in the face of mounting evidence, leaving Nixon to reluctantly appoint House Minority Leader Gerald R. Ford as vice president in October 1973, the first vice president to be appointed under the provisions of the Constitution's Twenty-Fifth Amendment. In addition to the continued festering of Watergate, further investigative scrutiny of the Nixon White House showed a web of deceit and trickery beyond the scandal at hand.

On July 23, as Watergate's cancerous tentacles continued to spread throughout the Nixon White House, Bush penned a long letter to his four sons—"Lads" he called them—elucidating his support of Nixon, despite the growing sentiment against him, including the clamor around the president's possible impeachment. "Dad helped inculcate into us a sense of public service [and] I'd like you boys to save some time in your lives for cranking something back in," he wrote. "It occurred to me your own idealism might be diminished if you felt your Dad condoned the excesses of men you knew to have been his friends or associates." After enumerating the strengths and weaknesses of Nixon as he saw them, he affirmed his belief that Nixon should be censured by the Senate but not impeached, stating, "I will never feel the same around the President after all this, but I hope he survives and finishes his term. I think it's best for the country in the long run." He also used the situation to impart a lesson to his boys: "Listen to your conscience," he urged.

Don't be afraid not to join the mob—if you feel inside it's wrong.

In judging your President, give him credit for enormous achievements . . . but understand too that the power accompanied by arrogance is very dangerous. It's particularly dangerous when men with no real experience have it—for they can abuse our great institutions.

Avoid self-righteously turning on a friend, but have your

friendship mean enough that you would be willing to share with your friend your judgment.

Don't assign away your judgment to achieve power.

George W. wrote later of his father's counsel, "He couldn't have realized it at the time, but his words set a standard that both Jeb and I would strive to follow when we held public office."

The day after Bush wrote the letter to his sons, the Supreme Court ruled unanimously that Nixon must turn over tapes to federal prosecutors, something Bush urged Nixon to do all along as the quickest route to absolution. Instead, the tapes damned Nixon. His house of cards fell in on August 5, as a "smoking gun" recording came to light revealing Nixon's complicity in covering up the Watergate break-in by ordering Bob Haldeman to block an FBI investigation six days after the burglary occurred. Already "battered and disillusioned" to that point, Nixon's betrayal devastated Bush. Though Nixon was only eleven years older than he, Nixon was a father figure who had taken a personal interest in his career. As Barbara explained it, "Both of our fathers were the most honest, decent people. [They] never lied to us. So, Nixon [lying] was a huge shock." Fed up, Bush told Nixon's chief of staff, Alexander Haig, "the whole goddamned thing has come undone and there [is] no way it can be resolved," and he had no compunctions about sharing his judgment with Nixon.

On August 6, Nixon convened his cabinet. Bush recalled the atmosphere hanging over the Cabinet Room was as "one of unreality." Refusing to acknowledge the scourge on his presidency, Nixon instead wished to address "the most important issue confronting the nation, and confronting the world, too: inflation." The men in the room were stunned.

Nixon continued to talk about the issue before suddenly, awkwardly shifting to Watergate, and maintaining there had been "no intentional breach of the law" and "no obstruction of justice." After a pained silence, Gerald Ford spoke up, suggesting that he would not have supported the president if he had known of the substance of the evidence against him. "I'm sure there will be impeachment in the House," he said. "I can't predict the Senate." He then said it would be his last word on the subject since he was "a party of interest." After Nixon tried to get back to economic matters, including the prospect of an economic summit between the White House and the Congress, his attorney general, William Saxbe, rejected the notion; "We ought to wait to see if you have the ability to govern," he said. As the momentum of the meeting shifted, Bush weighed in: Saxbe was right, he asserted, Watergate was a drag on the economy. If Nixon was intent on fighting for his presidency, Bush said, he should do so "expeditiously," so the country could move on.

The following day, Bush wrote the president a letter affirming his position. "It is my considered judgment that you should now resign . . ." he asserted. "Until this moment resignation has been no answer at all, but given the impact of the latest development, and it will be a lasting one, I now firmly feel resignation is best for this country, best for this President. I believe this view is held by most Republican leaders across the country." Nixon offered no response.

But a day later, on August 8, Richard Milhous Nixon became the first American president to resign the office. Bush was among those gathered on South Lawn the next morning to bid farewell to the fallen president. Nixon charged up the stairs to Marine One, incongruously flashed his trademark victory sign along with a forced smile before entering, and he was gone. The presidential helicopter ascended skyward, bound for Andrews Air Force Base, where Nixon would soon be on his way toward his home in Southern California, where an uncertain future awaited him. "I'm glad dad's not alive," Bush said of Nixon's resignation. "It would have killed him to see this happen. He thought that we were the party of virtue and all the bosses were Democrats." A little over an hour after Nixon's departure, Gerald Ford was sworn in as the thirty-eighth president. He took the oath of office in the White House East Room before a small audience—

Bush among them—who watched hopefully as the new president then offered not an inaugural address but "a little straight talk among friends." "My fellow Americans, our long national nightmare is over," he said, extending a promise of sorts. "Our Constitution works. Our great republic is a government of laws and not of men. Here the people rule."

Just as Prescott had in the 1950s, when he censured Joe McCarthy in an effort to save the Republican Party from drifting toward extremism, Bush had stepped in to help stop the bleeding, sparing his party—and America—further damage at the hands of a man who he dolefully concluded was amoral. As the nightmare of Watergate ebbed, Ford offered the nation a new beginning—just as he would, soon enough, to George H. W. Bush.

13
FAMILY VALUES

TO A LARGE EXTENT, THE Bush family fit the mold. The most distinctive members of the great American political families are generally characterized by an abiding sense of responsibility to live up to their heritage, to which they are bound by family ethos and tradition. They share pride and an air of confidence that comes from pervasive achievement throughout the bloodline and a fierce competitiveness spurred, in part, by their need to show their worthiness within it. They also have in common a belief in American exceptionalism and their own brimful confidence that they can further the nation's cause. They are individuals on a quest toward realizing their own rampant ambitions and making their own marks, but they recognize that they are a part of something greater.

It wasn't especially surprising to John Adams that his son John Quincy realized the presidency in his wake; he had bred his eldest boy for public service, prodding his character and forcing his education by enlisting him as a special envoy and shipping him off on foreign missions, first as his apprentice, then as a personal secretary to other American diplomats. The elder Adams instilled in his son a responsibility as an Adams, and John Quincy was meant to fill it. When John Quincy failed to meet his father's towering expectations, his father demanded reproachfully, "Can't you keep a steadier hand?"

Joe Kennedy had every expectation that he would make his oldest son, Joe Jr., the first Catholic president, and Joe Jr. accepted the charge. When Joe Jr. died in an explosion on board the plane he was piloting on a dangerous World War II bombing mission—another chance to prove himself—Joe Sr.'s plans had "all gone to smash." As he lamented to a friend, "You know how much I had tied up my whole life to his and what great things I saw in the future for him." The task then fell to Joe Jr.'s younger brother and rival, Jack, whose life had to that point been defined by his competition with Joe Jr., to fulfill the family destiny. "You know my father wanted his eldest son in politics," Jack Kennedy told a reporter. "'Wanted' isn't the right word—he *demanded* it." Later, when he was running for the presidency in 1960, Jack explained to a biographer, "For the Kennedys, it's either the outhouse or the castle, nothing in between."

But there was no contrived sense of destiny or paternal-driven agendas for the children of Pres and Dottie Bush or George and Barbara Bush. George's choice to follow Pres's path into politics was his alone. "My father was totally honest, totally committed to public service, per se. And that inspired me. So, if there's some parallel there, fine," he said. "We've never given anything to these myths of families that try to put a lot of propaganda about themselves . . . 'I gotta do this because my dad did, and I gotta do this because of my granddad.' That's not what motivates us." As Barbara explained of her children, "I never thought any of them would be president, and I don't think [George] did either. He just felt like I did, if they went to college and graduated they'd be a success." George W. had no sense of what his father's ambitions were for him at any point in his life. "There was no game plan," he said. "There wasn't some family council sitting around saying this is the way it's got to be." Any expectations George W. may have had by family obligation were inferred. The same was true of the other Bush children.

Yet there was something about the old man, a moral force eliciting a palpable reverence that the Bush kids all had for him, just as there had been between him and his own father. Even the thought of George H. W. Bush could bring the most phlegmatic of them to tears. "All five of them wanted George's approval," Barbara said in 2015, well after her children were on the far edge of middle age, "and they still do." He was, by the definition of George W. and Jeb, a "beacon" whose example would always be front and center, the family's Atticus Finch.

There were, in essence, two phases of the Bush family. The first was the early years, the salad days in which George and Barbara struck out from Yale into West Texas. Those years, in large measure, belonged to George W. as the dominant child in the household. The second phase was the latter years after George had struck it rich as an oil entrepreneur and the family moved to greener pastures in Houston, where George W. was soon sent to Andover, and George hurled himself into the rough-and-tumble of politics, eventually establishing a home in Washington. While George W. would come back home, igniting a spark in the household through his sheer magnetic presence, he thought of himself as, more or less, on his own. The latter years of the Bush family belonged to the balance of the Bush children: Jeb, Neil, Marvin, and Doro, all of whom were establishing their own identities as their older brother was out in the world.

Still, they were all "very close . . . some would say dysfunctionally close," Neil observed, bonded by family ethos, which George characterized succinctly as "love, loyalty, faith, friends." Friends were certainly an important part of it; the vast network of intimates the Bushes had in their orbit would come into play personally, politically, and professionally throughout their lives. But blood would always be thicker than water. Their greatest love, their greatest loyalty, was reserved for each other.

Doro, soft-spoken, diffident, and deferential to her protective older brothers, would go to the all-girls Miss Porter's School in Farmington, Connecticut, where her paternal grandmother had gone before her, then on to Boston College. Marvin, after attending Woodberry Forest School in Virginia, remained in the state to graduate from the

University of Virginia. "If they had had IQ tests when the kids were growing up, Marvin would have been the smartest," Barbara said of him in 2015. "He's not a big lover of politics, and he's not a lover of publicity, and he wouldn't like me saying that—but the truth is Marvin was the brightest." Neil, agreeable and ebullient, was given the family nickname "Mr. Perfect" by George W. Overcoming dyslexia in his teenage years, Neil went to St. Albans in Washington, D.C., then on to Tulane University.

Jeb, the most serious of the brood, grew up thinking of himself less as George W.'s younger brother—he was only nine when George W. left for Andover—than being the older brother of his three younger siblings. "My memories of my childhood are more of being the older brother of Neil and Marvin and Doro because George went off to school during those formative years," he related. In this manner, George W. and Jeb both held unspoken claim to being the de facto leader of the next generation, but Jeb's earnestness and early success made him the family's most likely to succeed.

Still, there was no outward rivalry between the two. All the Bush children were competitive; it was in the Bush DNA, or at least bred in them at birth, just as it had been for previous generations of the Bush-Walker clan. As in the household their father grew up in, where Dottie Bush rode herd over her five children, the home of George and Barbara Bush meant an endless series of competitions. The family instituted a system called "the rankings," in which one could only advance in a competition if they were next in line in the pecking order. Jim Baker, whom George partnered with as doubles tennis champions two years running at Houston Country Club and later dragged into Republican politics, recalled annual touch football games at the Bush home called the "Turkey Bowl," where passions ran high and George W. would "always be ready to give you the elbow." But the six years separating George W. and Jeb—an eternity in childhood years—meant that they weren't peers.

It was their father who was the immediate point of reference, anyway, not each other, and both handled any self-imposed pressure they may have felt around measuring up to him in different ways. If George W. had honored his dad through emulation, following in his footsteps and inviting direct comparisons, Jeb took a different direction entirely. Realizing early on that assessing himself by his father's standard would set himself up for failure, he, as he put it, "grinded myself to a place mentally to take it off the list of things to worry about." He set his own aspirations accordingly. "I actually lowered my expectations of my own goal in life to be a good husband and father to maybe half of what he was. It sounds strange but it made my life pretty easy in terms of not constantly worrying about this larger-than-life figure," he said.

Yet, inadvertently, his early years looked a lot like those of George H. W. Bush—far more so than his older brother's—at least after unpromising, unhappy sophomore and junior years at Andover. In his senior year, Jeb stepped up academically, adding his name to the honor roll by the end of his senior year while becoming captain of the tennis team. The catalyst for his change in effort and attitude was Columba Gallo, the diminutive sixteen-year-old girl he met in Ibarrilla, a small village in central Mexico, where he went with nine other Andover students on a ten-week high school exchange trip in the winter of 1971 when he was seventeen. Upon meeting her it was as though he had been struck by lightning, he recalled later. Love soon blossomed, despite Columba's reservations about her suitor. "She thought he was the son of a very rich man, and she thought that was all he was good for," Barbara recalled. Jeb returned to Andover determined to show Columba otherwise, applying himself while writing her nearly every day and visiting her in Ibarrilla biannually.

Barbara harbored her own reservations. "How I worry about Jeb and Columba. Does she love him?" she wrote in her diary when the two were engaged two years later. "I know when I meet her, I'll stop worrying." Eventually, she did. "It took me a while to realize this, but Columba honestly changed his life," she conceded much later.

After graduating from Andover, Jeb opted to forgo the family tradition of going on to Yale and instead attended the University of Texas at Austin, where he majored in Latin American studies and played varsity tennis. He sailed through college in two and a half years, graduating Phi Beta Kappa and magna cum laude. Afterward, he was "blowin' and goin'," throwing himself into a career in banking in an entry-level spot at Texas Commerce Bank in 1974. He married Columba the same year in a small ceremony in Austin with immediate family in tow, all meeting her for the first time. Three years later, he and Columba moved to Caracas, Venezuela, where he opened a new branch for the bank, serving as its manager and achieving the rank of vice president.

By the time he was thirty-one, he was a father of three and had embarked on a lucrative career in commercial real estate development in the Miami area, where he and Columba had settled in 1980. Just like his father, he had graduated from college with distinction in under three years, eager to get out into the world, married his first love in his early twenties and had a succession of children quickly thereafter, and eventually settled in a new state, unchartered territory for the Bush family, where he continued to pursue a successful, ultimately lucrative career. But any symmetry with his father's early path was purely chance. "It's not that I wasn't aspirational, but to try to measure myself against a near-perfect person would have been really hard. I would have felt inadequate," he explained.

It wasn't that easy for George W.—the firstborn, the namesake. As one family member said, "When they were growing up, everything fell on George. He felt the old man's disappointment the most. The others almost felt, 'That's W.'s job. We don't have to measure up. We're just living our lives here.'" Like George W., Jeb was aware of the scrutiny that comes with being the son of a man of great accomplishment, and the toll it can take. "You look at second-generation kids of successful people, it's not a great record; it's pretty ugly," he said. "It's not something I fretted about. I don't think George was all-consumed with it either, but I can imagine it being in the back of his mind and part of his drive, reacting to this and trying to figure out where he was going."

Where George W. Bush was going in his early and middle twenties was not clear. Much later he conceded, "I'm not good at psychoanalysis, but I suspect that, yeah, there was a certain sort of expectation. What's interesting is that my life somewhat paralleled [Dad's]—and on the other hand, I was able to, well, I charted my own course." It would take a while to see where his course was leading. As his father furthered his career on the East Coast, George W., having earned his diploma from Yale and his wings from the National Guard, was back in Texas faced with starting his own—and, at least on the surface, he didn't seem much concerned about it.

14

HORIZONS

WHILE GEORGE W. BUSH HAD eschewed much of the Aquarius zeitgeist that swept up many of the privileged youth of the 1960s and early '70s, he gave himself permission in his twenties to find himself, an indulgence not generally afforded the previous generation, who came back from the war, married, and went off to work, just as his father had. "The events of our life are somewhat affected by the cultures in which we live," he reflected. "And I didn't want to be rooted, didn't want any possessions. I kind of just moved around for a while." With no career path in mind, he drifted, if not aimlessly, then restlessly, peripatetically. In five years, he would hold three jobs and live in seven apartments in three different states. He later called them his "nomadic years."

His primary focus was on having a good time. Back in Houston after his thirteen-month flight training in Georgia, he fulfilled his two-year active duty commitment in June 1970, continuing his Guard duty on weekends piloting F-102 jet fighters with members of the 147th at Ellington Air Force Base. He moved into the Chateaux Dijon garden apartments in the southwest part of town, where he was surrounded by young, upwardly mobile tenants who made up most of the Chateaux's three hundred–some units. Bush raised hell with the best of them, leading a life of beer, blender drinks, and barbeques, often around one of the complexes' six pools, a procession of attractive dates on his arm. With his Triumph convertible, military buzz cut, and National Guard flight jacket, young George W. Bush exuded a classic top-gun cool, in contrast with the long hair and psychedelic bell-bottomed fashion trends of the day. "I was a badass back then," he later recalled to an aide. Don Ensenat, a Yale classmate and later one of Bush's Houston roommates, recalled, "He had a couple of girls that were more than one date, but nothing looked like a serious romance. Dates and the opposite sex were always high on the agenda . . . But we didn't do anything anyone else in their twenties didn't do."

George W.'s "active lifestyle," as Jeb put it, contrasted with his father's decidedly chaste view of the world during his young adulthood, another reflection that the times had changed. In a letter he dashed off to his mother during the war, George wrote that he and his older brother, Pressy, "share a view which few others, very few others even in Greenwich share. That's regarding sexual intercourse." He continued, "I would hate to find that my wife had known some other man, and it seems to me only fair to her that she be able to expect the same standards from me. Daddy has never discussed such things with us—of this I am very glad. But we have learned as the years went on by his character what is right and what is wrong." He signed the missive, "Pop professor 'sexology' Ph.D." But, like his own father, George didn't later foist his "sexology" views on his son. "Dad was shy. We never had 'the talk,'" George W. said. "He never told me to wear a 'raincoat' or anything."

It would be years before George W. would find a woman with whom he would settle down, as his father had with Barbara while still in his teens. When he did, at age thirty-one, he would discover that the object of his affection, Laura Welch, had lived at Chateaux Dijon—albeit on the opposite, more "sedate" side of the complex— at the same time as he had while she pursued a career as a librarian at a local public school.

George W.'s own career was not high on his list of priorities. Like his romances of the period, he flirted with a number of possibilities, but nothing quite stuck. In the fall of 1971, at the age of twenty-five, he was already considering entering the political arena, contemplating a run for a state Senate seat in Harris County against Democratic opponent, Jack C. Ogg. He dropped the idea early in January 1972, after talking it over with his father. "We hope he'll feel settled," Barbara recorded afterward in her diary.

He was hardly settled. He applied to law school at the University of Texas only to get rejected. Upon looking into the matter, Barbara discovered his grades were better than those of many accepted applicants. Concluding that his rejection may have been "a political thing," she and George offered to "do something about it." George W. declined. "No," he told his parents, "I never really wanted to be a lawyer, I just sort of thought I had to do something."

His great-uncle Herb Walker secured a job for him as a management trainee for Stratford of Texas, a Houston-based agribusiness company, but it didn't stir his passion, either. He left his position after just nine months, dismissing it as a "dull coat and tie job." An unpaid job working for an antipoverty program for troubled teens in Houston's largely African American Third Ward called the Professional United Leadership League—PULL for short—was a better fit. Arranged by his father through a connection with its executive director, "Big John" White, a former tight end for the Houston Oilers, the post satisfied Bush's desire to do "some good" for a cause he cared about. And given its philanthropic nature, it was an accepted family alternative to starting down a path toward wealth and success. At least it would buy him some time till he figured out a logical career path. Despite a background that contrasted dramatically with those of his young at-risk charges, Bush fit in as well as he could. "He was the first white boy the kids really loved," recalled one coworker. Bush said later that it was "an eye-opening experience," though his mother recalled that while he found it "very rewarding," he also found his nine-month stint "very frustrating" due to the high rate of recidivism among the teens with whom he worked.

When asked in 2013 if he ever worried about his son's lack of direction during those early years, George H. W. Bush replied simply, "No." An entry in his diary in May 1971 reflects his faith in him. While the elder Bush was immersed in a new life in New York as UN ambassador, he expressed concern for Jeb, newly enamored with Columba after their initial meeting, and his belief that George W., with whom Jeb would be living at W.'s Houston apartment during his summer break, would be "a good influence" on him:

> I worry about the family situation. The boys are in good shape. Jebby is going to need some help I am sure. He is a free and independent spirit and I don't want him to get totally out of touch with family. He doesn't want to be . . . Jeb will be staying with George and this is good. George will be a good influence on him.

Still, there was an edge to George W., "Bombastic Bushkin," as his friends would eventually call him, which could be unsettling. When he impertinently asked the wife of Gerry Bemiss at a Kennebunkport dinner party, "What's sex like after fifty?" that was one thing. It was in keeping with his flippant nature, a waggishness that had marked him since his boyhood. But another incident, legendary in Bush lore, suggests an uneasiness and disquiet in George W. during the "nomadic years." During a Christmastime visit to his parents' Washington home in 1973, George W., twenty-seven years of age, paid a visit to Jimmy Allison, a longtime friend and Bush political aide, as his kid brother Marvin, ten years his junior, tagged along. A round of tennis was followed by several rounds of drinks. Afterward, George W. drove home the worse for the alcohol. "I got drunk, and I was a free-spirited lad," he recalled, "and I ran over the neighbor's garbage can." The aluminum can lodged itself cacophonously under the front wheel of the car as he approached his parents' home. As he and his inebriated younger brother emerged from the car, Barbara castigated George W. for his "disgraceful behavior" and informed him that his father, who was upstairs reading, wanted to see him. George W. stormed upstairs truculently inviting a confrontation. "You wanted to see me!" he demanded. Some accounts of the incident have George W. going further, challenging his father. "You want to go mano a mano, right here?" he is alleged to have demanded beligerently, though Barbara maintained later, "He never said that. I don't remember him threatening to hit him or, 'Do you want to fight?'" Neither George nor George W. remembered it going that far, either.

"Mano a mano" or not, it was, as Jeb later put it, a "holy shit" moment. None of the Bush brood had ever approached their father like that. "Instead of saying the obvious, which was 'Holy cow! I've made a huge mistake; I'm really sorry,' which was the smart thing to do, he got defiant with my dad," Jeb remembered. "That's the first time I'd ever seen anybody do that. I was like, 'I may just step back and see what happens with this one.'"

Not much, as it turned out. George W.'s drunken bluster was met anticlimactically with his father's dispassion. The elder Bush merely looked up from his book, removed his reading glasses, and glanced disapprovingly at him, before putting the glasses back on and resuming his reading. "He didn't have to tell me he was disappointed," George W. said, it was all there in the look on his face. Defused, George W. "slinked out of the room," thinking *Oh, God, I made a complete ass out of myself.*

The incident may have reflected George W.'s pent up frustration that he wasn't quite living up to his father's standing. Jim Baker described him during the period as "a juvenile delinquent, damn near," but saw George W.'s pugnacity during his twenties as a rite of passage. "I don't know if he was trying to prove something or not," he said, adding, "I have four sons and every damn one of them at some point had to separate himself from his father, had to prove that he was his own man, had to show that he could do the stuff himself." George W. believed that the episode created "a myth" that his defiance of the moment defined his relationship with his father, maintaining instead that the fact that the confrontation didn't escalate said as much about his feelings toward his father as his initial belligerence. If George W. was resentful toward him or trying to prove something, it was fleeting. What resounded was his shame at disappointing his father and tacit resolution to do better down the line. "That was better than caning him," Barbara observed. "He was so mad that he had let his father down."

Even more so, the incident owed to Bush's drinking. "I don't think George [W.] could drink," a family friend said. Alcohol made him a little edgier, more acerbic. His uncle Prescott, his father's eldest brother, said of George W. during the period, "he was becoming a real boozer." His parents, along with other members of the Bush and Walker families, were active social drinkers, nothing more, but their son's routine partying, in which alcohol was a staple, largely escaped their attention.

Tellingly, the elder Bush never made mention of the trash can incident the next day. It was no different from the time when George W. was a boy and his father had discovered that toy soldiers he was playing with in their dusty Midland backyard had been stolen. George drove his son to the store where they had been lifted, where George W. walked in alone and returned the loot to the manager along with an apology. Afterward, the episode was forgotten. No estrangement or disaffection ensued between father and son; both moved on. George's faith in his son remained intact; so did George W.'s respect for his father. That was the way it worked. If his son was off the path, George would gently, unobtrusively help him find his way.

Though no transgression preceded it, George had helped him forge a new path a year earlier when George W. had unexpectedly been admitted to Harvard Business School after a friend from Yale suggested that he apply. George W. had filled out the application, including writing an essay on the jets he flew in the National Guard as his "biggest accomplishment," and mailed it off without giving it much more thought, much as he had to the University of Texas School of Law a couple of years earlier. His family knew nothing about it.

Not that he had any desire to go back to the Northeast. Seven years—three at Andover and four at Yale—had been enough. Nor was he determined to beat a path to the business world. "I wasn't sure what I wanted to do, truly," he said later. But actually getting into Harvard Business School at a time when he didn't have a lot to show for his fledgling career might prove something—to himself, to his parents, and to his friends, at least one of whom had applied to Harvard Business School and been rejected. Yet when he was unexpectedly accepted, it was not he who told his father but Jeb. The subject came up when the three of them had dinner, and the elder Bush asked George W. if he was going to enroll.

"No, I'm not really sure I want to go," his son replied.

"You really ought to think about it," his father said. "It'll broaden your horizons."

George W. appreciated his father's soft touch—not "You must go to Harvard Business School," but "Broaden your horizons," the tried-and-true Bush family maxim. At the time, as he explained it, "I had no possessions and really didn't want any. I was kind of floating around, and I had a variety of jobs . . . [Dad] really didn't interfere, but I'm pretty confident that he was perplexed at what I was doing." The idea wasn't something George W. came to immediately, but his father had "planted a seed" by, in essence, saying "don't stagnate," setting "an expectation to live life to the fullest."

John White, his boss at PULL, the antipoverty program in Houston's Fifth Ward where Bush was working, added his own two cents. "You know something, George, there's no money in poverty," White said. "Why don't you go up there and make a whole bunch of money and then come back here and help the folks in our program." Eventually, Bush warmed to the notion. In the fall, after fulfilling his commitment for the National Guard, he headed once again, improbably, for New England.

While George H. W. Bush hadn't preceded him at Harvard Business School (HBS) as he had at Andover and Yale, there were few who didn't know who the father of George W. Bush was, though, as always, it wasn't something George W. went out of his way to advertise, even to friends. It could well have made the difference in his acceptance in the first place. Despite Harvard's outward egalitarianism, the son of a prominent Washington insider would only add to its prestige, and it likely didn't escape the notice of the HBS admissions office. The students recognized it, too. "At a place like Harvard Business School, you always [knew] who the sons and daughters— but mostly the sons—of famous people were," said a fellow classmate. "And then there were the rest of us." The student body was no less liberal and steeped in intellectual snobbery than it had been at Yale. Under 8 percent of Cambridge's fifty thousand voters were registered Republicans at a time when George W.'s father was the head of the party, which was reeling from the slowly unwinding Watergate scandal and the unrelated resignation of Spiro Agnew, Nixon's corrupt vice president. Anti–Vietnam War protests had turned to those around a Nixon impeachment, though less thunderous.

George W. didn't exactly look the part of the Ivy League–educated son of an establishment Republican. On the contrary, his regular garb—blue jeans, cowboy boots, and National Guard flight jacket—along with the wad of Copenhagen chewing tobacco lodged in his cheek sent an *anti*-antiestablishment message. So did his favorite watering hole, the Hillbilly Ranch, the only local bar where authentic country music was a staple. Like the hit Barbara Mandrell song that would come at the beginning of the next decade, Bush was "country when country wasn't cool," and he wore it proudly, just as he did his last name.

Even if Cambridge was not an indigenous environment for Bush, he fit in easily, whether in the classroom, barroom, or playing intramural basketball or baseball. He formed lasting friendships and was often the one his peers looked to for leadership, despite not necessarily being seen as the smartest in the room. "He was not a star academic performer here," noted Michael Porter, an HBS professor who would later become an economic adviser to his former student during the latter's 2000 presidential bid, "but he was very good at getting along with people and getting things done." One of his classmates, Mitch Kurz, observed, "George's leadership was not based on raw brain power. He had an intangible quality about him, that there wasn't a problem that couldn't be solved—not necessarily by him, but somehow."

Bush considered his two-year experience at HBS a "turning point," where he gleaned "the tools and the vocabulary of the business world." It also gave him the confidence to know that he wanted to be his own boss—and that he could do it while circumventing the traditional corporate route. Upon earning his master's degree in business administration in the spring of 1975—becoming, in time, the first MBA president—he resisted the tide that pulled many of his fellow graduates toward a career on Wall Street. A spring-break road trip to Arizona to visit a friend a couple of months earlier included a stopover in Midland to visit Jimmy Allison, who had moved to West Texas to get into the oil business. Afterward, Bush was convinced that his immediate future lay in the Permian Basin. Midland, with a population of over sixty thousand, had grown nearly threefold since his boyhood, and he, like Allison, drew inspiration from the "energy and entrepreneurship of the oil patch," which continued to boom as the prices of oil had skyrocketed in the wake of the Middle East oil embargo perpetrated by the Organization of Petroleum Exporting Countries—OPEC. Bush "could smell something happening," and he wanted to be a part of it.

Twenty-seven years after George H. W. Bush had journeyed from Kennebunkport to Odessa in his new 1947 red Studebaker to begin his career in the oil business, George W. would climb in a blue

1970 Cutlass and drive from Cambridge to Midland toward similar aspirations. But unlike his father, whose destination represented a new frontier, George W. was going home. Midland was in his blood, capturing the essence of who he had become as well as where he had come of age while his father built his career and, eventually, amassed riches. It was where business was still done on a handshake and where the Bush name would open doors, along with the $20,000 in seed money he had cobbled together from Bush-Walker relatives who had placed their bets on him. Going back was an easy decision. "Cambridge, Massachusetts is a very heavy environment compared to Midland," he wrote later, though he might as well have been referring not just to Cambridge but the Northeast at large. "People there don't realize that horizons can be broadened. There is no growth potential there." In Midland, on the other hand, he wrote, "your horizons suddenly expand."

But before heading back to West Texas, he, along with Neil, Marvin, and Doro, set off on a six-week summer adventure halfway across the world where their father had accepted a mission that would broaden his own horizons: the People's Republic of China.

15

GO EAST

GERALD FORD HAD A CHOICE to make. The first man to ascend to the office not by a national vote but by the provisions of the Twenty-Fifth Amendment to the Constitution, which allows for the presidential appointment of a vice president when the post is unoccupied, Ford had a vice presidential vacancy of his own to fill. During the first week of his presidency in the dog days of August 1974, as the dark cloud of Watergate began to dissipate over a weary land, the new president turned his attention toward whom he should tap as his second in command. Bryce Harlow, an adviser to Ford, was tasked with compiling a list of potential candidates evaluated based on national stature, executive experience, and an ability to broaden Ford's base of political support. A roster of sixteen was narrowed to five, each one quantified with a total score measuring the candidate's standing against the selection criteria, and submitted to Ford. The first name to appear, with a corresponding score of 42, was that of George H. W. Bush.

On August 11, Ford's second full day in office, Bush himself made a pitch for the job in a half-hour Oval Office meeting with the new president. Ford asked Bush about other possibilities including Republican heavyweights Nelson Rockefeller, the moderate former New York governor, who ranked fifth on Harlow's list with a score of 35, and Barry Goldwater, the conservative stalwart. Bush expressed worry about the backlash Rockefeller would cause with the party's right wing, which had stridently thrown Rockefeller over for Barry

Goldwater as the party's presidential nominee a decade earlier, and about Goldwater as anathema for moderates. By implication, Bush was the safe choice, firmly in the middle. Harlow agreed, contending in his assessment for Ford that for the sake of party harmony, "plainly [the choice] should be Bush," but added as a downside that it might be "construed as a partisan act, foretelling a Presidential hesitancy to move boldly in the face of known controversy." Ford culled his own short list to three final candidates: Bush, Rockefeller, and Donald Rumsfeld, the brash forty-two-year-old NATO ambassador and former Illinois congressman who, like Bush, was considered a comer in the party.

Among those who wondered at Bush's prospects was George W., who was in Fairbanks, Alaska, working for Alaska International Air during a summer break from Harvard Business School. He called his father after learning in a local newspaper that he was on Ford's vice presidential short list. "Well, there are some who think I could do a good job, but I wouldn't make too much out of it," Bush told W., perhaps managing his own expectations as well as his son's.

Indeed, there were drawbacks for Bush. Ford's advisers regarded the fifty-year-old former congressman and former UN ambassador as "not yet ready to handle the rough challenges of the Oval Office," a view Ford felt was unfair. More damning was a story circulating that Bush's 1970 Senate campaign had received roughly $100,000 from an illegal secret slush fund controlled by the Nixon White House called the "Townhouse Operation." Bush, who immediately made Ford aware of the revelation as a potential liability, would eventually be cleared from any wrongdoing in the matter, but the association with the disgraced Richard Nixon didn't help his cause.

On August 20, just before announcing his decision, Ford called Bush who had retreated to the late-summer tranquility of Kennebunkport. Nelson Rockefeller, the president told him, would be his nominee. As Ford explained to Dick Cheney, Rockefeller was his choice because he was "nationally and internationally known," a counterbalance to Ford himself, who only months earlier had been a congressman from Michigan's Fifth District with little name recognition around the world. When asked for his reaction to the breaking news by a local television crew that had reached the front porch of his Kennebunkport home, Bush's first response was raw. "It hurts so much," he said.

The vice presidency now spoken for, Ford was open to hearing Bush's thoughts on other job possibilities later the same month. Bush inquired about becoming secretary of commerce or White House chief of staff. Ford was lukewarm on both. Instead, he volunteered the two most prized diplomatic posts—the ambassadorships of either Great Britain or France—and a third: chief of the United States Liaison Office in China, with which the U.S. had not yet established full diplomatic relations in the wake of Nixon's historic trip toward rapprochement two years earlier. Each opportunity was consistent with Bush's future ambitions. As he wrote in his diary, "I indicated that way down the line, maybe 1980, if I stayed involved in foreign affairs, I conceivably could qualify for Secretary of State. The President seemed to agree." After consulting with his own secretary of state, Henry Kissinger, Ford told Bush he believed China was his best option. Bush agreed to take the post, likening his decision afterward with that of his going west to Odessa in 1948. "An important, coveted position like London or Paris would be good for the resume," he wrote later, "but Beijing was a challenge, a journey into the unknown."

The journey began late the following month. "Sure the place is different, but that's what I wanted," Bush told the *Washington Post* before he and Barbara left in late September for Beijing, then called Peking. Different it was. The intensity and demands of Washington half a world away, Bush found himself, as he explained in a letter to his children a month after his arrival at the Liaison Office, "cut off from the day to day news," keeping tabs on the western world through a short-wave radio that hissed and wheezed like a Depression-era radio set. He was, he related, enjoying a "pace in my life less hectic than I have known in many, many years," adding for emphasis that his phone hadn't rung in over a week. "The difference between our two countries is immense—and yet there's a feeling that the people would like to be friends," he observed. It represented an opening for Bush, for whom personal diplomacy was both a strong suit and an effective tactic, he had found, toward strengthening national ties.

Friendship with the Chinese was not of import to Henry Kissinger, who believed that power trumped ideological commonality and amity in advancing America's position in the world. "It doesn't matter whether they like you or not," Bush was told by Kissinger, who would travel to China in two high-profile visits with the nation's leader and storied founder and chairman, Mao Tse-tung, and Deng Xiaoping, the first vice premier, during Bush's tenure in China. As he did during his time at the UN, Bush found himself an outsider with the State Department. Though respectful of Bush, Kissinger, intellectually brilliant but secretive and imperious, kept him at a distance. When the last ignominious gasp of the Vietnam War came on April 30, 1975, with the fall of Saigon, which had been seized by North Vietnamese forces, Bush was kept out of the loop. He learned the news at a national day reception for the Netherlands in Beijing, which was illuminated by a triumphal fireworks display launched from the North Vietnamese embassy later that same evening.

As always, the Bushes approached the assignment as a shared adventure. Embracing the culture around them, they took Chinese lessons (though proficiency with the language eluded them) and spent considerable hours absorbing their surroundings. Often they would walk through Beijing with their cocker spaniel, C. Fred, in tow, or join the hordes of drably, monotonously attired Chinese who bicycled through the city as their sole means of transport, George often donning a People's Liberation Army cap. Like her husband, Barbara got the most she could out of the experience, relating to her hosts in a way she didn't expect. "We found them very like us. They loved family; we were told they didn't," she recalled. "We were very excited about going there—by what we learned."

At Christmastime, she returned stateside to be with the Bush children for the holiday season, leaving her husband behind in China to entertain his mother during her visit the same month. After a call home to Barbara on December 4, George mused in his diary at how the family had risen to the occasion around his overseas mission:

> Great talk with Bar on the phone. The kids are all doing fine. It is as if each one of these five kids, recognizing that the family was undergoing a different experience, are pulling together much more. There are no longer those juvenile battles and each one comes through strong, vibrant, full of humor and different, full of life and we are awfully lucky. It is right that Bar be there but boy [do] I miss her . . .

In June of 1975, the Bush children—minus Jeb—arrived in Beijing for an extended Chinese visit that would include sojourns to Nanking, Shanghai, Wuxi, and other family highlights. The Bushes hosted a Fourth of July party for the diplomatic community complete with red-white-and-blue flourish and American picnic fare: hot dogs, hamburgers, potato chips, and beer. ("The Great Hot Dog Crisis," as Bush called it, was averted when the State Department complied with his last-minute request to ship seven hundred much-needed hot dog rolls.) And, in a small ceremony at Beijing's nondenominational Bible school, sixteen-year-old Doro became the first American to be baptized in the world's most populous nation since it had fallen to Communism in 1949, a bit of unfinished business the family hadn't quite gotten to back home through the years.

George W.'s first trip outside the U.S. left a deep impression on him. As he reflected in his 1999 autobiography, *A Charge to Keep*, "My visit underscored my belief in the power and promise of the marketplace, and deepened my belief that by introducing capitalism and the marketplace, China will free her people to dream and risk." It also made him appreciate his good fortune in being an American—and his desire to achieve his own dreams in West Texas. His father reflected hopefully on George W.'s next chapter while acknowledging his slow start in a diary entry on July 6, 1975, his oldest son's twenty-ninth birthday. "He is off to Midland," he wrote, "starting life a little later than I did, but nevertheless starting out on what I hope will be a challenging new life for him. He is able. If he gets his teeth into something semi-permanent or permanent, he will do just fine . . ."

As the father pursued his career in the Far East and his son began his own seven thousand miles away in Midland, each would accrue experience that would mold them as leaders and eventually help define their presidencies. George H. W. Bush, the would-be U.S. senator and might-have-been vice presidential appointee, had accepted his positions at the UN and in China as consolation prizes and made the best of them. At the UN, he had seen what the force of friendship could do to deepen relationships with foreign dignitaries and, by extension, the countries they represented. In China, he would build on that ability, never missing an opportunity to forge a new bond, however ostensibly unimportant—his first visitor at the Liaison's Office was the head of the delegation from Kuwait, a small Middle Eastern nation of little concern at the time. After Saigon had fallen, he would see U.S.-aligned South Pacific nations calling on the Chinese to bolster relations in the event that America's loss in Vietnam portended a weaker U.S. presence and diminished support in the region. Countries of little strategic significance could become consequential with a shift in world events; strong ongoing diplomatic relations could make a crucial difference. The diplomatic expertise would come dramatically to bear in Bush's presidency, as he built an unprecedented coalition of thirty-nine nations around the liberation of Kuwait from Iraqi invasion, and in his judicious responses to the collapse of the Soviet Union and the massacre of protesters in Beijing's Tiananmen Square, where his relationship with China's paramount leader Deng Xiaoping allowed him to express his views "with a frankness reserved for respected friends."

George W. Bush was going on to a career where he was determined from the start to be his own boss, not in a midlevel corporate position with an eye toward working his way up the ladder like many of his fellow Harvard Business School graduates. He would go with his instincts, answering to no one above him—launching his own company in Midland before going on to become co-owner and co-managing general partner of the Texas Rangers and then governor of Texas—preternaturally confident in his abilities. He trusted his gut on decisions professional and personal and moved ahead without hand-wringing or second-guessing, as he would when running for the Texas State Capitol despite profuse counsel that the popular incumbent Democratic governor, Ann Richards, was unbeatable. He would bring the same sensibility to the White House. The self-proclaimed "decider," Bush's presidency would be marked by brazen executive decisions, often in the face of fierce doubt or withering criticism, most notably to wage war and nation build in Iraq, and when the war was going awry, ordering a surge in troops.

But all of that remained to be seen in the summer of 1975, as W. struck out for Midland. One thing, though, would soon become clear: He was driven to succeed. "He was focused to prove himself to his dad," recalled Joe O'Neill, a friend of W.'s since their baseball-playing boyhoods in Midland. It applied not only to the oil industry but broader ambitions. "Right away," O'Neill said, "he started talking about running for Congress."

16
COMING HOME

THOUGH HE HAD LEFT THE Northeast resolutely behind, returning to his hometown of Midland in July of 1975, George W. Bush had not escaped Harvard completely—at least nominally. In the well-heeled section of west Midland, he settled into 2006 Harvard Street, a brick two-room alley-side guest cottage over the garage of a business friend of his father. There, he lived no more grandiosely than his parents had in their move to West Texas in 1947. His five-hundred-square-foot laundry-strewn quarters, dubbed a "toxic waste dump" by his friend Charlie Younger—a spare necktie held up one corner of his dilapidated bed—bespoke his frugality as he sought to learn the oil business. While he had $15,000 left over from an education trust established for him by his parents, the money, representing his own investment capital and sole means of support, would have to last.

He set his sights on being a land man, a studious function necessitating long hours in county courthouses sorting through records to ascertain the ownership of mineral rights below the surface of Permian Basin property. It also required assessing the property's availability for leasing and, often, carving out leasing deals with the identified owners. His father's old friends lined up to help George W. cut his teeth in the industry, but there was also resistance among some new to the business for whom he was just an Ivy League stranger. Former director of the Permian Basin Petroleum Association, Ed Thompson recalled, "Harvard, at first, was a hindrance" for Bush, who kept his head down and, according to Thompson, "did more listening than talking." Bush took nothing for granted, working hard not only to understand the business but to form key relationships. Eventually, as Thompson observed, Bush "began to fit in."

In fact, as his uncle Bucky Bush saw it, Harvard Business School was instrumental in his nephew's humble approach to the business. The "swagger" he had seen in him in his younger days "was replaced by a much more intelligent approach to things," he said. "And I'm not sure what it was there, it may have been the peer pressure—that 'Jesus, they're a lotta smart guys in the world and if I'm gonna compete against 'em, I can't bluff it, I've really gotta know my stuff.'" George W. knew that what he did in Midland mattered in a way that scholastic achievement or athletic accomplishment earlier in his life did not. He was out in the real world now, not exactly on his own—the Bush-Walker clan was a reliable safety net for contacts and capital—but putting himself out there in view of a family that placed real value in business success. Money was tangible evidence that one had achieved it, which, in the Permian Basin, was all about how to extract crude oil from the parched, dusty earth.

Soon Bush was actively freelancing as a land man for a hundred dollars a day, sharing an office with F. H. "Buzz" Mills, one of Midland's many larger-than-life characters with a pithy nickname, tall tales to tell, and a canny knowledge of the industry on whose back he had ridden through boom times and bust. The two shared office space on the fourth floor of the Midland National Bank Building, where Bush eschewed furniture for milk crates, making small investments in the wells Mills and his partner, Ralph Way, had commissioned and making some of his own. "I was collecting the crumbs," he wrote later. "But I was making a decent living and learning a lot." In a nod to the shadow Bush's father still cast in Midland, Mills gave his officemate a nickname of his own: "Bush Boy."

Close friendships were key to Bush's happiness in Midland. He drew on a triumvirate of friends—Joe O'Neill, Charlie Younger, and Don Evans—who would be with him for the long haul, all of whom, like him, had arrived in 1975. Younger, a few years Bush's senior, had come back home to Midland from East Texas to continue his career as an orthopedic surgeon. But O'Neill and Evans, like Bush, were among the throngs of young men who flocked to Midland to get in on the boom times during which, from 1973 through 1981, the price of oil would swell by 800 percent. Adhering to his father's directive, O'Neill had returned home to run his family's oil business after learning the industry at a corporation in San Francisco. Evans, a Houston native, arrived in Midland with his wife, Susie, who had attended Sam Houston Elementary School with Bush, prepared to learn the business from the ground up. Starting out as a roughneck at Tom Brown, Inc., an independent oil and gas concern, Evans would climb the ranks to become its CEO a decade later.

The friends, en masse or in various combinations, were a familiar sight in Midland, on daily jogs of three or four miles at the high school, in the bleachers at Midland Angels AA baseball games, on the golf course, knocking back whiskey or beer at local watering holes, and grilling at backyard barbeques. Bush's friends became a solid support system. Lacking proper attire for business meetings, Bush gladly took several dress shirts given to him by O'Neill, augmenting a spotty wardrobe that included golf shirts, tattered khakis, a pair of loafers tenuously held together by Scotch tape, and from China, a pair of black slippers and an ill-fitting custom-made suit his father had ordered for him in Beijing. (Outfitting George W., Neil, and Marvin in custom suits cost their father a cool seventy dollars.) "He didn't dress down, which intentionally looks bad," explained O'Neill. "He just dressed bad. He dressed in the dark." What little clothing he had he laundered at Don and Susie Evans's house. While the oil business consumed much of their time and talk, Evans, O'Neill, and Younger saw clearly Bush's greater ambitions.

"He wasn't obsessed with politics, but it was always there," Younger said. It was just a matter of time, they knew, until Bush would play his hand.

On Saturday, November 1, 1975, George H. W. Bush was tracked down by a messenger as he and Barbara rode their bicycles through Beijing. Bush was handed an "eyes only" cable from Henry Kissinger who was writing on behalf of President Ford. The president, Kissinger related, was making some "major personnel shifts" in Washington; would Bush consent to Ford nominating him as director of the CIA?

The request came to Bush as "a total shock." Like the assignment as chairman of the Republican National Committee as the Watergate scandal played out, heading up the CIA would be no easy task. The agency, under the leadership of its outgoing director, William Colby, was beleaguered after continued revelations of abuse of power and authority and intelligence leaks to the press, as its credibility and reputation ebbed along with morale at Langley. Six directors had come and gone in a decade. There was also the matter of the Senate confirmation process, which could be contentious. Bush would almost certainly be walking into a hornet's nest.

As the Bushes considered the potential assignment, one of their first reactions was, "The CIA! What would our children think?" They called George W. in Midland to see if he would survey his siblings on whether their father should accept. As Barbara explained it later, "He called back and said, 'Come home,'" though she wondered if he actually solicited their views or "waited a reasonable period of time and called back with his opinion." Shortly afterward, W. expressed his support in writing: "I look forward to the opportunities to hold my head high and declare ever so proudly that, yes, George Bush, super spook, is my Dad and that yes I am damn glad for my country that he is head of the agency." Regardless of his son's view,

Bush would almost certainly have taken the post due to the influence of his own father. As he explained in a cable to Kissinger the following day:

> Henry, you did not know my father. The President did. My Dad inculcated into his sons a set of values that have served me well in my own short public life. One of these values quite simply is that one should serve his country and his President. And so if this is what the President wants me to do the answer is a firm "YES."

However, he enumerated several concerns, one of which read, "I do not have politics out of my system entirely and I see this as the total end of any political future." He was more blunt in a letter to his brothers and sister, calling the job a "graveyard for politics." One thing was certain: Accepting the position would preclude him from being Ford's running mate in 1976. Though the incumbent vice president, Nelson Rockefeller, had taken himself out of the running voluntarily when Ford expressed his view that Rockefeller would be a liability among the GOP's increasingly volatile right wing, Ford had smoothed the politicized Senate confirmation process by ensuring that Bush would not be among those he would consider.

Rumor had it that Don Rumsfeld planned it that way. In the shake-up in the Ford administration, Rumsfeld had moved from being Ford's chief of staff to becoming, at forty-three, the youngest secretary of defense in U.S. history, leaving his former deputy, Dick Cheney, to take up the chief of staff role. Rumsfeld and Bush, along with Rockefeller, had been on Ford's short list for the vice presidency a year earlier, and word around Washington was that Rumsfeld had not only engineered the secretary post for himself to heighten his profile and set himself up as a natural vice presidential contender, but also had Bush relegated to Langley and out of consideration. Rogers Morton, Bush's friend and the outgoing secretary of commerce, told

Bush flatly, "Man, Rummy got your ass." Addressing the speculation head-on, Ford and Rumsfeld both had face-to-face meetings with Bush, denying that Rumsfeld had anything to do with the orchestration. Reflecting on it later, Bush said, "There was some feeling that [Rumsfeld] was not anxious to see me go forward. He denied that when [I] asked [him] about it. We just weren't close, put it that way."

Though they were outwardly civil—upon taking his China post, Bush loaned Rumsfeld his purple denim-appointed AMC Gremlin—there would always be bad blood between the two. From the beginning, Rumsfeld and Bush, both esteemed by Ford, regarded each other warily. David Hume Kennerly, Ford's White House photographer, observed that the men were "like two cocks in the barnyard."

Before the end of the 1975 calendar year, the Bushes had packed up and moved back to Washington, where Bush was confirmed by the Senate as CIA chief. He was sworn into the office in late January. Charged with improving the agency's strained relations with Congress, a job for which he was well suited given the many friendships he had established with its members, he would be called upon to testify before Congress fifty-one times. Ultimately, he would succeed not only in defending the agency but also in increasing its budget.

It was not a happy time for Barbara Bush, who had come home from Beijing an empty nester—all the children were off in school or making their way in the world—and the wife of the director of one of the world's most secretive organizations. Unlike George's past posts, his job at Langley was not a shared adventure. She was no longer the recipient of pillow talk about her husband's day or a frequent partner in personal diplomacy. As her husband kept the nation's secrets, she harbored one of her own: "I had a husband who I admired, the world's greatest children, more friends than I could ever see—and I was severely depressed," she wrote later. "I hid it from everyone, including my closest friends." George was the only one who knew, and he spent many nights trying in vain to console her. After six months of loneliness and despondence, the veil of depression lifted as inexplicably as it had descended on her.

While Bush considered his directorship at the CIA the most fulfilling of his appointments since leaving Congress after his Senate loss in 1970, the job would last only a year before the White House turned over. During the course of Bush's tenure in Langley, Jerry Ford was battling to earn the presidency in his own right after being lifted into the Oval Office upon Nixon's resignation. Ford had staved off a close primary challenge by conservative darling and former B-movie actor Ronald Reagan, whose star had risen in the party following his powerful defense of the GOP's ill-fated presidential contender Barry Goldwater in 1964, leading to two successful terms for Reagan as California governor.

Bypassing Rumsfeld and choosing as his running mate Kansas's senator Bob Dole, a concession to the party's right wing, Ford squared off in the fall against Jimmy Carter, a one-term governor of Georgia whose grassroots effort to secure the Democratic nomination shook the party establishment. Carter relentlessly reminded voters of the ills of the "Nixon-Ford administration," painting the disgraced former president and the incumbent president Ford with the same brush, promising never to lie to the American people. On Election Day, Carter and his running mate, Senator Walter Mondale from Minnesota, walked away the victors with 49.9 percent of the popular vote versus the Ford-Dole ticket's 47.9 percent, winning the Electoral College by 57 votes, which made it the closest vote since Woodrow Wilson squeaked to reelection victory sixty years earlier. Ford's "full and unconditional" pardon of Richard Nixon just a month into his fledgling administration lingered in the minds of voters. The maelstrom of controversy and indignation that came from the media and the American public had not fully subsided when the polls opened over two years later; nearly two-thirds of Americans disapproved of the pardon immediately after Ford issued it in September 1974—and 7 percent of voters had cast their ballots for Carter as a direct result.

On November 19, during the last of regular intelligence briefings he had given Carter since early July, Bush offered to remain on as the director of national intelligence. Carter indicated that his services were no longer required. As the new calendar year opened and a new president took office, Bush returned to Houston, his political future once again uncertain, just as it had been seven years earlier when he had lost his second bid for the U.S. Senate. Back in Texas, he would await his next chance to enter the political arena. As he did, his oldest son would seize an opportunity of his own.

17

"DEFEAT ISN'T DEFEAT"

"IT WAS GREAT TO BE back in Texas," George H. W. Bush wrote later of his return to his adopted state. Indeed, in some ways it *was* good to be back home. Jeb and Columba were living in Houston, offering George and Barbara ample opportunity to spend time with their first two grandchildren, George P. and Noelle, born in 1976 and 1977 respectively. They were able to catch up with old friends. George sat on several corporate boards, made a little money on the speaking circuit, and taught a course on government as an adjunct professor at Rice University. They were comfortable. Comfortable enough that he could turn down lucrative opportunities that came his way. Among them was an overture from Texas billionaire H. Ross Perot to run his oil business in Dallas. "Well, this is your big mistake," Bush recalled Perot telling him when he politely declined the offer, presaging difficulties with Perot in the years ahead. "You don't say no to Ross Perot."

But within weeks, Bush was restless. He had been gone for a decade—in Washington, New York, China—rubbing elbows with those who were shaping the world and, though he had the means, was not content to settle into a sedate country club life. There was too much to learn, too much to do. As he wrote to his old friend Gerry Bemiss in early March, "There is a missing of stimulating talk. I just get bored silly about whose daughter is a Pi Phi or even bored about who's banging old Joe's wife. I don't want to slip into that 3 or 4 martini late late dinner rich social thing." Wrestling with his innate humility, he provided Bemiss a glimpse of what might come next as he set his ambitions—tentatively—on the presidency: "I think I want to run or at least be in a position to run in '80—but it seems so overwhelmingly presumptuous or egotistical, yet I'll think some on that."

By early summer, in West Texas, George W. also had some thinking to do. On July 6, W.'s thirty-first birthday, conservative Democrat George Mahon, Midland's congressman of forty-three years, unexpectedly announced he would be stepping down, leaving a vacancy in Texas's Nineteenth District—and an opportunity. By Bush standards, George W. hadn't made much of a mark in his thirty-one years. A couple of Ivy League degrees, a pair of National Guard wings, a few jobs here and there, followed by his foray into the oil business, which had yet to produce much; in the beginning of 1977, he had launched his own small company, Arbusto, Spanish for Bush, whose first oil well had come up dry. Though he hadn't achieved the business success that was an unspoken prerequisite in the Bush family for moving on to a political career, Mahon's congressional seat was a chance to get into the game now and to do it in a big way.

To be sure, much of Bush's motivation was political. He had quickly become disillusioned with Jimmy Carter, whose energy policy didn't sit well with him. "I had just gotten out of Harvard three years earlier, and there [in Midland] in the oil business watching risk/reward ratios stacked or changed by government policy—if you drilled a well at five thousand feet, you paid less [in taxes] than if you drilled a well at ten thousand feet," he recalled. Moreover, he believed Carter's attempt to control the price of natural gas was tantamount to European-style socialism. But there was his father, too—an instinct to enter the political ring to prove himself within the family. "Was it [Carter's] policy alone that motivated me?" he asked of himself after leaving the White House, flirting with introspection that was largely foreign to him. "Obviously, there had to be—watching Dad, I was introduced—it was logical to assume that he influenced me because he introduced me to politics." He was also mindful of the Bush creed that had been passed on by his father: "You can't win if you don't run."

But his odds for securing the seat were long, and he knew it. First he would have to win the Republican primary against Jim Reese, a former sportscaster and mayor of Odessa, who had run against Mahon two years earlier garnering 45 percent of the vote and was seen by the party establishment as its best hope to pick up a district that had never before sent a Republican congressman to Washington. Then he would have to face off against the Democratic nominee, Kent Hance, a Lubbock-based Texas state senator with higher ambitions.

Barbara Bush expressed concern that her son's run for Congress would compromise her husband's political aspirations. "You know," she wrote her friend Millie Kerr in September, "I wonder if George Bush Jr. understands just how difficult it is to have two people thinking about running for office in one family? It reminds me of the time that George took all the heat when his father was in the U.S. Senate and voted against the oil industry." She also worried about his chances of winning. So did George. Regardless, he gave his son no advice on whether to get into the race but simply suggested that he go see Allan Shivers, the conservative former two-term governor of Texas. George W. dutifully called on Shivers, not knowing what had transpired between Shivers and his father, who set up the meeting. He knew only the pronouncement that Shivers gave him—bluntly: "You're not going to win the race. The district was created for Kent Hance."

"I mean if you're an expert in psychoanalysis you can figure that out yourself," George W. said of his father's guidance.

But, he didn't say, "Don't run." He never tried to dampen my ambition. He never said why, and I never asked him why he made me go see Shivers, which in itself is interesting to just accept it, rather than challenge [it]. On the other hand, he didn't want to snuff out the candle and yet, he wanted to make sure there was a little wind blowing across the candle to give a little perspective.

His father was just managing George W.'s expectations, indirectly giving him an objective sense of the inherent risk he would be taking without dissuading him. "And I ran anyway, that never deterred me," George W. said well over three decades later with a lingering hint of defiance not toward his father but the circumstance.

Less than two weeks after Mahon's announcement, George W. held news conferences in Midland and Lubbock, the southern- and northernmost cities in the Nineteenth District, to declare his candidacy. "[O]ut of nowhere—and I mean nowhere, even after Reese has shown so much strength—comes George Bush," recalled V. Lance Tarrance, a pollster who worked for Reese, the presumptive nominee. Despite his son's long odds, George H. W. Bush was pleased. His son was honoring the Bush family through the "passing on of tradition, passing on legacy." It's what the Bushes did. The following month, on July 25, he wrote his friend, GOP political consultant Eddie Mahe Jr.: "George is off and running in Midland. The Party pooh-bahs are not thrilled, but the guy is energetic, attractive, he grew up in Midland and has lots of friends. I'd say he's an underdog now but he'll acquit himself well I'm sure. I'm tickled pink about this." Alluding to his own political future, he added, "Other political churning going on—nothing definitive yet."

The summer of 1977 was a crowded hour for George W. Bush. Ten days after he declared his candidacy, he met Laura Welch, a Midland native and Southern Methodist University graduate, at a backyard barbeque dinner for four hosted by mutual friends Joe and Jan O'Neill. The O'Neills had tried for a year or more to put the two together, though they didn't have much hope for a match. George W. and Laura were so different: "Bombastic Bushkin," gregarious, out there, the life of the party; and Laura Welch, the quiet, bookish Austin school librarian who was decidedly uninterested in politics. At age thirty, Laura had eluded marriage, just as Bush had. "He was at the age where it was getting awkward to be a bachelor, but I don't think he thought about it," Joe O'Neill said. "He wasn't exactly presidential timber yet. It took some coaching for us to get the girls to go out with him." But the two clicked. Bush was taken by the blue-eyed brunette, and the feeling was mutual. "At that particular moment, on that warm summer night, both of us were hoping to find someone," Laura wrote in her memoir. "We were ready to build an enduring future."

In many ways, the coupling made sense. They had friends in common in addition to shared West Texas roots. She had grown up the only child of Harold Welch, a World War II veteran and Midland home builder and developer, and Jenna, a bespectacled homemaker and bookkeeper for her husband's business. Laura and George W. had even attended seventh grade at San Jacinto Junior High School, though they had never crossed paths that they could recall. Still, she felt as though she had known him her whole life. They were, Laura found, "two complementary souls." Death had cast a shadow over both of their homes at early ages as they had borne witness to life's fragility and precariousness. Bush had lost his sister at age six. Laura, desperate for siblings, saw her parents' pain after a series of stillborn births, which they euphemistically referred to as "late miscarriages," befell her mother. Much later, in her senior year in high school, Laura ran a stop sign resulting in an accident that killed her friend Mike Douglas, the driver of an oncoming car.

Both George W. and Laura were, in their own ways, free spirits—

products of a small, conservative town in the 1950s, but uncompelled toward conformity. And there was another commonality, too, an especially important one as Bush set his sights on political office. "He has a steadfastness that I love in him and I have a steadfastness that he likes," Laura said later. "Both of us are really tough; I mean, we are really strong. You can't live in the White House without [being strong]."

A brisk courtship took hold. The night after their meeting at the O'Neills' the foursome met up again for an evening round of miniature golf. As cicadas chirred, they laughed well into the hot, dry summer night, just as they had the previous evening. Six weeks later, after Bush traveled to Austin for successive weekend visits with Laura, they were engaged. Bush had met his match and acted quickly.

Laura met her future in-laws in early October, the weekend after her engagement, when she and George W. drove to Houston for the baptism of Jeb and Columba's second child, Noelle. George W. introduced Laura to his parents along with the news that they were to be married. "They are perfect for each other!!!" Barbara enthusiastically scribbled in her diary afterward. After the baptism, George W. and Laura visited the elder Bush's Houston home, where George pulled out a pocket calendar to ensure no conflicts with his busy schedule, and a wedding date was set for just three weeks later. The couple was married on November 5 at the Welch family's church, First United Methodist Church, before a group of some seventy-five family members and close friends. Along with other members of the Bush family, Marvin, meeting Laura for the first time, likened Laura's joining the Bush family to Audrey Hepburn's arriving at "Animal House." The bride, a day after her thirty-first birthday, wore an off-the-shelf white silk skirt and blouse and gardenias in her hair, and the groom was uncharacteristically dapper in a gray suit and tie, a gardenia pinned on his lapel. Knowing tears were inevitable, George H. W. Bush refrained from toasting the couple at the reception, ceding the task to Barbara. As Laura wrote much later of her husband and father-in-law, "In years to come, to others, the cool remove of television would frequently obscure the depth of their caring, how much and how deeply their own hearts open."

Laura complemented her husband, playing a vital supporting role. She knew what battles to fight, when to gently rein him in, as she did when he invoked "Wanted Dead or Alive" cowboy rhetoric in his administration's effort to hunt down Osama bin Laden after the attacks on 9/11. George W., in turn, pushed Laura beyond her natural diffidence, violating his premarital promise that she would never have to make a public speech almost as soon as they finished exchanging their vows. A honeymoon for the newlyweds was not in the cards. Laura was quickly swept up not only into married life, leaving behind her job as a school librarian in Austin and moving back to Midland, but also into the political campaign on which her husband had embarked just before their introduction. She accompanied him on the campaign trail, driving the same West Texas roads she had traveled as a girl when she and her parents visited her maternal grandmother in El Paso on weekends. The car rides brought the two newlyweds closer as they rode the long miles between campaign events and talked about their lives, their plans, their goals. "We spent nearly a year on the road," wrote Laura later, "and in many ways the bonds of our marriage were cemented in the front seat of that Oldsmobile Cutlass."

From the beginning of the race, Bush distanced himself from his father, running as his own man, just as he would in subsequent political contests. "We don't need Dad in this race. We don't need anyone in this race but the people in this district," he explained to the *Midland Reporter-Telegram*. Reese, however, put the old man front and center. Referring to his opponent as "Junior," Reese aligned him with the Bush patriarch who was still looked upon skeptically by Texans as a Rockefeller Republican, by implication a liberal-minded northeasterner.

By April of 1978, as the campaign intensified a month before the primary election, Reese garnered the support of Ronald Reagan, the former California governor, whose political action committee gave a total of $3,000 to Reese's campaign through two separate contributions. Reagan also offered his endorsement of Reese in a campaign commercial. The primary election attracted national attention as its dueling candidates became proxies for two Republican heavyweights—Reagan and the elder Bush—as possible candidates for the Republican presidential nomination in 1980. "I'm not interested in getting into an argument with Reagan," George H. W. Bush groused to the *Washington Post*'s David Broder the day before the primary. "But I'm surprised about what he's doing here in my state. They are making a real effort to defeat George." Earlier the elder Bush had appealed directly to Reagan to stay out of the fray, and Reagan assured him that his PAC would refrain from future contributions. "Governor, we're supporting a guy who supported you," Lyn Nofziger, a Reagan aide, counseled his boss about Reese, suggesting that Reagan owed nothing to either George H. W. Bush or his son. Though Reagan didn't take to the stump for Reese himself, his view was promulgated by Clarence Warner, a Republican activist from Oklahoma and a Reagan surrogate, who eviscerated the elder Bush—and by implication George W.—for the elder Bush's membership in the Trilateral Commission, a group of leaders from the U.S., Europe, Canada, Mexico, and Japan who convened regularly to discuss international issues.

Mindful of the weight he put on his son's candidacy, George nonetheless remained sanguine about his son's chances in the primary, writing Jeb, who had recently uprooted from Houston to Venezuela, that reports from West Texas "are pretty darn good." His optimism was well founded. Though George W. won only his home county of Midland, one of the district's seventeen counties, he edged Reese out by 1,400 ballots by plain outworking Reese, riding into the general election against Kent Hance with a war chest of $400,000 raised from the Bush family's abundant network of friends and family.

Charles Guy, editor of the *Lubbock Avalanche-Journal*, wrote George that his son had shown "a lot of class" in the race, to which George replied, "You know how close our family is; and therefore, you know of the pride I have in my son. He kept his head high in the face of some tough, bitter, mean little ultra-right attacks during the primary."

The issue of George W. Bush as a carpetbagging outsider in his father's Ivy League mold didn't go away in the fall. A natural on the stump, Hance spun a folksy yarn about working in a field in Lubbock's farm country when a Mercedes drove up. "The guy rolled down the window and wanted to know how to get to a certain ranch," Hance explained, directing him to drive down the road and take a right after the cattle guard, a metal grid found commonly on rural roads to keep livestock from straying. "Then he asked," Hance continued, "'What color uniform will that cattle guard be wearing?'" As the audience twittered with laughter, Hance added, "The car had Connecticut license plates." So it went.

"Daddy and granddad were farmers," Hance boasted. "They didn't have anything to do with the mess we're in right now, but Bush's father has been in politics his whole life." The race may have turned with a "Bush Bash" thrown by a Bush volunteer at Texas Tech at which free beer was served to students. Though the party was neither attended nor endorsed by Bush himself, Hance jumped on it as a chance to contrast Bush's values with those of "fellow Christians" in the district, telling a local reporter, "Maybe it's a cool thing to do at Harvard or Yale," but not at Texas Tech. The Bush campaign learned that Hance owned a Lubbock property near the Texas Tech campus that was leased to Fat Dogs, a local bar frequented by students, offering a chance to show his hypocrisy on the issue. But Bush insisted on taking the high road, refusing to counterattack. "Kent lives here. I'm not going to ruin the guy in his home town," he told his campaign aides, including Neil Bush, newly graduated from Tulane University, who served as his brother's campaign manager in Lubbock. It was a decision he would later regret. "When someone attacks your integrity," he wrote later, "you have to respond." In Texas terms, Bush had been out-churched and out-country'ed. They were not mistakes he would make again.

On Election Day, Allan Shivers's forecast was on the mark. Though Bush once again carried Midland County, in addition to two contiguous counties, he came up short against Hance, who won the race by four percentage points, 53 to Bush's 47. Bush's friends later saw his defeat as a blessing in disguise; the Harvard MBA's executive skills were conducive to a larger management role—a CEO or governor, say—where his decisions could effect immediate change, not a role where his voice and vote would be just one among a body of 434 others.

Horizons broadened, Bush took the lessons of the loss to heart, using them to see the possibilities that might be in his future, something he chalked up to his father's influence. "One of the things he teaches—and I didn't realize it at the time—is defeat isn't defeat if you've got big perspective. I think that's what he taught me," he reflected later.

Six months after the door closed on George W.'s immediate political prospects, putting his ambitions on hold for the next sixteen years, his father opened up a door decisively toward the ultimate political prize.

PART III

Turning Points

18

"THE BIG MO"

RONALD REAGAN HAD BEEN GIVING the same speech more or less for fifteen years. Since catching fire with the Republican right wing with his 1964 television address, "A Time for Choosing," supporting the doomed presidential run of Barry Goldwater, Reagan had been sounding the same conservative messages: America was under siege by a federal government that was big and getting bigger, grossly excessive in its ambitions and infringing on individual liberties; taxes were too high, adding to a bloated bureaucracy; entitlements were too great. "The Speech," as it would become known in the Reagan myth, culminated with Reagan suggesting that the choice Americans had was not between left or right, but between "up and down. Up to a man's age-old dream—the maximum of individual freedom consistent with law and order—or down to the ant heap of totalitarianism." This was Reagan's clarion call.

It had catapulted the dulcet-voiced, camera-ready former B-movie actor to an A-list Republican, who after two terms as California governor had come within a whisper of capturing the GOP presidential nomination from the earnest-but-lackluster moderate Gerald Ford, the incumbent president, in 1976. In a party that often rewarded the runner-up in the previous presidential primary with the nomination four years later, Reagan quickly emerged as the favorite to become the party's standard-bearer in 1980. He made his candidacy official in November of 1979, adding his name to a crowded field that included Kansas senator Bob Dole; former Texas governor and secretary of the treasury John Connally; two Illinois congressmen, John Anderson and Phil Crane—and George Herbert Walker Bush.

Bush had thrown his own hat in the ring six months earlier, notwithstanding national name recognition, as George W. recalled later, "so low that in many early presidential polls he didn't register enough support to be included in the results." He launched his dark horse candidacy on May 1, 1979, in a speech at Washington's National Press Club, as "a lifelong Republican who has worked throughout his career, in business and in public office, on behalf of the principles of Lincoln, Theodore Roosevelt, and Dwight Eisenhower."

The reference to Eisenhower was a portent: Bush would be running for president as a moderate in the Eisenhower mold—much like his father had been—offering a viable alternative to the conservative Reagan, a former Democrat who had defected to the GOP after supporting Franklin Roosevelt and Harry Truman. Eisenhower had ridden into the presidency in 1952 on a wave of popularity implied in his iconic campaign slogan, "I Like Ike." The hero of D-day, Eisenhower had edged out his conservative GOP rival, Ohio senator Robert Taft, at the Republican National Convention in 1952, but it had more to do with the force of Ike's personality and towering stature than it did with his moderate leanings. The old notion of Republicanism dominated by the northeasterners like Prescott Bush—nonideological; probusiness; fiscally conservative; moderate, even progressive, on social views—was waning even then. Despite a plank in the 1960 party platform espousing "progressive Republican policies" like "liberal pay," the GOP continued to drift rightward. By the mid-1960s, the party's Southern right wing—composed partly of former Democrats alienated by what they saw as the federal overreach of Lyndon Johnson's Great Society, including civil rights—became a greater force in the party. It became evident in 1964 when Goldwater, who became the party's nominee after a divisive convention floor battle, upset Nelson Rockefeller, an emblem of northeastern Republicanism. Since then, the moderate and conservative wings, the latter growing in numbers bolstered by evangelicals and harder-edged right-wingers, had been battling for domination of the party. The year 1980 would be no exception.

With Bush's chances slim, his campaign went all out in winning the first battleground on the road to the nomination, the Iowa caucus, slated for January 1980. The strategy had worked for Jimmy Carter in securing the Democratic presidential nod in 1976, launching the unknown former one-term Georgia governor from "Jimmy who?" to the lead horse out of the gate and garnering all the media attention that went along with it. Jim Baker—chairman of Bush's political action committee, the Fund for Limited Government, and now chairman of his campaign—consciously emulated Carter's approach. He had the right candidate in his former doubles partner, whose natural enthusiasm and feverish energy were conducive to outworking his competitors, especially Reagan, whose advanced age for the time, sixty-eight, was an Achilles' heel. "The age thing is going to get him," Bush told his young campaign aide, Karl Rove, who had worked on George W.'s congressional bid the previous year. Bush, thirteen years Reagan's junior, emphasized his relative vitality by vigorously working all Iowa's ninety-nine precincts, frequently donning running gear after campaign events for daily jogs and additional time with local constituents. The retail politics of Iowa suited Bush's strengths. "George Bush had the knowledge that grassroots politics win primaries," George W. said of his father, "having working knowledge of how politics work, who the players are, their positions, what trigger points are to their mentalities." Less an ideologue than a decent, competent man whom one would want to see in charge, he was more likely to make a friend who would support him at the polls than he was to make a point or impart a vision that would sway a voter to come to his side.

The campaign fast became a Bush family crusade, with members expressing their love and loyalty by taking time out of their lives to devote themselves to the cause. Bush-Walker relatives, mostly in the

Northeast, did everything from helping to raise money to licking envelopes and answering phones in local campaign offices. Barbara, as always, was supportive though it meant endless grueling days on the campaign trail, often without her husband who maintained his own hectic travel schedule as a means of dividing and conquering. As she described it later, "We campaigned in backyards, ballrooms, courthouses, bowling alleys, at a cider and doughnut farm, hot dog plants, high schools and colleges, retirement homes, and hospitals. You name it, we were there."

So were the Bush children, who threw themselves into the campaign as a way of not only showing their devotion to their father but also paying him back. "I look back many years later and I see friends of mine who work at the family business and contribute to the family company's success," recalled Marvin, who took off a semester at the University of Virginia to stump in Iowa in a beat-up Chevy Blazer sporting a "George Bush" vanity plate. "For me, it was really a neat feeling because I'd always been a taker from Dad." It wasn't that he agreed with him on every issue—he openly opposed his father's pro-life stance—but he loved him and he felt he owed him. Neil, having gained grassroots experience in his brother's campaign, was stationed in New Hampshire, the next crucial campaign front after Iowa. Doro, a student at Boston College, attended nine months of secretarial school to offer administrative help to her father in the campaign's Boston office. Jeb made a bigger sacrifice, leaving his banking post in Venezuela to throw himself into his father's campaign in Florida, where his fluency in Spanish helped to win over Cuban American voters. His natural ability and work ethic impressed his dad, who noted in his diary that whenever Marvin, also a volunteer in the campaign, talked of returning to UVA to resume his studies, "Jeb bawl[ed] him out."

George W.—the oldest of the brood, possessing his father's name and having shown his political chops in his congressional race the year before—was the candidate's most plausible surrogate, a role he relished. From his home base in Midland, he made as many as seventy calls a day to members of the oil community in Texas and beyond to grow the campaign's war chest while tending to his own oil business, which had lain fallow for over a year during his own campaign. And at crucial moments, he was dropped into key areas to take to the stump, including barnstorming in the northwestern part of Iowa with a local congressman several weeks before the caucuses.

The campaign's intensive Iowa effort proved a prudent investment. On January 21, 1980, Bush walked away the victor in the Iowa caucus with 32 percent of the vote, upsetting Reagan whose token effort in the state yielded 30 percent; Howard Baker placed a distant third at 15 percent, with the remainder of the field pulling down single digits. Bush had broken through. "I've got the 'Big Mo,'" he boasted to reporters of his momentum as the campaign shifted its focus to the New Hampshire primary.

If so, it stopped dead in the gymnasium of Nashua High School a month later. With Reagan and Bush running neck and neck out of Iowa, the Reagan camp had earlier approached Jim Baker with a proposal to participate in a New Hampshire debate limited to just the two front-runners—Reagan vs. Bush—leaving the remainder of the candidates out of the proceedings. Baker eagerly accepted the chance to create the impression that the primary was a two-man race. The *Nashua Telegraph* agreed to sponsor the debate at Nashua High School. Then, after the Federal Election Committee ruled that the *Telegraph* couldn't do so without including the other GOP candidates, the Reagan campaign agreed to put up the funds with the *Telegraph* playing host to the one-on-one debate, but not its sponsor.

What the *Telegraph* and the Bush campaign didn't know was that the Reagan camp had reached out to the other candidates in advance to invite them to come to the debate anyway, promising that they would be allowed to participate in some fashion—and banking on it creating confusion and discord for Bush. All the candidates agreed except Connally, who picked up on Reagan's plan to outflank

Bush, his rival and fellow Texan. "Brilliant strategy," he responded, "but I ain't coming. Fuck him over once for me."

On the winter evening of February 23, three days before the New Hampshire primary, an audience of twenty-four hundred packed into the stuffy gymnasium, including John Anderson, Howard Baker, Phil Crane, and Bob Dole, who ostensibly showed up to protest their exclusion as the crowd quickly came to their aid with jeers and boos. Just beforehand, Baker was told by Reagan emissaries that Reagan was going to expand the debate to include the other candidates. After consulting with Bush, Baker's response was adamant: "Goddamn it, you guys are not going to fuck this up! This is going to be a two-man debate and you are not going to do anything to change it!"

Reagan felt otherwise, leading the other candidates out on the stage. A fracas ensued leading up to the beginning of the event, with both the *Telegraph* and the Bush campaign balking at changing the format and insisting that they hold to the original agreement. The *Telegraph*'s publisher, J. Herman Pouliot, conceded that the candidates could make only closing remarks. As a bespectacled Bush sat mutely unengaged, Reagan picked up his microphone to protest; presently, Jon Breen, the *Telegraph*'s editor and the debate's moderator, ordered that it be turned off. Flushed with anger, Reagan tapped his mic, "Is this on?" he asked, then erupted, "I am paying for this microphone, Mr. Green!"

Mr. *Breen*, actually. The man's name was Breen. But it hardly mattered. In the moment, Reagan came off as strong, commanding, and in control. His "mad as hell" posture reflected the attitudes of a significant cross section of voters at a time when America's position in the world had waned with the sustained capture of fifty-two American hostages in Iran, a flagging economy marked by high inflation and unemployment, and sapped national pride. The audience thundered its approval as Anderson, Baker, Crane, and Dole added their own applause to the din. Always at his best when indignant— "Mr. Gorbachev, tear *down* this wall!"—Reagan, as he wrote later, "may have won the debate, the primary—and the nomination—right there."

Bush's posture reflected his penchant for playing by the rules, instilled in him early on by his parents. He was not a revolutionary, someone who colored outside the lines. While driven with a fierce desire to win, he wanted to do so fair and square. "It cost him in . . . New Hampshire," George W. said of his father. "Instead of saying, 'Okay, I'll break the rules. Let's change the rules. You guys can debate, too,' he said, 'I agree with the newspaper publishers that this is the way it's going to be.'"

The debate eventually proceeded, with Reagan pitted against Bush as originally planned, only after Reagan escorted the other four candidates willingly offstage, but the damage had been done. Reagan was the hero; Bush the heavy. Reflecting the views of all four of the nondebating candidates, Dole complained that they were treated "like second-class citizens." "Bush stiffed us, with the help of the paper," he maintained. Reagan said he didn't know why Bush refused to meet with his other opponents, "but that's something he will have to explain." Bush was eviscerated not only by his rivals but by the press. If it stung Bush, it was worse for his family. "Looking back at it, the New Hampshire primary was the first time I experienced the unique pain that the child of a public figure feels," George W. wrote. "I was used to hearing my father get criticized in his Texas campaigns and in his Washington jobs. This was different. The stage was bigger, the stakes were higher, and the barbs were more personal."

Three days later, Reagan took the New Hampshire primary over Bush by more than double, 50 percent to 23 percent respectively. Bush never recovered. While he would go on to eke out northeastern wins in Pennsylvania, Connecticut, and Massachusetts, the bulk of the country, especially the party's conservative base in the South and West, lined up behind Reagan. When Reagan took Texas with 53 percent of the vote versus Bush's 47 percent, Bush knew his hand had been played. He folded on Memorial Day weekend, after assurances from Reagan's campaign that he could speak at the Republican National Convention in Detroit in mid-July.

"What went wrong?" *Time* magazine asked of Bush's campaign in a June feature story after Reagan had wrapped up the nomination, providing its own answer: "Many things. Though Bush had broad experience as a former Congressman, ambassador and CIA director, his surprise victory in Iowa caught him somewhat unprepared for the national spotlight, and he failed to develop any issues . . . He could be seen, finally, as a decent, likable man but not a political power. Of such perceptions, some would say, are excellent running mates made."

That became the next question: Would Reagan tap Bush to round out his ticket?

19

THE CALL FROM REAGAN

BUSH RETREATED TO HOUSTON IN the wake of his campaign. "What's it going to be like?" he pondered in a diary entry. "Driving a car, being lonely around the house?" He talked to George W. in Midland, who encouraged him to think about accepting the vice presidency if it were offered to him, but he was tapped out, with little interest in talking about anything relating to his political future. Besides, the idea of hitching his wagon to Reagan was also hard to swallow, not just because Reagan had beat him out of the nomination but because Reagan had stood in the way of George W.'s quest for Congress three years earlier. "[T]his guy had campaigned against his son," said Bush family friend Doug Hannah. "I know for a fact that if George Bush could have killed Ronald Reagan, he would have done it. He couldn't stand the man."

The matter of the vice presidency was still undecided when the GOP convened in Detroit to crown Reagan as its nominee. Going into the convention, there had been much speculation on whom Reagan would choose. Along with Bush, the names raised were Howard Baker, New York congressman Jack Kemp, and Nevada senator Paul Laxalt. Bush was a logical choice. He had been a worthy runner-up and would balance the ticket as a moderate. But Reagan was resistant. There was still lingering enmity between the two former rivals, who had gone after each other tooth and nail during the campaign; Reagan had taken a particularly hard hit when Bush leveled his specious economic plan, calling it "voodoo economics," and expressed concern with his position on abortion. "I have strong reservations about Bush," Reagan told a friend. Baker labeled Reagan's VP strategy "A.B.B."—Anyone but Bush. Still, Bush, with most of his family in tow, showed up at the Hotel Pontchartrain in Detroit on July 13, knowing that Reagan's decision represented his only immediate political lifeline—one more shot, however remote, at securing a national office and setting himself up for another run at the presidency. But it looked less likely as the convention kicked off.

As the delegates arrived in Detroit, a buzz arose about a "dream ticket" consisting of Reagan and former president Gerald Ford as his running mate. Though Ford had been run off from the White House by Jimmy Carter nearly four years earlier, he remained popular, more so as Carter struggled in the presidency. Moreover, Ford would add weight to the ticket as a respected former president, and as a moderate would be a bigger-name alternative to Bush. The notion of a Reagan-Ford ticket captured the imagination of party leaders—including Reagan.

If no love was lost between Reagan and Bush, the same was true for Reagan and Ford. Ford had been less than pleased four years earlier when he narrowly beat out Reagan for the nomination and Reagan refrained to offer him his wholehearted support, including declining to stump for Ford in the swing state of Texas, where Reagan had shellacked Ford in the 1976 GOP primary. Ford ultimately lost Texas to Carter, and with it went the presidency. "They needed a putty knife to get him off the ceiling," a friend recalled of Ford's apoplexy. Nonetheless, Ford remained distantly open to the possibility of rounding out the ticket. At least, that's what Reagan hoped, even though Ford rejected the idea when Reagan privately proposed it to Ford on July 14, the convention's first night, just as he had in an earlier meeting at Ford's home in Rancho Mirage, California.

Reagan wasn't wrong. Ford had held the presidency for just two years and five months, the only president unelected on a national ballot, and it gnawed at him. His narrow loss to Carter had cut deep.

Coming back as Reagan's running mate was a chance to get back in the game, and to do it in a big way. Yet Ford had no desire to take up the traditional role of VP, which he had held for an interminable nine months under Richard Nixon during Watergate's darkest days. If the role were "expanded" on the other hand—with members of the White House staff reporting to him and him reporting to Reagan, and with a veto vote on cabinet secretaries and other enhanced duties—Ford told advisers that he might consider it. With a narrow window of opportunity open, negotiations transpired between the Reagan and Ford camps who tried to strike a balance between how much power it would take for Ford to agree and how much Reagan would be willing to give up.

The following night, July 15, after addressing the convention delegates at Detroit's Joe Louis Arena, Ford was interviewed by CBS's Walter Cronkite, who raised the question of Ford's receptivity to being on the ticket. Ford confirmed that he would entertain it if Reagan were willing to share some of the presidential duties. "It's got to be something like a co-presidency?" Cronkite suggested. Ford replied that the arrangement would be problematic but didn't dismiss the idea out of hand as an inappropriate division of presidential responsibilities. It gave Reagan pause. *Wait a minute,* he thought as he watched the interview in his suite at the Detroit Plaza Hotel, *he's really talking about two presidencies here.* "As far as Ronnie was concerned, that did it," Nancy Reagan recalled of her husband's reaction. But the matter wasn't closed.

Bush watched Ford's interview with great interest and then went to the arena himself to address the delegates, where he was told by a backstage worker erroneously that Reagan had chosen Ford. The matter ostensibly closed, Bush delivered his speech, offering his unqualified support of Reagan, then retreated to his suite at Hotel Pontchartrain. After a couple of beers, he was prepared to call it a night and then head back to Houston with his political career behind him. Jeb, who worked on the campaign as hard as any of his siblings, reflected the somber mood as his parents began packing their bags. "It's not fair," he lamented to his father. "It's not fair to you."

"What do you mean it's not fair?" Bush interjected, devoid of self-pity. "'Course it's fair. We lost, and we have no reason to feel entitled to something."

At the same time, across town at the Detroit Plaza Hotel, another conversation took place between Reagan and Ford. Henry Kissinger, negotiating on Ford's behalf, held out hope that he could work something out the following day. But at eleven thirty, Ford went to see Reagan at his suite to officially withdraw his name from consideration, convinced that his initial instinct was right: A Reagan-Ford ticket just wouldn't work. The notion of a co-presidency had collapsed of its own weight. "It was a dream ticket, that threatened a political morass and a constitutional nightmare," *Newsweek* later opined.

Ford's withdrawal left Reagan with few options. Richard Allen, Reagan's foreign-policy adviser and later national security adviser, suggested that Reagan reconsider Bush. He was the safe choice: experienced, a nod to moderates, the runner-up in primary votes. Reagan, a pragmatist with his back against the wall and an election to win, agreed.

It was a quick decision. At 11:37, the phone in Bush's suite rang. Baker answered. It was Reagan, who asked to speak to Bush. Bush took the receiver. "Hello, George," Reagan boomed. "This is Ron Reagan. I'd like to go over to the convention and announce that you're my choice for vice president . . . if that's all right with you."

"I'd be honored, Governor," Bush said, whose thumbs-up to the room reflected his twist of good fortune to the immense relief of his family. After assuring Reagan that he would support his platform, Reagan announced his choice.

George W., in New York City raising capital for Arbusto, learned the news from a maître d' at the 21 Club, who wheeled out a portable TV set so he could see the scene play out in Detroit. He had reluctantly concluded during the campaign that it was Reagan's time, not his father's, and had developed grudging admiration for his father's opponent, whom he dubbed the "Western" man. When Reagan fired his campaign manager, John Sears, after *winning* the New Hampshire primary, replacing him with Bill Casey, George W. judged him "a man of decisive action," a high compliment from one who saw himself similarly. But, though George W. had watched the drama over Reagan's vice presidential pick with great interest, he thought his father stood "little chance" of getting the nod. "Surprised and thrilled" at his dad's change in fate, he and Laura quickly hopped a plane to Detroit to join his family to watch his father formally accept the nomination at Joe Louis Arena. "If anyone wants to know why Ronald Reagan is a winner, you can refer him to me," George H. W. Bush declared self-effacingly to the convention delegates. "He's a winner because he's our leader, because he has traveled this country and understands its people. His message is clear." Bush was back in the game; the Reagan-Bush ticket was off and running.

Bush had practiced what he preached to his children. He had taken a big risk and raced fervidly toward his ambition, leaving everything on the field. But he had once again come up short. And just when the light was about to go out of his political career, "out of the clear blue sky," as he said at the time, came the call from Reagan. If the Gipper's "Morning in America" was in the offing, its first glorious rays were cast on George Bush. Indeed, defeat wasn't defeat.

20

A HEARTBEAT AWAY

REAGAN-BUSH COASTED TO AN EASY victory in November, taking nearly 51 percent of the vote to Carter-Mondale's 41 percent, with John Anderson, running as an Independent, yielding just shy of 7 percent. Any hopes Jimmy Carter had to retain the White House had largely dissipated seven months earlier, on April 24, with his failed mission to rescue the fifty-two American hostages who had been held captive in Iran since the fall of 1979 when Iranian revolutionary "students" stormed the U.S. embassy. The aborted offensive, seen by many as a metaphor for Carter's well-intentioned but luckless administration, ended in tragedy as a helicopter exploded when it hit a transport plane, resulting in the deaths of eight American troops. Reagan offered a marked alternative. In contrast with the malaise of the Carter administration, he radiated the promise that America's best days were ahead; with his victory, the electorate turned the page to a decidedly new chapter, one they hoped would be as sanguine and sunny as the messenger's twinkly eyed disposition.

Preparing for their own next chapter, George and Barbara sold their Houston home for $843,273, rented a suite at the Houstonian Hotel, and moved into the official vice president's residence, the rambling three-story, 9,150-square-foot white brick home on the grounds of the Naval Observatory less than a quarter mile from the White House. Walter Mondale was the first vice president to call the mansion home, which had been designated for the vice president a few years earlier, though Mondale's predecessor, Nelson

Rockefeller, in keeping with all that his last name implied, chose instead to live in his own much larger Washington mansion.

Any lingering animosity between Reagan and Bush had long melted away on Inauguration Day, January 20, 1981. Bush wasn't one to look back and harbor grudges, nor was Reagan, who appreciated that Bush had held to his pledge to "work, work, work" for him during the campaign. The new administration began serendipitously with the release of American hostages who boarded a plane that left Iranian air space during Reagan's inauguration address, the fruition of a sleepless negotiation by Carter in the waning hours of his presidency, but a symbol of the new beginning as America and its new president looked forward.

Reagan's propensity for forgiveness extended not only to Bush but also to his campaign manager, Jim Baker, as well. Impressed by Baker's performance, Reagan tapped him to be his chief of staff. Baker would prove himself an adept administrator. Along with deputy chief of staff, Mike Deaver, and counselor to the president, Ed Meese, holdovers from Reagan's gubernatorial days, he rounded out the "troika" that would become instrumental in the smooth running and overall effectiveness of Reagan's first term in the White House, keeping a spit shine on Reagan's favorable public image. Baker and Bush occupied neighboring offices in the West Wing, steps away from the Oval Office. "I know I satisfied the Reagans that I was totally loyal to them," Baker said, "and yet I was in a position to see that the vice president was included in meetings."

Bush's instinctive loyalty flowed naturally to Reagan, who, after all, had rescued him from political obscurity and with whom Bush now marched in lockstep. In a 1982 press conference, Bush went as far as to deny that he had ever called Reagan's economic policy "voodoo economics" during the campaign, challenging the media to find evidence. (NBC News did, subsequently airing a clip of Bush making the accusation at an April 1980 campaign stop at Carnegie Mellon University.)

Two Texans had held the vice presidency within a half century of Bush taking the office, and both had earthily expressed their disdain for the office. Franklin Roosevelt's first VP, John Nance Garner, famously likened the office to "a warm bucket of piss," while Lyndon Johnson, Kennedy's number two, compared it to being "a stuck pig at a screwing match." It was better for Bush, thanks in part to Carter. Intent on strengthening the VP role, Carter had met regularly with Walter Mondale, including weekly one-on-one lunches. Reagan continued the practice with Bush. Every Thursday the two were in town, they met for Mexican fare in Reagan's private West Wing dining room, where Bush had the president's ear and Reagan was able to talk freely. As Bush said later, "Presidents don't get to do that very much." Reagan appreciated his vice president's ability to keep their conversations in confidence. "He would be forceful with Ronald Reagan, but not in front of a whole bunch of people trying to grandstand," George W. said of his father, whom he would emulate when acting as his father's adviser during the elder Bush's presidential run in 1988. "Reagan expected Bush to be a younger Nelson Rockefeller, and Bush thought Reagan would be a better dressed, better stated John Bircher," said Chase Untermeyer, Bush's executive assistant. "But they were naturally bound to be friends based on their humanity and the value they placed on good relations."

To be sure, there were personal as well as political differences between Bush and Reagan. Bush's Waspy northeastern background contrasted sharply with that of Reagan, the Irish son of Jack Reagan, an alcoholic shoe salesman whose peripatetic career took the Reagan family of four from town to town across Illinois just ahead of his sagging reputation. Bush revered his own father, the sterling character of whom made him his role model. Jack Reagan, on the other hand, was hardly an exemplar of moral rectitude. Ronald Reagan, in a rare moment of emotional openness, recalled as an eleven-year-old boy discovering his father flat on his back from a whiskey binge, "dead to the world" on the snowy front porch of their home where he managed, reluctantly, to drag him out of the wrath of winter, inside and into bed. While George H. W. Bush was the same kind of hero to his children that his father had been to him, Reagan maintained a distant relationship with his own children, two from his first marriage to actress Jane Wyman and two from his marriage to Nancy, his closest confidante and chief protector, whom he married in 1952. When Reagan expressed envy for Bush's close family and gaggle of grandchildren, Bush gently reminded him that he had grandchildren of his own. But the pair found common ground and developed a warm camaraderie. "There was no doubt who was the senior partner in our relationship," George said, "but our trust and friendship had grown with each passing day."

Bush actually *liked* being vice president. "I guess every Vice President had to endure the annual rounds of 'whatever happened to VP _____ stories,'" he wrote Nixon. "They don't bother me a bit. I like my job. I have plenty to do, and I believe I can be helpful to the President. So what else is there?" Actually, there was something else—something big. Bush bargained on the vice presidency offering a launching pad for the Oval Office after Reagan's turn, which alone may have been enough for Bush to withstand the inherent indignities of the job.

His ascent to the presidency nearly happened accidentally. Just seventy days into Reagan's administration, on March 30, Reagan made a speech to the Construction Trades Council at the Hilton Hotel, a mile and a half from the White House. As Reagan left the hotel through a side entrance at 2:27 p.m., raising his right arm to acknowledge a small crowd, he heard what sounded like "a small fluttering sound—*pop, pop, pop*." It was followed by a Secret Service agent pouncing on him and thrusting him into an awaiting limousine, which at the Secret Service's direction sped to George Washington University Hospital, not the White House, when blood began flowing from Reagan's mouth. Reagan thought the "excruciating pain" he felt was a broken rib, the result of the agent's force against him. He would discover later that one of three .22-caliber bullets from the gun of John Hinckley Jr., a deranged twenty-five-year-old from Colorado, had struck him in the chest—less than an inch from his heart. The decision to take the president immediately to the hospital likely saved his life. Reagan's press secretary (Jim Brady), a Secret Service agent, and a police officer were also hit in Hinckley's assault.

Bush was at Carswell Air Force Base in Fort Worth, on board Air Force Two awaiting a short hop to Austin, when he was alerted to the news of the shooting at 1:43 p.m. central time. The details were blurred. Initial reports from Washington indicated that Reagan was unharmed. Bush awaited direction and further details from the White House as he watched news coverage on a small color television set in Air Force Two's sixteen-square-foot stateroom, whose walls resounded with history; it was the same room in which Lyndon Johnson had taken the presidential oath of office on a runway at Dallas's Love Field after the assassination of John F. Kennedy, when the 707 aircraft served as Air Force One. Given the plane's nonsecure phone lines, Bush learned through a telex from Secretary of State Alexander Haig that Reagan had been struck in the back and was in serious condition. "Recommend that you return to D.C. at the earliest possible moment," Haig wrote. After a necessary stop for refueling in Austin, Air Force Two bound for Washington.

While Bush was airborne, Haig, a former four-star general who had been the second highest ranking officer in the army, asserted command in the White House Situation Room before bursting in on a press conference about the president's condition. Asked by a member of the press corps who was making decisions for the government, Haig replied erroneously, "Constitutionally, gentlemen, you have the president, the vice president, and the secretary of state. I am in control here at the White House, pending the return of the vice president."

Bush was conspicuously less zealous during the crisis. Upon his arrival back in Washington at 6:30 p.m. eastern time, he struck the right balance of urgency and understatement, rejecting a proposed plan for him to fly directly to the White House via Marine Two, the vice presidential helicopter. "Something about it didn't sit well with me," he wrote later. "It may well have made for great TV, but I thought it sent the wrong message to the nation and the world." As he explained to a military aide, "Only the president lands on South Lawn." Instead, he was transported by helicopter to his residence at the Naval Observatory, then by motorcade to the White House. There he entered "with perfect equanimity," recalled a White House aide, and confidently assumed the head of the table in the Situation Room. After reading a short statement to the press, he adjourned to his office, called members of the congressional leadership, and met with the wives of the police officer and the Secret Service agent who had been shot. Before leaving for the evening, he visited a shaken Nancy Reagan, whom he described to an aide as "tiny and afraid." Reagan recovered slowly, returning to the White House on April 11, but the attempt on his life was far more serious than the public realized.

The vice president won plaudits for his cool-headed restraint during the crisis, as well as respect among Reagan acolytes in the West Wing. The president did better, achieving hero status nearly overnight. He had faced the attempt on his life with cinematic stoicism and aplomb, walking into the emergency room at George Washington University Hospital before collapsing, and tossing off one-liners embossed immediately on the Reagan legend. When Nancy Reagan rushed to him at the hospital, he greeted her with a borrowed line from the boxer Jack Dempsey, who, after taking a pounding from rival Gene Tunney, remarked to his own wife, "Honey, I forgot to duck." When a medical team prepared the president for emergency surgery, he quipped, "I hope you're all Republicans," and asked by a nurse afterward how he was doing, he replied with a quote from W. C. Fields: "All things considered, I'd rather be in Philadelphia." How could you not like the guy?

It translated into legislative advantage for Reagan, whose political capital surged as his approval numbers, which reached a high of 68 in May, soared. Even the canny Tip O'Neill, presiding over a majority of Democrats in the House, knew he was beat. Congress, he said, would go with the will of the people, and "the will of the people will go along with President Reagan." The White House sped Reagan's conservative agenda through Congress, including a 25 percent across-the-board tax cut over three years and sharp increases in defense spending, which would swell by 35 percent throughout the course of his presidency.

There were setbacks. In 1982, the budget deficit surged to $110.7 billion despite Reagan's pledge to reduce federal spending, as the economy fell to its lowest levels since the Great Depression, with unemployment rising into double digits. Reagan's poll numbers plunged to 35 percent in 1983. The same year the country lost a barracks of 241 marines on a peacekeeping tour of duty in Beirut, Lebanon, to a bombing traced to Hezbollah, a militant Islamic terrorist group, prompting a change in Reagan's Middle East policy. But as the presidential election year of 1984 approached, beckoning Reagan's reelection, the economy righted itself enough that Reagan once again rode high.

21

ARBUSTO OR BUST

THE BUSH FATHER AND HIS eldest son lived separate lives during the bulk of George's vice presidential years. As Laura wrote later of herself and George W. during her father-in-law's first term as vice president, "[W]e were outliers on the family curve. We were not with them for Christmas, only for summer visits in Maine, with all the other cousins and meals for a small army of Bushes and wet beach towels strewn about the Kennebunkport house." When Barbara made a rare visit to Midland in 1984 as part of the Reagan-Bush reelection campaign she was greeted by her granddaughters as "Ganny," the traditional name for Bush grandmothers, for which Barbara expressed gratitude to Laura. "Thank you for teaching your girls to know me," she said.

In early 1983, George brought all his sons together for Washington's annual Alfalfa Club dinner, stag at the time, after which he wrote to them wistfully, "My heart was filled with pride when I got to introduce you to my friends at Alfalfa . . . I'm getting a little older. I'm not sure what the future holds. I don't worry about that. Win or lose, older or younger, we have our family." Otherwise, George and George W. kept in touch through frequent phone calls. "George would call young George for advice," said Shelley Bush Jansing, a Bush cousin. "He always appreciated George because he knows George's strength. He would call him when he had to make tough decisions. Reagan made George do some hatchet jobs that George didn't want to do. He didn't like to do them, so he used to call young

George [W.] for bolstering up. They've always had that kind of relationship." When Reagan was shot, George W.'s secretary saw her boss uncharacteristically shaky as he told her to hold his calls in the event his father wanted to connect with him. George was itinerant for much of his two terms under Reagan, traveling more than a million miles to seventy-four countries, including manifold funerals for heads of state and other VIPs, which prompted Jim Baker to offer an unofficial vice presidential motto, "You die, I fly."

As Vice President Bush pursued his duties, George W. was settled into life in Midland, where he and Laura lived in a $200,000 one-story brick home on the tree-lined Country Club Drive. After looking into adoption when they had difficulty conceiving, the couple discovered in the summer of 1981 that Laura was expecting twins. It wasn't an easy pregnancy. In her last trimester, she was diagnosed with toxemia, an affliction with symptoms that can include spikes in blood pressure, blurred vision, and pronounced swelling. She was admitted to Dallas's Baylor Hospital, a five-hour drive from Midland, where her condition could be better monitored. When it posed the threat of kidney failure, she underwent a cesarean section five weeks before her due date. Barbara and Jenna, named for their grandmothers, were born on November 25, 1981. The media delighted in the news of the arrival of the vice president's twin granddaughters, who appeared at their first press conference two hours after their births. The vice president himself would meet them during a visit to Midland at the end of the calendar year six weeks later.

"Laura and I didn't know anything about babies, and suddenly we had two," George W. wrote later. Both rose to the challenge. Wrote Laura later of her husband, "He changed diapers. He got up at night to help feed them their bottles. He would come home and think of adventures for them." Coached by their mother, the first word spoken by both girls was "Daddy." Bush quickly recognized fatherhood as the "most important responsibility" he would ever have, and as it had for George H. W. Bush, it came easily to him.

Business success did not. After his failed congressional push, George W. recommitted himself to building Arbusto, the company he established in 1977, and began actively operating in early 1979. Politics, though, was never far from his mind. "Running came up a lot," said Mark Owen, a geologist who went to work for Bush as the company grew. "After '78, he consciously decided to concentrate on his business and family, but he was always a politician. When his dad was vice president, he didn't want to get in on that because the 'dad's coattails' thing would come up. He was waiting for the right time." Despite no immediate outlet for his political ambitions, Bush's talent was put to use as he sought to raise capital to build his erstwhile one-man operation and enhance his small company's drilling capacity. "The politician was in him. He was a great promoter and a great money raiser," recalled Jim McAninch, who ran Bush's drilling operation. The Bush family network helped. George W.'s uncle Jonathan Bush, who had followed his father, Prescott, to Wall Street where he became a trusted stockbroker and money manager, teed up prominent East Coast investors who poured money into the company, now a full-fledged operating company called Bush Exploration Company, which garnered $4.67 million from limited investors. "George was an easy sale," Jonathan Bush said of his nephew. "I mean the people who met him would say, right away, 'I want to drill with this guy.'"

But in spite of George W.'s hope of finding a "company maker," a gusher that would propel Bush Exploration Company into the big time just as his father had at Zapata, the company never rose above an average performance. "I had some success in the oil business," Bush said later. "Not much." Of the ninety-five wells Bush drilled in the company's first five years of operation, about half yielded petroleum, nothing more than standard production in a capricious industry where fortunes are often made from rolls of the dice and blind luck. It was a far cry from his father's run of seventy-one consecutive profitable well drillings during Midland's fabled 1950s oil boom.

George H. W. Bush's ascent to the vice presidency was a mixed blessing for George W., eliciting pride in what his father had accomplished and pressure because the bar had been set higher—expectations were greater. He was "totally inebriated in hitting the big one," but it eluded him. While he lacked ambition earlier in his life, his company—mocked by some Midland oil insiders as "Arbust-oh"—was his concerted effort to prove himself in the business world, just as he had hoped to launch himself in the political world in 1978. It wasn't the money or the prospect of material gain that motivated him. Bush was hardly avaricious, as his shoddy wardrobe would attest, eschewing the conspicuous trappings of wealth even when he had the means. During what she described as Midland's "doodah days," Jan O'Neill recalled, "a lot of people couldn't resist—jets, boats, cars. George didn't go for that." Tangible achievement was what he wanted, proof that he had done something with all that was given to him. It was part of being a Bush. Andy Card, who would work for both of the Bush presidents, observed, "The competitive nature of the family is an underlying burden they all carry."

Now George W. worried about poor yields for his investors, which returned only $1.55 million in distributions, and what it would do to his standing in the industry—in addition to any future he hoped to have in politics. "Let's face it, George [W.] was not real happy," recalled Joe O'Neill of Bush. "It's the first-son syndrome. You want to live up to the very high expectations set by your father, but at the same time you want to go your own way, so you end up going kicking and screaming down the same path your father made."

Success proved more elusive as the industry went into bust mode in 1982, with an ominous dip in oil prices. A dim harbinger came the following year with the failure of Midland National Bank, awash in red ink from bad loans. Bush's prospects were bolstered in 1984 as he merged the Bush Exploration Company with Spectrum 7, a Cincinnati-based oil-drilling company run by supporters of his father, doubling the size of his operation. As Spectrum 7's president, Bush would oversee the drilling of 180 wells during his three years with the company. With each came the hope of a turnaround. "Stay alive till '85," went the saying among the oil community in Midland—but all the signs pointed to tougher times ahead.

22

"WHY DON'T YOU COME TO WASHINGTON?"

WHEN AMERICANS RETURNED TO THE polls in 1984, Reagan and Bush did more than stay alive. Reagan's masterful "Morning in America" campaign ad boasted, "Under President Reagan's leadership, America is prouder, and stronger, and better." The nation overwhelmingly agreed. A challenge by the Democratic ticket topped by Walter Mondale and his running mate, New York congressman Geraldine Ferraro, went down in a landslide as Reagan and Bush eased to reelection with 59 percent of the popular vote and an unprecedented 525 electoral votes, propelled by a gust of economic uptick and swelled national pride.

The question of Reagan's advanced age—seventy-three—came up in the campaign, just as it had in 1980, prompted not only by the fact that Reagan was four years older but also by an incoherent, meandering closing statement he made in the first of two debates with Mondale, seventeen years Reagan's junior. Afterward, Reagan's lead in the polls slid from 18 percent to 7. But in a second debate, Reagan deftly put the issue at bay. When asked about it directly, he replied, "I won't let age become a factor in this campaign. I will not exploit for political purposes my opponent's youth and inexperience." Even Mondale laughed.

But the creaks of age, and perhaps the Alzheimer's disease that would be diagnosed after he left the White House, began to show in Reagan's second term, which, while triumphant in many respects, suffered at times from Reagan's disengagement. When Jim Baker and Secretary of the Treasury Don Regan approached Reagan with the notion of swapping jobs, with Baker heading up Treasury and Regan taking over as chief of staff, Reagan blithely approved. It would soon prove an ill-fated move for Reagan's White House. Exploiting the autonomy as chief of staff given to him by the president, who readily delegated authority, Regan hoarded power and expanded the role to become a sort of prime minister, drawing the wrath of Nancy Reagan, her husband's chief protector. The drama that unfolded was a disruptive influence in the West Wing throughout much of the second term.

The Iran-Contra affair was another complication. In 1986, it was revealed that the White House had sold weapons, mostly antitank missiles, to Iran's Islamic government to help secure the release of American hostages held by terrorist groups in the Middle East, a breach of Reagan's policy of neutrality in the Iran-Iraq War. Proceeds from the sale were, in turn, illegally diverted to the contras, a revolutionary force in Nicaragua fighting to overthrow the Marxist government of Daniel Ortega Saavedra. Reagan's public denial—"We did not, repeat, did not, trade weapons or anything else for hostages, nor will we"—rang hollow. A prolonged congressional investigation into the matter, implicating members of his National Security Council, tainted Reagan, whose knowledge about what transpired was put into question. "A few months ago, I told the American people I did not trade arms for hostages," Reagan said in a nationally televised address. "My heart and my best intentions tell me that's true, but the facts and the evidence tell me it's not."

Both the Regan and Iran-Contra matters clouded the balance of Reagan's tenure in the White House—and both would touch his vice president.

With Reagan's second term in motion, the path was clear for George H. W. Bush to begin plans to seek the office himself in 1988. The dawn of his presidential campaign came on April 27, 1985, at a meeting around a long table at Camp David, the presidential retreat in Maryland's Catoctin Mountain Park, which had been loaned to Bush by the president for the occasion. On one side sat prospective hired guns—consultants for every aspect of the effort—along with Lee Atwater, the brash thirty-four-year-old political strategist from South Carolina, whom Bush had selected to lead his election effort. On the other was the Bush family: the vice president flanked by his wife, brothers and sister, and children. There was some practicality to the seating arrangement, with the possible campaign team poised to pitch the stakeholders across the table. But it underscored a constitutional truth: For the Bushes, family was family, and blood would always be thicker than water. If there were any doubt, the conversation that ensued would dispel any notion otherwise.

The Bush sons, led by George W. and Jeb, had come to the meeting leery of Atwater. George W. was particularly concerned about Atwater's close relationship with Charlie Black, the campaign architect for Jack Kemp, the former Buffalo Bills quarterback and New York congressman, a presumptive GOP presidential candidate who George W. believed would be a "formidable" competitor to his father. Atwater was a rising star among Republican political operators, a scrappy no-holds-barred whiz kid with a big personality and a Southern drawl as thick as South Carolina pluff mud. But would he be a Bush loyalist focused solely on victory for George H. W. Bush or a soldier of fortune trying to make a bigger name by hitching himself to the vice president's wagon?

"Why should we trust you," George W. snapped at Atwater, getting straight to the matter at hand.

It was a question he would ask repeatedly throughout the campaign of others trying to get close to his father. He could be brutal when protecting family—especially the old man. For the Bushes, loyalty was an outward manifestation of love. The two—love and loyalty—were fused together intrinsically. Politics, the Bushes knew all too well, is a contact sport. As in any campaign, there were those around the candidate who set strategy for him to win the fight, but few beyond family who thought about the resulting blood, bruises, and scars and could be counted on to subordinate their own ambitions. Early on, George W. designated himself the campaign's "loyalty yardstick," aggressively taking measure of those who wanted access to his dad.

"Let me rephrase what he's saying to you," Jeb interjected. "If there is a hand grenade rolling around the candidate, we want you diving on it first."

Atwater responded, ostensibly unfazed by the grilling, "I've known George Bush since I was in the Republican National Committee. I was the YR (Young Republican) chairman during the Republican National Committee when your dad was chairman. I've respected him and grown to like him and know that he is the kind of guy who needs to be our president."

George W. was satisfied with Atwater's answer, though he remained cautious. But the exchange had actually rattled Atwater, who approached George W. and Jeb afterward. "That was some tough questioning," Atwater said.

"Yes, but you've got to understand, Lee, I don't know what you've been dealing with in the past, but we are motivated by love in this family. We love this guy," George W. replied.

Atwater looked at the two Bush sons. "If you're so damned worried about my loyalty," he asked, "why don't you come to Washington?"

Two months later, during a Washington dinner with George W. and Jeb, Atwater raised the question again, this time with an entreaty. "I want y'all up here," he said.

Jeb, wary of the "palace intrigue" of national campaigns, would eventually opt to support his father's efforts in Florida, where he had settled in Miami and was fast becoming a player in the city's burgeoning commercial real estate market. George W., though, chewed on the proposition as he returned to Midland. Though he recognized that his father "did not want his life's course to interrupt my ventures," it was a chance he didn't want to miss. "At the time, the oil business in Texas had gone into the tank," he explained. "I had merged my company; I wanted to get out of Midland. The campaign was the perfect way for me to get a sabbatical out of the business world and do something really fun and meaningful for George Bush." But first there was the matter of how to extricate himself gracefully from Spectrum 7.

Indeed any hope of the energy business staying alive in '85 was dashed with the steady slide of oil prices, which toppled from eighteen dollars a barrel to ten in a bleak six-month span and sunk to nine dollars as the calendar year turned. As George W. wrote, "Midland was suffering through its version of the stock market crash of 1929." Drilling halted, banks failed, businesses went under, jobs were lost. The future of Midland, the mecca of the American oil industry, was uncertain even as the bulk of the country was knee-deep in red-white-and-blue Reagan optimism. "I don't know where the hell this is all going," Bush conceded to one of his fifteen employees in 1986, after taking a pay cut of 25 percent and cutting the salaries of his staff by smaller amounts. What he *did* know was that the only hope he had of saving the company was to sell it.

Serendipity and relief came from a merger Bush struck with Dallas's Harken Energy Corporation. The deal would turn Spectrum 7's business operations over to Harken, which would assume the company's $3 million debt, in exchange for $2 million in stock. Bush himself was given two thousand shares and a seat on the company's board—an A-list of big-time investors including George Soros—at an annual salary of $120,000. Not bad considering the state of the industry and the fact that Spectrum 7 had lost $406,000 in the two previous quarters. Little doubt that the deal was more attractive to Harken given that the name of the son of the vice president and presumptive '88 presidential candidate was attached to it, along with George W.'s Rolodex of contacts. Harken's founder, Phil Kendrick, who sold the company in 1983 but remained on as a consultant, said, "His name was George Bush. That was worth the money they paid him."

While the merger offered Bush an elegant exit from the oil business, he made sure his employees landed on their feet, remaining in Midland through the fall of 1985 to ensure that all of them had jobs after Spectrum 7 closed its doors. He spent much of the following year commuting back and forth from Midland to Dallas as he helped transition the business to Harken, though he had little to do with day-to-day operations. By October, he collected his shares of Harken stock amounting to $312,000, a nest egg that would yield a cool $835,000 when he sold them in 1990 to put up collateral as a minority owner of the Texas Rangers.

Liberated from Midland, George W. turned his attention toward his father's campaign. As he considered Atwater's proposed move to Washington to enlist in the effort full-time, the question became what role and title he would assume, accompanied, as he explained, by "enormous worry that I never fit in a niche." The problem was resolved during a visit to Washington when his father in the predawn hours slid a handwritten note under the door of his guest bedroom as he slept at the vice president's residence. It read:

> In politics, access is the key to power. You don't need a title, because you'll have all the access you need. If you really want to come up here, if it's right for you to do it, I'd love to have you.

Early in 1987, after a yard sale in front of their Midland home at which they sold off extraneous belongings, George W. and Laura packed up a small moving van and, with their five-year-old twins in tow, headed northeast to Washington in their Pontiac sedan and Chevy station wagon. The family moved into a modest town house in the Westover Place Apartments on New Mexico Avenue, a short walk to the vice president's residence on Massachusetts Avenue—the first and only time that George W. and Laura would live in the same city as his parents. In the ensuing months, George W. Bush would take up the most important mission of his forty-one years: making his father the forty-first president.

23

AWAKENING

"WHEN I WAS A CHILD, I spake as a child, I understood as a child, I thought as a child," reads one of Corinthians' oft-repeated verses, "but when I became a man, I put away childish things." Well before leaving Midland, coinciding with a religious awakening that had taken hold within him, George W. Bush resolved to put his drinking behind him and, in essence, put away childish things. There was a discernible change in the man who showed up in Washington to aid in his father's campaign. He was sharper, more mature and confident, almost serene, and sober—the result of a resolute effort toward self-improvement.

George W. realized that his life needed changing as he neared his forties. Fate had not tested him the way it had his father, who before the age of thirty, had faced the most agonizing of crucibles—the ugliness and desolation of war and the devastating loss of a child—that imbued him with maturity, grace, and equanimity. Neither had George W. realized his father's early promise or prominence. A Bush cousin put it bluntly: George W. "was on the road to nowhere at age forty." But as he drifted toward middle age, a confluence of circumstance had come to bear, driving him toward needed reform in his life: evening drinking that he conceded had become a problem; a struggling business in the doomed climate of Midland; a quest for answers and spiritual enlightenment; and a feeling of responsibility to clean up his act now that his father had a real shot at the White House. His crossroads came in 1986.

On Monday, July 28, 1986, George W.'s passage toward sobriety began after he awakened with a hangover in his room at the Broadmoor Hotel in Colorado Springs. Quitting was a decision that was long in coming. He knew he was drinking too much, and it gnawed at him. Since his college days at Yale, he had been avidly bending the elbow, which had resulted in a number of "mistakes," including his arrest for driving under the influence in Kennebunkport, Maine, at age thirty, in late summer of 1976. After moving back to Midland, where alcohol was a staple among the town's oil set, George W. frequently indulged in what he called the "four B's": beer, bourbon, and B&B, an after-dinner liqueur. Hangovers were a frequent consequence.

So Bush's booze-soaked dinner at the Broadmoor Hotel on the evening of July 27 was nothing new. He often partied hard as he did that night with Laura and their friends Joe and Jan O'Neill, Don and Susie Evans, and Penny Slade-Sawyer, who had all traveled from Midland for a quick Rocky Mountain getaway, joined by Neil Bush, also now in the oil business, who had driven from his home in Denver to meet them. As they celebrated the common fortieth birthday years of George W., Laura, Don, and Susie, the drinks flowed. Laura recalled later that she must have heard the same toast at least twenty times.

She had been asking her husband to quit drinking for some time, though it wasn't a threat to their marriage, or an ultimatum. "I wasn't going to leave George," she wrote. "But I was disappointed. And I let him know that I thought he could be a better man." Her husband's natural bombast was exacerbated by alcohol, and his impishness could turn to obnoxiousness or truculence, which became more unseemly as he got older. An illustration came in early April 1985 while Bush was at a Mexican restaurant in Dallas and spotted Al Hunt, the Washington bureau chief for the *Wall Street Journal*. Hunt had recently contributed to a story in the *Washingtonian* in which high-profile pundits speculated on what the presidential tickets for both parties would look like in 1988. Hunt had drawn Bush's rage by offering as his prophecy, "Kemp and Indiana Senator Richard Lugar against Hart and Robb," excluding any mention of George H. W. Bush, a logical front-runner for the GOP presidential nod. George W. descended on Hunt as he dined with his wife, Judy Woodruff, the chief Washington correspondent for PBS's *MacNeil/Lehrer NewsHour*, and their four-year-old son. "You no good fucking son of a bitch," George W. yelled at Hunt as he approached his table, "I will never fucking forget what you wrote!" Hunt, who barely knew George W. Bush, was thrown off by his reference to the *Washingtonian* article since he hadn't even mentioned his father. "[H]e was quite clearly lubricated," Hunt recalled of Bush. After the incident, Hunt concluded, "This is a guy who's got problems."

In a less direct way, Bush had come to the same conclusion. Alcohol had become "a priority"; the morning of July 28, 1986, was an "inflection point." After the fortieth birthday celebration, he aborted his regular three-mile morning jog around the Broadmoor Hotel, his head pounding, feeling worse than usual. It was then that he resolved to quit for good. He carried out his pledge quietly, ultimately irrevocably, initially telling only Laura and his close friends. There would be no repentance, twelve steps, or meetings in church basements. He stopped cold turkey. As he said in 2000, "I realized that alcohol was beginning to crowd out my energies, and could crowd, eventually, my affections for other people . . . When you're drinking, it can be an incredibly selfish act." Joe O'Neill, who had his own problems with drinking for which he would later check into the Betty Ford Center, saw a greater urgency in Bush's decision. "He looked in the mirror and said, 'Someday, I might embarrass my father. It might get my dad in trouble,'" O'Neill said. "And boy, that was it. That's how high a priority it was. And he never took another drink."

John Ellis, a Bush cousin, contended, "There was a strong feeling that he had become an embarrassment to the family, there had simply been too many incidents. I think several people, including George and Barbara, were just fed up." That may have been an over-statement. George W. recognized that his father was "frustrated" with him, while his mother, nearly thirty years after George W. put drinking aside, said, "We didn't know he was an alcoholic—talk about dumb parents." When asked whether his son's drinking at the time concerned him, George replied, "No." His actions in July 1986, the same month George W. committed to sobriety, spoke louder as he and Atwater agreed that George W. would be George's senior ad-viser in the 1988 presidential campaign, taking the lead role among all his siblings. "It was a big deal for me," George W. said later, a sign that he had his father's full faith and confidence.

George W.'s temperance was part of a quest toward self-betterment that began two years earlier with a search for religious mean-ing. He had long been an active churchgoer, attending services at First United Methodist Church with much of his friend set on any given Sunday. Still, his ability to seek guidance was limited. "He couldn't trust anybody," said Doug Wead, a campaign aide who worked closely with George W. as he wrote a 1988 biography of George H. W. Bush titled *Man of Integrity*, as a means of outreach to religious conservatives. "All through his life, his dad was a UN ambassador or was head of the CIA or head of the Republican Party, before he was vice president for eight years. So, he couldn't talk; he couldn't go to a counselor. He couldn't talk to a friend about what's going on in his life. But he's going every Sunday to this church. He's hearing this stuff, and he knows he has a drinking problem."

But in the spring of 1984, when he heard the aptly named Arthur Blessitt proselytizing on local radio, Bush was intrigued. At the time, oil prices had started to crater, jeopardizing his business, and Laura had made repeated pleas for him to stop drinking. Blessitt, a traveling preacher who gained renown as a sort of folk hero among the evangelical community for dragging a twelve-foot wooden cross across the country and throughout much of the world, was in town for a week of speeches at Midland College's Chaparral Center. Through a friend, Bush arranged to have a private meeting with Blessitt at his hotel.

Bush was forthcoming with Blessitt, who while pledging his discretion, implored him to follow the path of Jesus. "I encouraged him to tell his wife and his friends what was in his heart," Blessitt wrote of their encounter on his website. "I told him he needed now to grow in the Lord and to study the word of God and be open to his testimony. I told him I would not get up and announce in the meeting about him being saved." Whether Bush was actually saved by the meeting remained to be seen. He and Blessitt never spoke again, nor did Bush show any outward signs of religious conversion.

A different spiritual mentorship, though, began again a year later, in the frigid Maine waters not far from the family warmth and tradition of Walker's Point. It spoke volumes about George W.'s rarefied life that it came from Billy Graham—and perhaps it was prophetic. Graham, a friend of George and Barbara's, had offered spiritual and political counsel to every U.S. president since Harry Truman. He and Ruth, his wife of forty-two years, were guests of the vice president and second lady during the Bushes' annual August reunion in Kennebunkport, where they took part in a frenzy of activities—golf, tennis, swimming, and anything else that may have a competition attached—that typified a Bush summer get-together. One afternoon George W. and Graham took a walk on the beach, then ventured into the dark, heaving Atlantic Ocean. While it wasn't a baptism, it stirred something in Bush. Graham remembered little of what was said as they swam but recalled, "It was just two people. That was our beginning."

Their relationship developed during Graham's visit. One evening, in the expansive living room at Walker's Point as the waves steadily crashed into the rocky shore outside, Graham lectured the assembled

Bush clan in a private Bible study. George; his mother, Dorothy; and Graham sat on a couch. "They say you have to be born-again to go to heaven," the vice president challenged Graham. "My mom, who I'm confident will go to heaven, will say that she's not born-again. Billy, what do you say?"

"I had a born-again experience," Graham replied. "But I needed it, and your mom didn't."

George W., who stood in the back of the room sipping wine, thought, *Wow, that's a good answer.* It wasn't an epiphany, but he started listening more intently to Graham throughout the evening.

The following day, Graham "saddled up" to George W. during a walk and asked, "Do you mind if I send you a Bible?"

"No, I'd be honored," Bush said. As he recalled thirty years later, "There was no lecture, no grabbing of the shoulders. And I started reading it. It was the beginning of me reading the Bible. A religious walk began, and still goes on today."

Graham had the advantage of seeing George W. around the larger Bush family, providing insight into his character. "Billy Graham was watching the interplay of [George W.] and his family, and he asked, 'Do you have the right relationship with God?'" said Karl Rove. It was a question Bush carefully considered, taken as much by Graham's message as he was its messenger. "The man is powerful and humble," he said of Graham. "The combination of wisdom and humility was so inspiring to me individually that I took up the Bible in a more serious and meaningful way." Forever after, he remained grateful to Graham, who "planted a seed in my heart, and I began to change."

In addition to faithfully reading the Bible, Bush joined a Bible study group with Don Evans and taught Sunday school, while embracing sobriety, which gave him new energy and vitality. As Laura said, "George was always disciplined in a lot of ways, except drinking. And when he was able to stop, that gave him a lot of confidence and made him feel better about himself."

George W.'s open embrace of his Christian faith marked a difference between him and his father, a product of northeastern Presbyterianism, to whom faith was a private matter, not something to be gushed exuberantly. George bore an uneasy relationship with the Christian Right, which had coalesced as he took aim at the White House, galvanized by the Supreme Court's 1973 *Roe v. Wade* decision legalizing abortion and creating a potent force in the growing conservative wing of the party. Televangelists like Jerry Falwell, Jimmy Swaggart, Jim and Tammy Faye Bakker, and Pat Robertson, who would manifest his own presidential ambitions, wielded prodigious influence with a flock of socially conservative Americans whose litmus tests for candidates revolved around social issues like opposition to abortion and homosexuality, and support of school prayer. It compounded George's troubles with the right wing, which saw him as a pretender to the Reagan mantle. Though by necessity he wooed the Christian Right and the right wing as a whole, it was an uncomfortable courtship steeped in mutual circumspection. In 1987, he dismissed an invitation to address the Conservative Political Action Conference by telling his advisers, "Fuck 'em. You can't satisfy those people."

But his son could. In contrast to his father, George W. wore his faith easily, often effusively, engendering the trust of evangelicals and fitting more naturally with conservatives. Part of it was his own background as a product of Midland, where the unbridled display of faith is as common as sweet tea on a summer afternoon. He was one of them in a way his father would never be. Within the Bush family, he became "the religious guy." But to him, religion was a deeper and more profound experience of understanding and accepting the "unbelievable gift of grace."

It suited his father's campaign, where initially George W. played the role of a "semi-surrogate, special voter group person," charged, among other things, with creating an all-important bridge between his father and the right wing, including religious conservatives.

During his own presidential bid in 2000, when asked to name his favorite philosopher, he instinctively blurted, "Jesus Christ." Afterward, his father called him to say, "Don't worry, son. It's not going to hurt [your campaign]."

"No, I meant it," George W. replied, never imagining that it would be construed negatively.

As he gained his own prominence, George W. Bush as the Prodigal Son would become an irresistible and expedient narrative as the media told the story of the ne'er-do-well eldest son of George H. W. Bush who saw the error of his ways, found God, went on the wagon, and eventually made good by reaching the Texas governor's mansion and, ultimately, following in his father's prodigious footsteps to the White House. But Bush never saw himself that way. Sometimes he lived "a little too large," he admitted, but he was never the Prodigal Son because he "never felt disconnected" to his family. On the contrary, his transformation as he crossed into middle age was an attempt to be a worthier member within it—a better husband, a better father, a better son, and as Laura had hoped, a better man.

24

JUNIOR

FOR YEARS, GEORGE W. BUSH had been called "Junior." Never mind that he wasn't a junior, pointing out (just as his parents had) that he lacked the third of his father's four names—people called him Junior all the same. Often it was conjured as a jibe, as if to say, "You'd be nothing without your daddy's name"—"All hat and no cattle," in Texas parlance. He was dubbed Junior by his opponents in the congressional campaign of '78 and would be called the same—or worse—by Ann Richards, the incumbent Texas governor, when he challenged her for the state house in 1994. "Junior" was a means of marginalizing him.

But when he took up his job on his father's 1988 presidential campaign, the nickname Junior stuck, and that was fine with George W. Bush. This time it wasn't invoked as a pejorative. Rather, it pointed to his indispensability in the campaign as his father's progeny, namesake, and closest aide, the one who had total access. And this time, he wasn't trying to make a name for himself beyond his father's long shadow; he was trying to get his father elected to the nation's highest office. Mary Matalin, who worked for Bush as vice president and on his presidential campaign, and would go on to work in the administrations of both Bush presidents, said, "The only person that [George] Bush completely trusted on everything as his force, magnifier, his perfect translator—giving him information and translating it back with no personal agenda—was, as we called him, 'Junior.'"

The campaign marked a new beginning for George W., and the

stakes couldn't have been higher. "George W. didn't learn to channel his energy until middle age," Joe O'Neill said, "and he didn't feel real comfortable until he went to Washington. He hated Washington, but it charged him up." In November 1986, before Junior's arrival in Washington, anticipating his and Jeb's involvement in the campaign, George H. W. Bush wrote in his diary:

> I think George [W.] Bush coming up here will be very helpful and I think he will be a good insight to me. He is very levelheaded, and so is Jebby. I think some of our political people are thinking, "Oh, God, here come the Bush boys." But, you know where their loyalty is and they both have excellent judgment and they are both spending a bunch of time on this project.

Still, Junior was also aware that his standing in the "project" and with his father would have to be earned. "If you're the candidate's son and people respect your judgment [and] you're not in there as a loose cannon, but you're working as hard as I was, people eventually respect you," he said in 1990. "[Dad] is so anti-nepotistic that he bends over backwards to make sure—we really have to earn our spurs to get him to, in any way, interject his children above anybody of equal talent."

Atwater was one whom Junior won over quickly, though early on the relationship hit a snag when Atwater posed for a December 1986 *Esquire* article titled "Why Is Lee Atwater So Hungry?" featuring a photograph of the political wunderkind saluting the camera in red boxer shorts with his pants around his ankles. It fell to Junior to "chew his ass out" and admonish him for his immaturity. "You think this is bad?" he yelled. "Wait till Mother gets ahold of you!" The "fiercest of warriors," in his words, he could be "unpleasant" when he thought someone forgot that the campaign wasn't about them, necessitating a reminder, often vociferously delivered, "that they needed to serve George Bush, not the man, George Bush the cause, what he stood for."

Junior became the embodiment of George Bush the cause, including a self-appointed role as his father's media gatekeeper—"the guard dog at the gate," as one reporter called him—screening journalists by demanding of them, "Give me one good reason I should let you talk to George Bush!" When members of the media pushed back, calling him arrogant, he would say, "Just doing my job, protecting the old man." As the campaign progressed, Junior's role grew beyond the one as special interest surrogate that had originally been imagined. "My role evolved into helping Lee," he said, "and Lee became the closest person to me on the campaign. It was advice giving as well as caution; he would use me as a sounding board." They talked constantly, in their neighboring offices at the campaign's downtrodden headquarters in the Woodward Building on Fifteenth Street, and on long phone calls and daily jogs, forming a friendship and becoming an allied force on the campaign staff, which often battled against the vice president's White House staff, derided by Atwater and Bush as "the clerks."

Doubtless, the two saw the mutual benefits of a close relationship. George W. used Atwater as another set of eyes and ears on the campaign. Atwater, in turn, used George W. as an emissary to the candidate and a safeguard within the Bush family—including Barbara Bush, whose tender, grandmotherly public image belied that of the family enforcer behind the scenes. As Mary Matalin observed, "Even people that were doing a good job and loved her son and her husband, if they were in the middle of something that didn't go well for either of them, then she would look askance at them. She is a grizzly mama, protective of all [her] people." George W., too, saw his mother's influence. "[E]veryone in the campaign was afraid of Mother to a certain extent, because they knew that access was power," he said.

George W. incurred his mother's ire early in the campaign as rumors surfaced in *Newsweek* and *U.S. News & World Report* that his father had had an affair with his longtime aide, Jennifer Fitzgerald. The media reported on the allegations, which consumed the campaign as it focused on the upcoming Iowa caucus. When George W. approached his father directly about the rumors, his father replied, "They're just not true." That was all George W. needed to hear. When the lie "impugned the character of a man I loved and respected, I couldn't hold back," he wrote later. Atwater agreed, advocating an immediate response—"Lance the boil," he prescribed. The two men hatched a plan to meet with *Newsweek*'s Howard Fineman to offer a denial on behalf of the family. "The answer to the big A, is N-O," George W. told Fineman. Barbara feared that her son's response would add fuel to the fire, triggering another round of coverage. "How dare you disgrace your father by bringing this up?" she demanded of her son, who believed she also reflected the views of his father. Happily for all concerned, George W.'s quip, after drawing national attention, effectively extinguished the rumor and the story.

Together the Bush mother and son; Atwater; media consultant Roger Ailes; pollster Robert Teeter; Bush's chief of staff, Craig Fuller; and longtime friends Nicholas Brady, Bob Mosbacher, and Jim Baker became Bush's chief advisers in the campaign. George W. called them "one of the great campaign brain trusts there ever was."

The experience offered George W. a political education at the highest level, including the lesson that perception in politics is reality. "I would intentionally be at the vice president's house having a Coke with Mother and Dad prior to the arrival of an important meeting and leave, just so people would know I was around," he said. "At times, I think they would probably think I was discussing some important matter of foreign policy. In fact, we were probably talking about how my daughters were doing." It helped that his proximity to power inspired "a little fear" of him among those around the campaign.

He learned early on how to use his access to his father. When he relayed advice from family friends imploring him to "just be yourself," his father looked at him quizzically. "Who the hell else am I going to be?" he replied. Afterward, George W. more judiciously filtered the information he passed on knowing his father was being bombarded with suggestions from all sides—much of it useless—and the importance of "decompression" during his down times. "He was his father's confidant, is what I'd say," Jim Baker said. "His father would ask, 'What do you think is going on over there at the campaign? What do you see? Who's doing what to whom?' And W. was the one who would be there to tell him. And I think he served a very useful purpose for his father."

Junior quickly learned how far to push his father, and when to back off and defer to his judgment. When he arrived in Washington, there was a growing clamor in the White House over Donald Regan, who was becoming a disruptive force in the Reagan White House, as he prodded Reagan, recovering from surgery from abdominal cancer, to wage an aggressive counteroffensive on the Iran-Contra affair. At the urging of campaign advisers, George W. met with his father to suggest that he publicly rebuke Regan and call for his resignation. It was a short conversation. "When he knows he's right or thinks he's right and gets tired of a subject, he can be pretty tough," George W. recalled. "He just looked at me and said, 'That's wrong policy.'"

George W. later realized that his father's reasoning during their conversation was sound: The vice president, or any other administration official, publicly demanding the resignation of Regan would compromise the president. "A strong [leader] makes those decisions," George W. said later. "George Bush knew that Don Regan had to go. [But] the way he went about it was very private in order not to make Ronald Reagan [look] weak."

It was a tenet to which the elder Bush would adhere after his son, the one-time "Junior," had become the forty-third president, and the shoe was on the other foot. When pressure mounted for George W. to reject the influence of his bellicose vice president, Dick Cheney, and the neocons in his administration who had ostensibly steered George W. off course in the war in Iraq, George never interfered, letting his son find his own way. Any public suggestion of disapproval, even the impression of dissent, he knew, might weaken him.

Defining portraits of the two Bush presidents: George Herbert Walker Bush, a blend of prim, patrician New England, where he was reared, and rough-and-tumble Texas, his adopted state, where he came into his own; and George Walker Bush, New Haven–born and Ivy League–educated, but a cultural product of his boyhood home, Midland, Texas, with his Lone Star swagger and unvarnished charm. *(George Bush Presidential Library and Museum; George W. Bush Presidential Library and Museum)*

TOP LEFT: George H. W. Bush, a child of Brahmin privilege, nicknamed "Little Pop" or "Poppy" for his namesake and paternal grandfather, George Herbert Walker, and "Have-half" for his natural generosity. *(George Bush Presidential Library and Museum)*

TOP RIGHT: U.S. Navy pilot GHWB during World War II. Defying his father's wishes, Bush enlisted in the Navy on his eighteenth birthday. When his torpedo bomber was shot down over the Pacific, Bush lost two crewmates, emerging as the plane's lone survivor. "In my view, there's got to be some kind of destiny," he reflected later, "and I was being spared for something on earth." *(George Bush Presidential Library and Museum)*

FACING PAGE TOP: Barbara, George W., George, Dorothy, and Prescott Bush, Odessa, Texas, March 1949. George declined to take his father's path to the riches of Wall Street, opting instead to make it on his own in the West Texas oil business. "And I never looked back," he said. *(George Bush Presidential Library and Museum)*

FACING PAGE BOTTOM: George, George W., and Barbara, Rye, New York, 1955. "George [W.] aggravates the hell out of me at times," wrote the elder Bush to his father-in-law the same year. "But then at times I am so proud of him I could die." *(George Bush Presidential Library and Museum)*

TOP: The Bush family—Doro, George, Jeb, Marvin, George W., Neil, and Barbara—pose during George's 1966 congressional campaign, as the boys imitate their father. Campaigning would become a Bush family affair. "Being in politics either brings you together or sends you apart," Barbara said, "and in our case, it brought us all together." (*George Bush Presidential Library and Museum*)

BOTTOM: George W. and Laura Bush in the late 1970s. Married just four months after being introduced in 1977, they were, in Laura's view, "two complementary souls." Laura would play a vital supporting role for George W., just as Barbara played for George. (*George Bush Presidential Library and Museum*)

Vice President Bush and President Ronald Reagan at the 1984 GOP National Convention. Four years earlier, overcoming "strong reservations," Reagan had tapped Bush to be his running mate, rescuing his former rival from almost-certain political obscurity. The two men went on to forge a close working relationship in the Reagan White House.
(George Bush Presidential Library and Museum)

George W. Bush and his twin daughters, Barbara and Jenna, pose with Ruth and Billy Graham, Kennebunkport, Maine, summer 1983. Three years later, in a quest to be a better man, George W. would put heavy drinking behind him, and, inspired by Billy Graham's spiritual mentorship, begin a "religious walk that still goes on today."
(George Bush Presidential Library and Museum)

TOP: Lee Atwater, George W., and Jeb with Vice President Bush on Election Day, Houston, Texas, November 8, 1988. George W. devoted a year to his father's presidential campaign, becoming one of his key advisers, a sign that he had his father's full faith and confidence. *(George Bush Presidential Library and Museum)*

BOTTOM: Bush, Ronald Reagan, and Mikhail Gorbachev in New York with lower Manhattan in the background, a month before Bush's inauguration, December 1988. Reagan would receive disproportionate credit when the Soviet Union collapsed the following year, but it was Bush as president who would help to ensure a peaceful end to the Cold War through his quiet diplomatic facility and inherent humility. *(Ronald Reagan Presidential Library and Museum)*

TOP: At Blair House on the morning of his inauguration, January 20, 1989, President-elect Bush reviews his speech, a reminder of one of his grandchildren to his left. Family was a balm for Bush during his presidency, lessoning its burden. *(George Bush Presidential Library and Museum)*

BOTTOM: Bush meets with his advisers Brent Scowcroft, John Sununu, James Baker, and Dick Cheney after Iraq's invasion of Kuwait under the order of Iraqi dictator Saddam Hussein, August 1990. Initially, Bush thought there was little that the United States could do before proclaiming that Saddam's "naked aggression" against Kuwait "will not stand." *(George Bush Presidential Library and Museum)*

Bush, with Norman Schwarzkopf, visiting U.S. troops in Saudi Arabia, November 1990. Reports of the atrocities inflicted on the Kuwaiti population by their Iraqi invaders made the Gulf War a moral imperative for Bush, who resolved to liberate Kuwait regardless of congressional approval. "If I don't get the votes," he told an aide, "I'm going to do it anyway. And if I get impeached, so be it." *(George Bush Presidential Library and Museum)*

George P., Jeb's oldest child; George; and George W. at Opening Day at Camden Yards, Baltimore, Maryland, April 1992. By then, George W., who had struggled to find business success, had helped engineer the purchase of the Texas Rangers and the building of a new state-of-the-art stadium, increasing his visibility in Texas and paving the road toward his political rise. *(George Bush Presidential Library and Museum)*

TOP: Bush and Bill Clinton, just prior to their third presidential debate, Michigan, October 1992. Clinton successfully reinforced public impressions that Bush was out of touch and lacked domestic vision. Bush would be devastated by his defeat at Clinton's hands, convinced that Americans didn't know his "heartbeat." *(George Bush Presidential Library and Museum)*

BOTTOM: George W. and Laura, the governor and first lady of Texas, pose with the elder Bushes at a celebration of the former President Bush's seventy-fifth birthday. The political ambitions of George W. and Jeb gave George a sense of purpose in his post-presidency, helping him, from George W.'s perspective, "to recover" from his 1992 reelection loss. *(George Bush Presidential Library and Museum)*

TOP: Lynne and Dick Cheney, George W., Don Evans, Jeb, Laura, and George watch election returns at the Texas governor's mansion in the early morning hours of November 8, 2000. The presidential election seesawed dramatically between Bush and his Democratic opponent, Al Gore, who conceded defeat to Bush before his retraction as polling results in Florida later proved inconclusive. It would take a U.S. Supreme Court decision thirty-six days later to proclaim Bush the election's winner. *(David Hume Kennerly)*

BOTTOM: Inaugurated earlier in the day as the forty-third president, George W. Bush greets his first Oval Office guest: the forty-first president. Trusting him to find his own way in the role, "41" refrained from proffering unsolicited presidential advice to "43." *"We had our chance,"* was his view. *"Now it's his turn."* *(George W. Bush Presidential Library and Museum)*

TOP: On September 11, 2001, after learning that the second of the two World Trade Center towers has been struck by a plane, Bush gathers more information in a classroom at Emma E. Booker Elementary School in Sarasota, Florida, as White House Director of Communications Dan Bartlett points to television footage of the burning tower. The defining event of his presidency, 9/11 instantly made Bush a "wartime president," something he "never wanted." *(George W. Bush Presidential Library and Museum)*

BOTTOM: Bush receives a briefing from Vice President Dick Cheney and Secretary of State Condoleezza Rice, at the White House, September 12, 2001. Cheney's dominant neoconservative influence on foreign policy in Bush's first term began to wane during Bush's second term as the president began to rely more on Rice. *(George W. Bush Presidential Library and Museum)*

FACING PAGE TOP: With manifest pride, George pats George W.'s arm after the latter's speech at a post-9/11 National Day of Prayer and Remembrance at the National Cathedral in Washington, DC, September 14, 2001. The moment would be the most indelible memory each would hold of the other during 43's eight-year presidency. *(George W. Bush Presidential Library and Museum)*

FACING PAGE BOTTOM: Later the same day, atop the rubble of the World Trade Center, President Bush rallies first responders in lower Manhattan, with retired firefighter Bob Beckwith at his side. During the visit, the mother of a missing Port Authority officer pressed her son's police badge into Bush's hand, which he carried throughout his presidency as "a reminder of all that was lost." *(George W. Bush Presidential Library and Museum)*

TOP AND BOTTOM: Echoing his father's resolve and solemn walk around the South Lawn after ordering Operation Desert Storm in 1991, 43 walks the White House grounds reflectively after giving the go-ahead to begin Operation Iraqi Freedom, his most fateful presidential decision, March 19, 2003. Immediately afterward, he wrote his father, "I know what you went through." *(George Bush Presidential Library and Museum; George W. Bush Presidential Library and Museum)*

TOP: Peering out a window of Air Force One, President Bush surveys the destruction of Hurricane Katrina over New Orleans, creating the impression of detachment and indifference, August 2005. Striking as the War in Iraq was becoming increasingly unpopular, Katrina marked the low point of 43's presidency. *(George W. Bush Presidential Library and Museum)*

FACING PAGE
TOP: Five presidents—Bush 41, Barack Obama, Bush 43, Bill Clinton (obscured by Bush 43), and Jimmy Carter—gather to inaugurate the George W. Bush Presidential Library, Dallas, Texas, April 13, 2013. Forty-one attended the gathering against all odds, after battling a near-fatal bout with pneumonia just several months earlier. Forty-three believed it was his father's love for family that pulled him through. *(George W. Bush Presidential Center)*

BOTTOM: George W. and Laura join Jeb Bush in a last-ditch effort to save Jeb's 2016 presidential campaign, North Charleston, South Carolina, February 16, 2016. The Bush presidential dynasty was derailed after Jeb's loss to his populist challenger Donald Trump, a "blowhard" in 41's view, for whom neither 41 nor 43 voted. *(George W. Bush Presidential Center)*

Statues of the Bush presidents—only the second father and son to hold the nation's highest office—at the George W. Bush Presidential Library in Dallas. *(George W. Bush Presidential Center)*

25

"BE TOUGH"

DESPITE BEING A HEARTBEAT AWAY from the presidency, and engendering the trust of Reagan, George H. W. Bush's path to the Oval Office would be an uphill climb. History lined up against him; no sitting vice president had been elected to the office since Martin Van Buren was elected in 1836, succeeding the two-term Andrew Jackson. Two incumbent VPs, Richard Nixon and Hubert Humphrey, had tried within the previous twenty-eight years to win the White House from the perch of the vice presidency; both failed by narrow margins. Additionally, no party had retained the White House for more than two terms since Franklin Roosevelt and Harry Truman served a collective two decades from 1933 to 1953. Moreover, an impressive field of candidates stood in the way of his party's nomination, including Senate majority leader Bob Dole, New York congressman Jack Kemp, and televangelist Pat Robertson, all with bona fide conservative credentials that made them viable heirs to Reagan.

There was also the matter of Iran-Contra, which clung to Bush as he sought the presidency. The scandal and its resultant congressional investigation by the Tower Commission, a three-person committee appointed by Reagan and headed by Republican senator John Tower from Texas, had led to the resignations of national security adviser Robert "Bud" McFarlane and his successor, John Poindexter, as well as National Security Council aide Oliver North. Bush, however, denied having an "operational role" in the matter and claimed not to know enough to weigh in when Secretary of State George Shultz and

Secretary of Defense Caspar Weinberger argued against an arms-for-hostage deal in a White House meeting he attended with Reagan in early January 1986, one of at least several meetings Bush attended in which the subject was raised. In an August 1987 interview with the *Washington Post*'s David Broder, Bush said, "If I had sat there and heard George Schultz and Cap [Weinberger] express [opposition to the Iran arms deal] strongly, maybe I would have had a stronger view. But when you don't know something, it's hard to react . . . We were not in the loop." As opinion polls reflected, the public had its doubts about Bush's stated denial. Even if he was telling the truth, the affair tarnished him. Either he was an ineffectual vice president who hadn't been privy to information relating to significant policy meetings, or he was deliberately misleading the public on his knowledge and acquiescence in the matter.

In fact, Bush *was* aware that the administration was trading arms for hostages. On February 1, 1986, after a meeting when trading arms for hostages was discussed as a means of cultivating relations with Iranian moderates, newly appointed national security adviser John Poindexter recorded a note afterward that read, "Most importantly, president and VP are solid in taking the position that we must try." "I'm one of the few people that know fully the details, and there is a lot of flack and misinformation out there," Bush wrote in his diary nearly two months earlier, on November 5, 1986. "It's not a subject we can talk about." He dismissed the subject when George W., returning from a Notre Dame football game after the Iran-Contra story broke, called from a pay phone in Tulsa to register his concern.

"Dad, what is this Iran-Contra thing?" he asked.

"Don't worry about it," his father replied, perhaps hopefully. "It's not going to affect my campaign at all."

Ronald Reagan loomed over the election despite approval ratings that flagged over Iran-Contra. Bush's sons often won bets by proving that their father was taller than Reagan—Bush's six feet two versus Reagan's six feet one. But it belied a more salient point: In the eyes of many Republicans, George H. W. Bush didn't measure up to the Republican icon. Deference to Reagan hurt public perceptions of the vice president who was seen as lacking the moral conviction and strength to stand on his own. In his popular *Doonesbury* comic strip, left-leaning cartoonist Garry Trudeau lampooned Bush as putting his manhood in a "blind trust." Conservative columnist George Will landed a more devastating blow, writing that the "unpleasant sound" Bush emitted "as he traipses from one conservative gathering to another is a thin, tinny arf—the sound of a lapdog." In October 1987, *Newsweek* put the issue front and center. Its cover bore an image of the Bush patriarch blazing through the waters off Maine's coast, eyes on the horizon, at the helm of his boat *Fidelity*, blaring the headline, "Fighting the 'Wimp Factor.'" The attendant story read:

> Bush, who formally declares his candidacy this week, enters the nomination fight with enviable advantages—high name recognition and stronger voter ratings for experience and competence. Other candidates can spend an entire primary season trying to match those assets. Yet Bush suffers from a potentially crippling handicap—a perception that he isn't strong enough or tough enough for the challenges of the Oval Office. That he is, in a single mean word, a wimp.

"[T]hat was about as angry as I've heard him," George W. said of his father's reaction to the story. "I mean, he was just really mad about what he thought was unfair treatment—and it *was* unfair." In a diary entry after the story appeared, George wrote:

> That *Newsweek* story was the cheapest shot I've seen in my political life. The "wimp" cover, and then everybody reacts—

pick a fight—be tough—stand for something controversial, etc., etc. Maybe they're right, but this is a hell of a time in life to start being something I'm not. Let's hope the inner strength, conviction and hopefully, honor can come through . . .

Inner strength, conviction, and honor were those essential qualities under the surface that revealed Bush's character. It's what Bush wanted the electorate to see, just as his family and close friends did. And it's what he wanted to be judged on, not the superficial, often fatuous posturing and stagecraft of elective politics.

Bush combatted those challenges as the campaign drew closer to the first pivotal battleground, the Iowa caucus, which had catapulted him into front-runner status eight years earlier with his upset over Reagan. This time, though, things were different. Bob Dole, whom Bush had supplanted as Republican National Committee chairman fifteen years earlier and who had won out over Bush as Gerald Ford's running mate in 1976, had the edge going into the Hawkeye State. A native of neighboring Kansas, Dole resonated with the local electorate, which had a long history of fickleness and held a dim view of Ronald Reagan.

On February 8, Dole came out of Iowa with 37 percent of the vote, with Robertson trailing at 25 percent, and Bush at 19. That evening, in Nashua, New Hampshire, where the all-important New Hampshire primary would be held in eight days, Bush held a press conference. "This is the beginning," he said, managing expectations. "I can't say that I'm not disappointed, but I'm not down. I will guarantee you that." Then, after congratulating Dole and Robertson, he added, "I'm coming after them." Still, once again, Bush was staring at the political abyss. The same evening, the *New York Times* wrote, "Even before tonight's weak showing, Mr. Bush's lead looked very shaky. National surveys have shown that a large share of his supporters are still not firmly committed to him and might easily shift to Mr. Dole."

Bush dug in, determined to use New Hampshire to reverse his political fortunes, just as Reagan had done at his expense in 1980. A scheduled interview with CBS anchor Dan Rather provided an opportunity. Beforehand, the Bush campaign had gotten wind that Rather was planning to use the interview to ambush the vice president on Iran-Contra instead of pursing a line of questioning related to Bush's presidential campaign. When Bush insisted that he didn't remember details relating to his involvement in Iran-Contra after Rather's repeated questions, Rather insisted he didn't want to be argumentative.

"You do, Dan," replied Bush.

"No, sir, I don't," Rather insisted.

Bush then summoned a planned response suggested by Roger Ailes pointing to a 1987 episode in which Rather, angered by CBS's delay of his broadcast as the network wrapped up coverage of a tennis match, huffed off the set of a live news broadcast followed by several interminable minutes of dead air. "I don't think it's fair to judge my whole career by a rehash on Iran," he said. "How would you like it if I judged your career by those seven minutes when you walked off the set in New York?"

Afterward, seeking reassurance, Bush called George W. who judged that he had "knocked it out of the park." "Dad, this is awesome," he said. "You stood your ground, you didn't let him bully you, and the American people are going to appreciate this."

They did. The counterattack coupled with a foray against Dole as a flip-flopper on key issues—"Senator Straddler," a Bush ad labeled him—showed Bush in a different light: aggressive, more in command. Dole, battling his own image issues as a kind of cantankerous sexagenarian who might not throw a ball back if it landed in his yard, hurt his own cause by snarling, "Stop lying about my record," to Bush in a televised interview the night of the New Hampshire primary. Aided by a strong endorsement from New Hampshire's governor, John Sununu, who would later be rewarded with the chief of staff role in Bush's White House, Bush won the primary just as resoundingly as Dole had in Iowa. The vice president yielded 38 percent of the vote to Dole's 28 percent, with Robertson pulling just 9 percent. The momentum shifted to Bush, who parlayed his New Hampshire victory into a decisive showing on "Super Tuesday," on March 8, by racking up wins in sixteen of seventeen primaries in Southern states, including his own Texas, which he won statewide for the first time, all but assuring the delegates needed to secure his nomination.

The vicissitudes of political fortune had fallen Bush's way. On March 17, the *New York Times*, which had sent a message of foreboding after Bush's defeat in Iowa, now changed its tune. A story titled "Bush vs. Dole: Behind the Turnaround," read, "George Bush's success in transforming himself from a loser to an almost certain winner in just twenty-nine days is one of the remarkable stories of recent American political history." The nomination now well in hand, the Bush campaign turned its attention toward the general election and the largely unknown governor of Massachusetts, Michael Dukakis.

26
NOT SO KIND, NOT SO GENTLE

THE DEMOCRATIC RACE WAS NOT as conclusive. Super Tuesday had come and gone with no clear winner or standout in a field that included Michael Dukakis, Missouri congressman Richard Gephardt, U.S. senator from Tennessee and congressional legacy Al Gore, and civil rights leader Jesse Jackson, who had never held elective office. In a game of inches, Dukakis broke out of the pack a month later, in mid-April, with a clear primary victory in New York and the mother lode of delegates that came with it.

Exuding competence if not charisma, the fifty-five-year-old Massachusetts governor had presided over the "Massachusetts Miracle," an economic turnaround that brought the Bay State back from the economic gloom of the 1970s. Dukakis boasted of 800,000 new jobs in his state in a dozen years and a negligible unemployment rate of just under 3 percent—imagine, he suggested, what he could do for America. It was enough to intrigue voters, who gave Dukakis a thirteen-point lead over Bush in the polls by late May—a gap that owed as much to negative impressions of Bush and a waning confidence in the country's economic future as it did to favorable views of Dukakis. As a Dukakis strategist put it, "They are losing the election at this point; we're not winning it."

Bush's bare-knuckled toughness in the primaries continued as his campaign shifted to the general election. "What's fourteen inches long and swings in front of an asshole?" Bush joked privately to his inner circle, answering, "Dukakis's tie." Meanwhile, publicly he hammered away at his opponent's liberalism, such as his veto of a bill that mandated the Pledge of Allegiance in Massachusetts's public schools and his sanction of a program allowing for weekend furloughs for imprisoned felons, including those who had been sentenced for murder. "Let 'em stay where they belong," he declared at a rally. It was a theme he would come back to with great effect.

The Democrats struck back at their national convention in Atlanta in mid-July. Dukakis sailed into the convention with a credibility boost after announcing his choice as running mate, Lloyd Bentsen, the smooth, probusiness conservative senator from Texas, who had defeated Bush for a seat in the U.S. Senate in 1970. Harkening back to the John F. Kennedy–Lyndon Johnson Democratic ticket of 1960, which boasted a Boston-to-Austin regional alliance, the selection of Bentsen provided geographic and ideological balance to Dukakis's ticket while lending it an air of gravitas and establishment credibility.

Ann Richards, the quick-witted, silver-coifed treasurer of Texas, offered a pithy keynote address. "After listening to George Bush all of these years, I figured you needed to know what a real Texas accent sounds like," she drawled to the party faithful. "Poor George, he can't help himself. He was born with a silver foot in his mouth." The zinger scored, launching Richards into the national spotlight and poising her for a successful bid for Texas governor two years later. The only hiccup came when the forty-one-year-old governor of Arkansas, Bill Clinton, angling for his own shot at national attention, gave a rambling, uninspired thirty-three-minute policy speech—over twice the expected length—before endorsing Dukakis, which provoked boos and sparked rousing applause only when he spoke the words "In conclusion . . ." Otherwise, the Democrats staged a polished convention resulting in a bounce in the polls for the party's ticket cresting at seventeen points. The Dukakis campaign left the convention, as the party's presidential nominee put it, "with the wind at our backs." But as the *Washington Post* warned in an article recapping the triumphant proceedings, "Dukakis, a native of often stormy New England, knows that the wind can shift abruptly."

George H. W. Bush, another New England native, knew it, too. His turn came a month later as the GOP convened in New Orleans on August 15, braving the sultry summer heat to crown him as its standard-bearer. The vice president arrived in Louisiana with much speculation as to whom he would choose as his own number two. A number of names were floated, including former opponents Bob Dole and Jack Kemp, both logical picks to satisfy the party's skeptical conservative wing. Other names had been considered, as well. Donald Trump, the self-promoting, braggadocious New York real estate mogul and author of the 1987 bestselling book *The Art of the Deal*, made unsolicited overtures to the Bush camp to suggest that he was available to be Bush's running mate. Intrigued by Trump's unconventionality, Lee Atwater entertained the notion, going so far as to have a phone conversation with Trump; Bush, when presented with the option, summarily rejected it for the same reason.

Acting alone without consulting any member of his family or inner circle, Bush went a different and unexpected direction. His choice was Dan Quayle, the attractive forty-one-year-old conservative junior senator from Indiana who had won the seat eight years earlier, riding the momentum of the Reagan-Bush landslide in 1980 with an impressive win over respected veteran Democrat Birch Bayh. Quayle, Bush reasoned, brought youth to the ticket—ushering the baby boomer generation into national politics, which had been dominated by the World War II generation since the "torch" had been passed from Dwight Eisenhower to John F. Kennedy nearly three decades earlier.

Sensitive to how the suspense of his VP pick might be "degrading" to those on the short list—just as he had been earlier in his career—Bush decided to announce his decision at a scheduled rally at New Orleans's Spanish Plaza on the banks of the Mississippi River, after a cruise on a riverboat.

Effusive and overeager, Quayle giddily bounded on stage radiating a less-than-ready-for-prime-time impression. He grabbed Bush by the shoulders and punched him on the arm as he exclaimed to the crowd, "Let's go get 'em. All right? You got it?" The press's reaction was tepid. The *Los Angeles Times* wrote, "In naming the staunchly conservative Quayle, scion of a rich and powerful family in a traditionally Republican state, Bush passed over a flock of better-known contenders who might have provided the ticket with greater experience, broader credentials and perhaps more political clout." Alluding to the popular beer, critics soon dubbed the ticket Bush-Lite.

While Quayle had been unprepared for his national debut—Baker had called Quayle with the news only ninety minutes earlier, after which Bush called Quayle personally—so was the Bush advance team, which hadn't planned effectively for Quayle's appearance at the densely packed rally in Spanish Plaza. When no advance men found them, Quayle and his wife, Marilyn, tried to work their way to the stage through the crowd, successful only after the Secret Service steered them through the chaos.

Controversy came almost immediately after Quayle's announcement when questions about his military service arose. Had he dodged the draft, as some had speculated, or had he fulfilled his military obligation by serving in the National Guard? A crisis was averted when it was revealed that Quayle had fulfilled his military commitment through service in the National Guard, but likely as a means of preempting deployment to Vietnam. George W., in defending Quayle's service—and perhaps his own—told the Associated Press, "The thing that's important is he didn't go to Canada. Remember, Canada was an option. Let's keep it in generational perspective." Still, Quayle's shakiness amid media scrutiny reinforced the impression of him as being callow and out of his depth. Behind the scenes, George W. believed swapping out Quayle for another running mate could invigorate the campaign, going so far as to briefly mount an effort to dump him from the ticket. He aborted it as his father made it clear that he was sticking with him. "Quayle was a generational statement," George W. conceded in 1990, "a statement that said, 'I'm willing to reach into our [baby boomer] generation, I'm not afraid.' It didn't work."

A wave of nostalgia swept over the convention, which marked a curtain call for Ronald Reagan, who despite the doldrums of Iran-Contra and a daunting federal deficit, would see his approval rating climb to 63 percent by the end of his term. Through his enduring faith in America, "a shining city on a hill," as he called it, Reagan had restored American pride in the wake of Vietnam, Watergate, and the disquiet of the Carter years. After establishing a warm but firm relationship with his Soviet counterpart, Mikhail Gorbachev—and abandoning the appeasement strategies of earlier administrations—he negotiated an arms-control agreement dismantling all Soviet and American short- and medium-range nuclear missiles. The Cold War, which had defined world geopolitics since the end of World War II, was in its waning days, with the Soviet Union, in financial distress, nearing collapse and the liberation of the Eastern Bloc countries not far behind. As he addressed his party for the last time as president, he gave his vice president a directive: "George, go out and win one for the Gipper."

Bush used his own address to tread a delicate balance between praising Reagan, who had delivered peace and prosperity—themes he emphasized in his own campaign—and differentiating himself from him. Bush had learned much from the boss "about decency and honor and kindness, and those broad values." But there was a perhaps unintended callousness to the unbridled capitalism Reagan espoused. Eight years earlier, he told Americans, "government is not the solution to our problems; government *is* the problem." His rhetoric, often angry and indignant, appealed to a nation that had seen the economy and its place in the world soften and slip. Bush's vision for America, while nebulous in many respects, was rooted in community and compassion. "I don't hate government," he declared. "A government that remembers that the people are its master is a good and needed thing . . . [W]here is it written that we must act as if we do not care, as if we are not moved? Well, I am moved. I want a kinder and gentler nation."

At the same time, he offered red meat to the party's right wing by standing firm on taxes. Three years earlier, Reagan had scored big by co-opting a line from a Clint Eastwood "Dirty Harry" movie when lawmakers proposed a tax hike, challenging them by threatening, "Go ahead. Make my day." Bush drew on the same kind of pithy glibness. "The congress will push me to raise taxes and I'll say, 'No,'" he said. "And they'll push again and I'll say to them, 'Read my lips: No new taxes!'" The line stirred a din of applause, but it would prove a fateful pledge that helped boost his candidacy but sink his presidency.

Putting his natural humility and his mother's admonishments to guard against self-aggrandizement at bay, Bush also seized the chance to define himself to a nation that, in spite of his seven and a half years as vice president, had yet to fully understand or appreciate him. He talked of his life as a series of missions, like the one he embarked on in World War II, nearly two generations earlier. He had seen over the course of his years working with Reagan the issues that come across "that big desk." Who should be sitting there for the next four years, he asked. "My friend," he said confidently, "I am that man."

I say it without boast or bravado. I've fought for my country, I've served, I've built, and I'll go from the hills to the hollows, from the cities to the suburbs to the loneliest town on the quietest street to take our message of hope and growth for every American to every American. I will keep America moving forward, always forward—for a better America, for an endless, enduring dream and a thousand points of light.

This is my mission, and I will complete it.

It was the most revealing, most inspiring speech of Bush's political life. Afterward, as the convention disbanded, a CBS poll showed Bush-Quayle up six points over Dukakis-Bentsen. The political winds had shifted.

Still, the Bush camp and the Republican Party took nothing for granted, relentlessly depicting Dukakis as a liberal out of touch with mainstream American sensibilities, evidenced by his status as a "card carrying member" of the American Civil Liberties Union (ACLU), and a foreign-policy neophyte. Taking aim at Dukakis's weak stance on crime, an independent group, Americans for Bush, ran a series of three television ads beginning in early September that highlighted the story of Willie Horton, an African American convicted murderer, who had raped a white woman and assaulted her white fiancé in the course of a weekend furlough from prison. While none of the spots featured Horton's image, and they ran on a limited basis for a period of twenty-eight days, the campaign garnered national media attention—"earned media," it would later be dubbed—that called out its racial overtones while giving it far greater exposure.

The fall saw a series of three debates; two between the presidential nominees and one between their running mates. The latter came on October 5, when Quayle squared off against Bentsen in a one-sided contest that compounded Quayle's woes. When Quayle, combating perceptions that he was callow, likened his experience to that of John F. Kennedy when he ran for president, Bentsen—anticipating Quayle's claim, which he had made on the campaign trail—won the contest with a single knockout blow: "Senator," Bentsen replied, "I served with Jack Kennedy. I knew Jack Kennedy. Jack Kennedy was a friend of mine. Senator, you're no Jack Kennedy."

The first presidential debate, on September 25, was largely anticlimactic. Bush mostly played it safe, avoiding long answers and specificity that might arouse controversy. A *Saturday Night Live* skit mocking the debate had Dana Carvey as Bush, responding to a question about how he would achieve a "kinder, gentler nation," running out the clock by reiterating a stream of pat campaign phrases. When told he had additional time to offer a more fulsome response, Carvey redundantly replied, "Let me sum up: [Keep] on track. Stay the course. Thousand points of light."

When asked for a rebuttal, Jon Lovitz, playing Dukakis, deadpanned, "I can't believe I'm losing to this guy."

Prior to the debate, Dukakis provided his own caricature. In mid-September, as part of an effort to show himself as tough on defense and a worthy commander in chief, Dukakis made the mistake of riding in a sixty-eight-ton battle tank during a visit to a General Dynamics facility in Michigan. Outfitted in an army helmet with his name on it—and defying an inviolable politician's rule of not putting anything on your head, which has stood since Calvin Coolidge's unintendedly comical 1927 photo op in an oversized American Indian headdress—the scene instead made Dukakis look like an exuberant child playing army.

Another costly blunder came in the second Bush-Dukakis debate, on October 13, when Dukakis was asked if he would reconsider his position against the death penalty if his wife, Kitty, had been raped and murdered; Dukakis offered a reasonable though passionless response that reinforced perceptions of him as a robotic technocrat. A political cartoon the following day had Bentsen whispering in Dukakis's ear, "Frankly, governor, you're no Jack Kennedy either." The public agreed. The situation the candidates had found themselves in in the late spring had reversed: It wasn't so much that Bush was winning but that Dukakis was losing.

The polls pointed to a Bush-Quayle victory when George and Barbara Bush, after campaigning till the final hours, arrived in Houston on Air Force Two on Monday, November 7, the day before Election Day, where they were met on the tarmac by George W. and Laura. George W. had spent much of the summer and fall on the road, traveling, often alone, to "backwater towns" promoting his father's cause—Bakersfield, California; Marion, Ohio; Texarkana, Texas.

"Off markets," where his presence would generate big crowds as "the closest thing they'd ever come to a presidential candidate," and local media would offer a bigger bang for the buck. In doing so, he saw his importance as his father's surrogate. "People are really looking at offspring to find out, 'What are this guy's kids like, because I may see something in the man through his children,'" he said. Now, with the campaign in its last hours, he was back in Texas to greet his parents after completing his own mission.

The journey had bound the two men closer as comrades in arms, slogging toward the same goal. George W. got to know his father as a "warrior" with whom he was hunkered down "in the trenches during a tough political fight." His father saw it the same way. "It was a wonderful experience for both of us. He was very helpful to me," George observed, "and I think it toughened him for the real world." It was a telling turn of phrase; George W. had been in "the real world" for some time. He was a husband and father, had held a series of jobs, made a respectable run for Congress, and launched, run, and sold his own business. The "real world" to George H. W. Bush, in other words, was politics at the highest level. In this way, George W.'s involvement in his father's 1988 presidential campaign was an apprenticeship. "[Junior] was the first and last word on the toughest issues and not only because he was his son," said Mary Matalin of him, "[but] because he always had great judgment, and the daddy always [had] unconditional love for all of his kids."

George's unconditional love was never in doubt. But his respect for his son's instinctive political ability grew during the experience, just as it boosted George W.'s self-confidence as well as his place within the family. "If there was competition with his father," Laura said of her husband, "it was certainly gone by 1988. He had an opportunity most people never have—to work with his parent as adult to adult. They had time to work through any competition." Invoking the classic film *The Godfather*—which features a Mafia don's sons, the hotheaded Sonny Corleone and his cunning younger brother, Michael, who eventually emerges to lead the family to new heights of power—Joe O'Neill said, "George went up [to Washington] as Sonny Corleone and came back as Michael." The experience reignited his own political ambitions far beyond the seat in Congress he sought a decade earlier, especially as he looked toward his return to Texas and the future it might hold. But those were thoughts for another day. This was George H. W. Bush's time, as the results on Election Day made clear.

The Bushes, their children, and grandchildren, convened at the elder Bushes' suite at the Houstonian where they watched the returns yield 53 percent of the popular vote for their family's patriarch versus Dukakis's 45 percent. At 11:00 p.m., after the polls closed on the West Coast, Dukakis called Bush to concede defeat. The following morning, at 7:45 a.m., the family attended a service at Houston's Saint Martin's Episcopal Church, where George W., the family's "go-to prayer guy" since his religious embrace two years earlier, was asked to offer an invocation. "Many of us will begin a new challenge," he intoned.

Please give us strength to endure and the knowledge necessary to place our fellow man over self . . .

We pray that as we face new challenges, we understand that through you we can clear our minds and seek wisdom . . .

Please guide us and guard us on our journey, particularly watch over Dad and Mother.

Afterward, the president-elect, Barbara, George W., Laura, and Barbara and Jenna boarded Air Force One to return to Washington. If Barbara Bush had any glamorous illusions about her new station as first lady in waiting, they dissipated when she rolled up her sleeves to remove wads of toilet paper her seven-year-old twin grandchildren had mischievously jammed into one of the plane's toilets.

Once back in Washington, George W. once again rolled up his

own sleeves to aid his father in the transition to the White House. The loyalty enforcer in the campaign, he now headed up "Scrub Team," a secret group of Bush insiders who determined which members of the campaign staff would be offered positions in the White House based on their competence and the allegiance they showed George H. W. Bush. As his father prepared for a move to the White House, George W. wound down his time in Washington. A few days after his father's election win, he and Laura closed on a home in Dallas's manicured Preston Hollow neighborhood, north of the city among many of Texas's most famous sons. There, in Texas's most moneyed city, George W. would begin a next chapter of his own.

PART IV

41

27

"A NEW BREEZE"

INAUGURATION DAY, FRIDAY, JANUARY 20, 1989, was cold and overcast. At three minutes past noon, on the south side of the U.S. Capitol, Supreme Court chief justice William Rehnquist issued the oath of office to the forty-first president as Barbara Bush, in a signature periwinkle-blue winter coat, held the family Bible, along with that of George Washington, who was sworn in as the nation's first president two hundred years earlier. Behind them on the inaugural platform among a flock of dignitaries were Bush's predecessor and, though it wouldn't be known for a dozen years, his immediate Republican successor.

The latter, George W. Bush, bearing an unmistakable resemblance to the new president, stood watching with his siblings and paternal grandmother several rows behind, basking in the shared triumph for the Bush family. Five months earlier, as head of the Texas delegation at the GOP convention, he cast his state's votes for "the man who made me proud every single day of my life and a man who will make America proud," putting his father's nomination over the top. But his pride abounded no more than on this day. His father had prepared nearly his whole life for the moment at hand; all the hard work, the nights away from home, the loyal soldiering, the perseverance through loss that he hadn't let define him, had led to this: the realization of his ultimate ambition, the highest honor he could attain.

The former, Ronald Reagan, sat bundled against the winter chill in a black overcoat and white scarf, as his vice president and friend, thirteen years his junior, succeeded him. The fortieth president would be leaving Washington having waged a revolution that changed the city more than any man since Franklin Roosevelt, for whom he had cast his first presidential vote. Reagan's thatch of shellacked dark hair blew gently in the cold January air, revealing a slight touch of gray. His time had come.

But his shadow would hover. In succeeding Reagan, Bush joined the ranks of those presidents who follow a colossus whose myth, based on an enduring image more than a record, wells up in the popular imagination nearly as soon as the presidency leaves his grasp. Like John Adams, Harry Truman, and Lyndon Johnson, who succeeded George Washington, Franklin Roosevelt, and John F. Kennedy respectively, Bush would live with an impossible though inevitable standard of measurement cast by the man who came before him. In this manner, he would experience what George W. Bush had been subject to for his whole life with his father's example.

Still, the residual excesses of Reagan's administration fell squarely on Bush's plate. Reagan's economic policy, while spurring manic growth, had left many Americans behind, while ballooning the federal deficit. The Reagan years had seen an eightfold increase in the number of American millionaires while real wages remained flat. AIDS had taken fifty-five thousand lives, the homeless population had soared, and poverty among children rose by 20 percent. Bush used his trim twenty-minute inaugural address to sound an overture for the "kinder, gentler nation" he hoped to nurture in his tenure, just as he had envisioned in his convention address. "America is never wholly herself unless she is engaged in high moral principle," he said; there were many disenfranchised Americans who needed the nation's compassion. At the same time, he added, "we must bring the Federal budget into balance. And we must ensure that America stands before the world, united, strong, at peace, and fiscally sound." Characteristically, he extended an olive branch to the Democratic opposition, which would control both houses of Congress through the entirety of his administration. "To my friends—and yes, I do mean friends—in the loyal opposition—and yes, I do mean loyal: I put out my hand," Bush said. This, he pledged, would be "the age of the offered hand."

Portending the central events of his presidency, those that would define it, Bush went on to talk about the state of the world. "A new breeze is blowing," he observed, "and a world refreshed by freedom seems reborn; for in man's heart, if not in fact, the day of the dictator is over. The totalitarian era is passing, its old ideas blown away like leaves from an ancient lifeless tree."

It was a breeze that would blow strong enough to take down the Iron Curtain, which had veiled the Soviet Union and Soviet-dominated Eastern Europe since the end of the Second World War. Reagan would receive disproportionate credit for its collapse, but it would be Bush whose quiet diplomatic facility and inherent humility would ensure that the Cold War did not end with shots fired and bloodshed. And it would be Bush whose determined leadership would catalyze a gale force of international allies that would sweep Iraqi insurgent troops under the direction of Saddam Hussein out of Kuwait, a little-known oil-rich Middle Eastern emirate. If Bush's ascent to the presidency was long in coming, the moment in history in which he found himself in the White House suited him. Though, as domestic pressures mounted despite his best intentions for a more compassionate society and a less encumbered economy, it would pass sooner than he expected.

While George H. W. Bush had a standard by which to be measured as president in his immediate predecessor, Ronald Reagan, the standard by which George W. Bush had been measured throughout his life had grown considerably. Being the eldest son and namesake of the vice president was one thing; being the eldest son and namesake of the president was another entirely. American history is rife with the stories of presidential scions who not only failed to live up to the benchmarks set by their fathers, but stumbled ignominiously, falling into poverty, debt, substance abuse, and mental illness. "One of the worst things in the world is being the child of a President," observed Franklin Roosevelt, whose children had struggles of their own. "It's a terrible life they lead."

In fact, stories of unqualified success among presidential offspring are the exception not the rule—including the most noble of American political families. The Adamses, the only family to that point to produce two presidents, was no exception. Charles Adams, the second son of John Adams and younger brother of John Quincy Adams, failed to make a satisfactory living as a lawyer, abandoning his wife and family after he descended into alcoholism, succumbing to cirrhosis of the liver at age thirty. John Quincy Adams's eldest son, George Washington Adams, shamed for conceiving a child out of wedlock with a chambermaid who had blackmailed him, drowned after a night of heavy drinking shortly after his father's departure from the White House.

George W. was aware of the bleak fates of many of those who went before him. Just before departing Washington for Dallas, as he considered what he would do next in his career, he asked Doug Wead to write up a summary of perceptions of the offspring of U.S. presidents. The resulting report, "All the President's Children," spelled it out in eleven chapters and forty-four sobering pages. "Being related to a president may bring more problems than opportunities," Wead concluded. "Two things the media and the public won't allow?: Success or failure. Keep the business mediocre, maintain a personal low profile and you will be left alone."

Mediocrity, however, was not what George W. Bush was aiming for—nor was he interested in lowering his profile. On the contrary. In the years ahead, despite whatever obstacles his father's exalted position presented, he was determined to mine the opportunities—

and he had a goal in mind: to be the governor of Texas. As early as November 1988, just after his father's presidential victory, he had expressed as much to a friend and would make his ambition plain in the New Year. But he also realized he had to achieve something of significance first—something that came from standing on his own, not riding "Daddy's coattails." He knew his biggest liability at home would be a simple question: "What's the boy ever done?"

28

"COMFORTABLE IN THE JOB"

"I FEEL COMFORTABLE IN THE job," the forty-first president allowed in his diary the day after his inauguration. He had been around the White House on and off for nearly two decades, and as he said later, "knew where the keys to the men's room were." In 1980, his campaign slogan was "A President We Won't Have to Train." Indeed, nine years later, after adding eight years in the vice presidency to his résumé, the duties of the office came easily to him.

The role of first lady came easily to Barbara, too. The "Silver Fox," as she was nicknamed, was warm, self-deprecating, and comfortable in her skin. Her understated style was a change from the imperial image of her predecessor—and a refreshing departure for many women. According to the *Washington Post*, her immediate popularity was "in no small measure a byproduct of the appearance of a woman at center stage who dared to look her age." "My mail tells me that a lot of fat, white-haired, wrinkled ladies are tickled pink," she said. When her husband asked her if she was going to eat her dessert, she replied, "I have to eat for my fans."

Just as Barbara lessened the burden for her husband as she always had, so did the balance of the family, which surrounded him as a balm. "We approach everything as family," George H. W. Bush said. So it was in his presidency; he never understood some of his predecessors who whined about the office as "the loneliest job in the world." His first Oval Office visitor as president was his mother, Dorothy, who visited her second-born child on the first full day of his presidency. When asked by a reporter afterward if it was the most exciting moment of her eighty-seven years, she replied, "So far. So far." Marvin and Doro, who both lived in the Washington area, could be at the White House to have dinner or a beer with him, and George W., Jeb, and Neil, off in Dallas, Miami, and Denver respectively, visited on special occasions. Bush also put his family to work. In the first eighteen months of his term in office, a member of the Bush or Walker families was included in fifteen of forty-one delegations sent abroad for ceremonial occasions.

The first-family dynamic helped make the president and first lady more accessible even to the loftiest of visitors. When Queen Elizabeth and Prince Philip, Duke of Edinburgh, visited the White House in the spring of 1991, George W. and Laura were on hand with the other Bush progeny and their spouses. At a small luncheon they all attended in advance of a state dinner the same evening, Barbara jokingly told the Queen that she had seated George W. as far away from her at the dinner as possible. "Is he the black sheep in the family?" the Queen inquired. When George W. admitted he was, she replied, "Every family has one," then asked Barbara why she considered her eldest son to be so dangerous. Barbara allowed that it was his plainspokenness—and the fact that he had threatened to wear to the evening's black-tie affair one of his gaudy pairs of cowboy boots, either a pair embossed with an American flag or another with the words "God Bless Texas." Which pair was he planning on wearing, Her Majesty asked George W. "Neither," he replied. "Tonight's pair will say, 'God Save the Queen.'"

Just as George H. W. Bush found the job of president agreeable, he had an immediate comfort with those in his administration, which was composed heavily of friends, old hands, and Reagan White House veterans—many of whom would serve in his son's White House. Jim Baker, perhaps his closest friend, whom he had passed up as a vice presidential possibility, was tapped instead as secretary of state. The two men had known each other for thirty years, and Baker, seven years Bush's junior, owed Bush for getting him into politics. Despite a warning from his grandfather to "work hard, study, and keep out of politics," and the fact that he had been a lifelong Democrat, Baker took a plunge into the political world as campaign director in Bush's 1970 Senate race at the urging of Bush, who saw it as a means of occupying Baker after the loss of his first wife to breast cancer the same year.

Brent Scowcroft, who would become another of Bush's closest friends, stepped into the role of national security adviser, a post he had occupied in Gerald Ford's administration a dozen years earlier. Along with Scowcroft came two promising deputies, Condoleezza Rice and Robert Gates. Reagan's last national security adviser, Colin Powell, got Bush's nod as chairman of the Joint Chiefs of Staff. Dick Cheney, the Wyoming congressman who had served as Gerald Ford's chief of staff, was appointed as secretary of defense after Bush's initial choice for the post, longtime senator from Texas John Tower, fell victim to a Democratic-controlled Senate, which exacted revenge for the rough campaign and a three-term Republican lock on the presidency. Tower was rejected due to a long history of boozing and womanizing, making it the first time a cabinet official had been rejected by Congress in thirty years.

John Sununu, who had helped to deliver the all-important New Hampshire primary to Bush, was given the post of chief of staff, while Lee Atwater was rewarded for his own efforts in the campaign with an appointment as chairman of the Republican Party before succumbing to an aggressive form of brain cancer at the age of forty in 1991.

The new president's leadership style, devoid of his predecessor's presidential majesty and theatrical flair, was marked by pragmatism, prudence, and restraint as a series of momentous international events began rolling out in his first year. His first major test on the world stage came in the spring. The desire to break free from totalitarianism, as Bush had presaged in his inaugural address, hung thick throughout the globe as the decade drew to a close. By early April 1989, it spread to China, where student protesters began gathering by the thousands in Beijing's Tiananmen Square to demand democratic reform. The dissident movement grew after a two-day state visit by Mikhail Gorbachev, whose call for glasnost and perestroika— greater openness and political and cultural reform—had given hope to those in the Soviet Union and Eastern Bloc nations. A week after Gorbachev's departure in mid-May, Chinese authorities instituted martial law to put down the protesters, whose ranks had escalated to almost a million. Tanks were mobilized only to be met by human blockades.

By early June, Deng Xiaoping, China's paramount leader, had had enough. In the dead of night on June 3, a fifty-tank convoy rolled into the square as soldiers armed with machine guns put down any protesters that stood in their way, prompting a massacre that would leave three thousand dead and another ten thousand wounded. The movement, symbolized by the image of a lone Chinese dissident standing defiantly in front of a line of four tanks, in Bush's words, "captured the imagination of the entire world."

Bush publicly condemned the attacks and called for sanctions, but began with a personal appeal through an "anguished letter to an old friend," Deng Xiaoping, emphasizing the importance of restraint. Deng responded the following day, accepting Bush's overture to dispatch an emissary to meet with the Chinese leadership. Brent Scowcroft was dispatched to "convey to the Chinese how serious the divide was between us but also how much we respected our friendship," as Bush put it later. "It kept the door open."

Elsewhere in the world, a "door of freedom," as Bush had foretold in his inaugural address, opened not long afterward. On November 9, East German officials called a press conference to announce the relaxing of travel restrictions from East Germany to West Germany. A confluence of circumstances had led to the announcement: Gorbachev's introduction of glasnost and perestroika, a burgeoning

East German resistance movement, and a decaying East German regime under the weakening thumb of the Soviet state. The announcement by a hapless member of the Politburo, who read it for the first time on air, indicated that travel from East Germany to other nations would be "possible for every citizen" and would begin "right away, immediately." But what was meant to be a limited and controlled opportunity for travel between nations was widely construed by the East German people and the international media as an unimpeded open door to freedom.

After the announcement, unchallenged by the East German police, East and West German citizens by the thousands descended euphorically on the Berlin Wall, where they hacked away at it with pickaxes. Twenty-eight years earlier, in August 1961, the wall had been erected ominously, marking the rise of Cold War tensions between East and West. Now the wall was being smashed into rubble by the very people it separated, symbolizing the Cold War's imminent end.

During an informal meeting with the press, Bush, seated behind the Resolute Desk in the Oval Office, greeted the news with subdued reservation. CBS's Lesley Stahl asked him why he was not more demonstrably satisfied. "I'm not an emotional guy, but I'm pleased," he replied dispassionately.

Hardly. As Jim Baker put it, Bush "could cry at the drop of a hat." That very month, after his beloved cocker spaniel, C. Fred Chambers, died, Bush left it to Marvin to offer a eulogy knowing that he would become a puddle of emotions if he did it himself. But Bush's muted reaction to the fall of the Berlin Wall was part of a conscious strategy. In Hungary in 1956 and in Czechoslovakia in 1968, the world had seen uprisings for liberty and calls for reform met with violent crackdowns. Bush knew the fragility of the situation, refraining from bravado that might embolden East German and Soviet hard-liners. He was also careful not to put Gorbachev in a compromising position, jeopardizing future negotiations.

Bush's response drew criticism among GOP conservatives, who flagged it as a departure from the probable approach of Reagan, who almost certainly would have used the milestone to assert the triumph of American democratic ideals. Bush and his advisers, on the other hand, saw restraint the most judicious course. "The worst thing, we thought, would be for the President to gloat that we'd won," Brent Scowcroft reflected, "because what we wanted was for this momentum to keep going. Most people advocated that the President ought to go to Berlin to dance on the wall. But I think the President had exactly the right approach. What he was trying to say was, 'Look, nobody lost here; we both win with the end of the Cold War.'"

In early December, a month after the Berlin Wall's collapse, Bush ordered an invasion of Panama and the capture of its dictator, General Manuel Noriega. Noriega had been a thorn in the side of America for years as a drug trafficker for Columbia's Medellín cartel, contributing to America's growing drug problem, while accepting money as a paid CIA agent. "I've got Bush by the balls," Noriega bragged shortly after Bush took office. Bush would prove otherwise. As early as October, after the harassment of American service members in the Panama Canal Zone, he had concluded that "a grab of Noriega" would be "more acceptable certainly at home, maybe abroad." The last straw came with the murder of an American soldier and the torture of another, prompting Bush to deploy fourteen thousand American troops, who joined the thirteen thousand troops already in Panama, for Operation Just Cause, with the mission of removing Noriega from power and bringing him to justice. After a manhunt lasting a few days, Noriega was apprehended and jailed in Miami as a prisoner of war, where he would later be found guilty of eight counts of drug trafficking, money laundering, and racketeering, and sentenced to forty years in prison. While largely considered successful, the incursion claimed the lives of eleven American soldiers, while wounding 325.

One of the "hardest days of George's presidency," Barbara recalled, was a New Year's Eve visit the first couple made to two San Antonio military hospitals to console some of the wounded and their families. But the commander in chief left bolstered by their patriotism and spirit. Many expressed pride in him and gratitude for his leadership. Another, an African American accompanied by his Filipina wife and their two sons, assured him, "I couldn't have better care if I had all Donald Trump's money."

"It's been some year," Bush wrote in his diary the same day, the last day of the decade. "I end the year with more confidence, and end the year with the real gratitude of our team . . . I'm certainly not seen as a visionary, but I hope I'm seen as steady and prudent and able." His 80 percent approval rating, which had climbed steadily throughout the year, suggested that he was.

He was grateful, too, for his family, which grounded him and gave him solace. In the same diary entry, he reflected on "one of the greatest highlights" of the year: The week prior, on the day after Christmas, before repairing to the office to exercise his duties as leader of the free world, his three-year-old granddaughter, Ellie LeBlond, Doro's second-born child, summoned him to a bathroom and pointed to an unflushed toilet. "Did you leave that poo-poo?" she asked.

29

A WHOLE NEW BALL GAME

EVEN BEFORE HIS FATHER TOOK the presidency, speculation about George W. Bush's political future was the subject of the Washington punditry. On December 29, Washington's widely read columnists Rowland Evans and Robert Novak cited George W. as "the increasingly likely Republican candidate for governor" in Texas, adding that he was "engaging and articulate, more conservative than the president-elect and the family member with the purest Texas accent." Soon after arriving back in Texas, Bush barnstormed the state to test the waters for a run at the governor's mansion in 1990, beguiling Republican leaders and potential donors with his zeal and self-deprecating wit.

Inevitable questions arose from the press. How did he differ from his father? "He went to Greenwich Country Day School, and I went to San Jacinto Junior High," Bush answered and often reliably repeated. Was he part of a political dynasty? No, because it implied that he had "inherited something." Were the Bushes analogous to the Kennedys? Again, no. "They never had to work. They never had to have a job," he said.

Politics was a calling, Bush explained, not something he was pursuing because it was expected of him. "I want to affect the lives of people. I want to make life better. I think politics is an arena where you can do that," he said in early 1989, sounding very much like his father and grandfather. Still, he saw the advantages of his name, adding, "If I run, I'll be most electable. Absolutely. No question in my

mind. In a big media state like Texas, name identification is important. I've got it." But his challenge, as made clear to *Texas Monthly*,
which profiled the possible gubernatorial contender in April 1989, was
"how to take advantage of his name and still establish himself as a
politician in his own right." Critics and doubters had already taken
to calling Bush, "the shrub."

To be sure, George W. was no stranger to hard work. It just hadn't
amounted to an achievement that registered on a large scale, especially in Texas with its fixation on bigness. As he weighed the governorship, he also considered another opportunity, one that could help
build his profile in the state as a businessman, establish himself apart
from his father—and better position him for a political run. Just
before his father's election as president, he got a call from Bill De-
Witt Jr., a friend from the oil business whose father had owned the
Cincinnati Reds in the 1960s. The Texas Rangers were up for sale,
DeWitt told him, adding, "This could be a natural for you. I know
you want to get back to Texas, and you've always loved baseball."

True enough, Bush had a passion for both Texas and baseball,
which had a special significance in the Bush and Walker families.
Even in West Texas, where football reigned supreme, it was baseball
that captured George W.'s imagination as he dreamed not of being
president but of being Willie Mays. Bush's father, the president, kept
his Yale first baseman's glove in the top drawer of his Oval Office
desk, and his great-uncle Herbie Walker had been the second-largest
shareholder when the New York Mets launched as a franchise in 1962.

Bush was taken by the idea of acquiring the Rangers, a fixture in
Arlington, Texas, between Dallas and Fort Worth, since the erstwhile Washington Senators moved west in 1971, later to be renamed
something suitably Texan. But though he had $606,000 in invest-
ment capital squirreled away from the Harken Energy deal that he
could use to borrow against for collateral, he "had more determination than money." Undaunted, he pursued the purchase "like a pit
bull on the pant leg of opportunity," partnering with DeWitt to raise

the $86 million at which the team was valued. As DeWitt lined up investors in Ohio, dubbed the "Cincinnati group," Bush mined his own contacts, including Yale buddies and Bush family friends. Concurrently, Bush wooed Rangers owner Eddie Chiles, another family friend, to ensure his willingness to sell to the group.

Bush's prospects hit a snag when Major League Baseball commissioner Peter Ueberroth blocked the sale to the group out of concern that the franchise would be purchased by a group composed mainly of non-Texans who might move the franchise out of state. Instead, Ueberroth himself sought out a potential buyer in Richard Rainwater, a prominent Fort Worth financier who considered the investment with Dallas businessman Rusty Rose. After Rainwater and Rose demurred due to Rainwater's disinclination to manage day-to-day operations and deal with the media, Ueberroth suggested that they meet with Bush, who had expressed an interest in taking on those responsibilities. The baseball commissioner's outreach had as much to do with Bush's standing as a member of the first family as it did with his viability as a partner. "There is no question that Rainwater and Rose were the primary investment group and I asked them to consider taking George in," said Ueberroth later. "He was an asset because his father's career was going up and reaching the top. We just brought the young man over somewhat out of respect for his father."

By late April, a deal was closed. An investment group comprising Rainwater and Rose, and the thirty-nine investors rounded up by and including Bush and DeWitt, purchased 86 percent of the Rangers for $75 million. Over time they would buy the team in its entirety. Bush, as the deal's front man, would receive a disproportionate amount of the team's proceeds, 10 percent of the profits once all the investors got back their original investments plus 2 percent.

After the sale, Bush remained the face of the Rangers organization, acting as co-managing general partner for the team, charged, along with Rose, with the management of day-to-day operations, at an annual salary of $200,000. Reacting to questions about whether he had the skills to act and speak on behalf of the investment group, Bush told Dallas reporters, "I've seen this before. I'm a guy whose father was called a 'wimp' on the cover of *Newsweek* the day he declared his race for the presidency. If it gets too bad, I'll let you know."

He didn't have to. He flourished in the role with the team, which suited not only his love for the game but his gregarious nature. No imperious part-owner behind the glass of a luxury box, the "first son" became a regular in the stands, mixing it up with the fans—posing for photographs, signing George W. Bush baseball cards, lining up at the concession stands, and "peeing in the same urinals." It wasn't bad for business, either. As he told a reporter, "being the president's son puts you in the limelight. While in the limelight, you might as well sell tickets." Gross revenue more than doubled, and the team's record improved, from a season of 71 wins and 91 losses in 1988 to a winning season in 1989, 83 wins and 79 losses, with the team's legendary pitcher forty-one-year-old veteran Nolan Ryan striking out over 300 batters on his way to 16 wins.

With the acquisition of the Rangers in hand and Bush immersed in his new responsibilities, his immediate political prospects dimmed. One of his investors had committed to the purchase with the condition, "I don't want to make the investment if you plan to run [for governor] in two years." Barbara Bush was just as opposed to the idea, believing that the political misfortunes of one Bush in office could adversely affect the other. Ever protective of her husband, she wanted nothing to get in the way of his presidency, helping to close the matter with a comment to White House reporters. "When you make a major commitment like that," she said of her son's Rangers venture, "I think maybe you won't be running for governor."

By August, Bush opted to put his ambitions on hold. But not for long. While it was his father's name that helped him cinch the Rangers deal, it was his involvement with the franchise all on his own in the next several years that would help pave the way to the governor's mansion by the middle of the following decade.

30
COMMANDER IN CHIEF

RANGERS OF A DIFFERENT KIND were on the mind of George H. W. Bush in the first weeks of 1990. During his visit to military hospitals in San Antonio on the last day of the previous year, he was given a small American flag by an army ranger who had lost his leg during military operations in Panama. The flag, the young man said, represented the men in his unit, and he added, "We're proud of you, and we back you." It would remain on Bush's Oval Office desk for the balance of his tenure, a reminder of the cost of war and the solemn duty he carried as commander in chief, and a harbinger of the difficult decisions ahead. On January 17, a few days after a Cincinnati meeting with the mother of another army ranger, who had died during the Panama incursion, Bush wrote his children, "I thought—I sent her son into this battle and here she is telling me with love about her son and what he stood for. She said, 'You did the right thing.'"

The inordinate weight of the office would be with him throughout the year. Melancholy visited him in mid-April. With Barbara traveling in the Northeast for a college speech, and a host of issues nagging at him, he recorded in his diary, "Bar is gone . . . I sit here at the White House . . . a little gloomy given the magnitude of the problems. It might be the first day I've really felt an accumulation of problems . . . but that's what I get paid for."

Among the problems Bush grappled with was a slowing economy, hampered by the crippling deficit as job creation slowed and unemployment climbed. A Democratic-controlled Congress refused to accept a budget that relied solely on budget cuts. Federal revenue, they contended, would have to be raised, which meant increasing tax revenue, an express violation of Bush's red-meat *"Read my lips: No new taxes!"* campaign pledge. With his back against the wall and in need of a compromise, Bush capitulated in late June. "It is clear to me that both the size of the deficit problem and the need for a package that can be enacted require . . . tax revenue increases," Bush said, fatefully grazing the political third rail. The *Los Angeles Times* noted that Bush may have "given up what many Republican strategists see as the party's most important issue—low taxes." Already wary of Bush as a right-wing pretender, GOP conservatives led by the second-year House minority whip, Georgia's Newt Gingrich, howled believing higher deficits to be a lesser evil than a boost in taxes.

In the fall, House Republicans went into open revolt as the 1990 budget deal Bush had brokered with Democratic leaders came down to a vote. The forty-eight-year-old Gingrich, a take-no-prisoners firebrand, had introduced a new brand of Republican political leadership that he referred to as "reform populist conservatism"; the opposition was viewed as the enemy and compromise as ideological capitulation, supplanting the pragmatism practiced routinely by Republican elders. In private, Gingrich voiced his support of the budget deal. In public, though, he not only distanced himself from it but also derided it as antigrowth. Much of the GOP abandoned the president, siding with Gingrich. The budget deal passed with the support of 218 of 246 Democrats, but only 32 of 168 Republicans—less than one-fifth of the party's ranks.

Compounding Bush's economic woes was the steady collapse of the savings and loan industry. After sweeping deregulation in the late 1970s, S&Ls were drawn to riskier investments, including the housing market. As the economy softened, the housing market went with it, pushing many S&Ls into insolvency. Bush settled with Congress on a taxpayer-funded bailout of $100 million to alleviate the crisis, a further drag on the sluggish economy.

The S&L failure hit uncomfortably close to home. Among the eighteen Colorado S&Ls that required the federal bailout was the Denver-based Silverado Banking, Savings and Loan Association, on whose board of directors Neil Bush had served, a position he had taken in 1985 at age thirty. Hardly an expert on banking or finance—his only experience was a summer clerical job at a Dallas bank—Neil had no illusions about why he had been tapped. "I would be naïve to think that the Bush name didn't have something to do with it," he said in a 1990 interview. But as the savings and loan crisis played out in 1990, he would also see the downside of his last name, becoming the subject of a congressional investigation charged with looking into the matter. Mirroring the intensive press coverage, a *Washington Post* headline read, "Neil Bush Stands in the Shattered Ruins of Denver Thrift."

Neil took the allegations of wrongdoing hard, obsessively monitoring the coverage and worrying what it would mean to his future and to his father's. His plight weighed just as heavily on his parents. Barbara wrote in her diary, "His whole problem is that he is our son." The president blamed himself for his son's pain to the point of contemplating not running for reelection in 1992. As Bush dictated to his diary on July 11, 1990, "[I'm] . . . worried about Neil . . . wondering in my heart of hearts, given what's happened to Neil, whether I really want to do this after I serve this term."

In late July, Bush retreated to the tranquility of Walker's Point, where the family joined him for the usual spate of manic summer activities—fishing, boating, tennis, and rounds of "aerobic golf" pursued with such alacrity that they constituted a higher form of exercise. George W. was there with Laura and the girls and recognized that after a difficult six months, his father's "spirits needed a lift."

The elder Bush shared his sentiments about Neil with George W., his closest adviser among his children, during a fishing outing. "I'm thinking of not running again," he told George W.

His son was taken aback. "Why, Dad?" he asked.

"Because of what Neil is going through," Bush replied.

"I know it's tough," George W. said, "but you've got work to do, and the country needs you."

The world, too, as it happened. Just a few days into Bush's Maine respite, he was called back to Washington. Ominous reports in the Middle East, his advisers told him, required his attention.

A t 8:20 p.m. on Wednesday, August 1, Brent Scowcroft tracked down Bush in the basement of the White House as Bush was getting a deep-heat treatment for a sore shoulder. "Mr. President," Scowcroft told his friend, "it looks very bad. Iraq may be about to invade Kuwait." It confirmed the president's "worst fears" over the ambitions of Iraq's truculent dictator, Saddam Hussein; 120,000 Iraqi troops along with armored units had amassed on Iraq's south-eastern border poised to invade its oil-rich neighbor. Seizing Kuwait, which Saddam had accused of stealing oil in the disputed Rumaila oil fields, would extinguish the debt Iraq had incurred from Kuwait to help fund its long-standing war with Iran, while providing unfet-tered access to the Persian Gulf. Moreover, it would double Iraq's oil supply, giving Saddam control of one-fifth of the world's oil reserves and making him a more significant figure on the world stage.

There was little doubt that Saddam was capable of large-scale military aggression. Saddam had been Iraq's dictator since seizing power in 1979, leading a totalitarian regime that was notoriously brutal. Political enemies and other antagonists were routinely shot in the name of patriotism as "traitors." In 1988, he had murdered thousands of his own citizens—75 percent of whom were report-edly woman and children—when he ordered poisonous gas to be dropped on the northeastern city of Halabja, to crush a mounting resistance movement by Kurdish rebels. A June issue of *U.S. News & World Report* featured a menacing Saddam on its cover under the headline "The Most Dangerous Man in the World."

By 10:00 p.m. the same evening, Scowcroft confirmed what both he and Bush had suspected. Iraqi forces had crossed the Kuwaiti border. Kuwait was under siege.

The following day, August 2, the news of the Iraqi invasion made its way around the world. The Middle East was shaken. Iraqi troops had taken the nation's oil fields and its capital city, Kuwait City, killing and wounding hundreds of Kuwaitis while driving the nation's sheik into exile. The magnitude of the situation, though, had yet to hit the president. "We are not discussing intervention," he replied when asked by a White House reporter how the U.S. would respond. But he added paradoxically, "I would not discuss any military options even if we agreed upon them." Naturally cautious, he hoped to convey that he was keeping his options open. Instead he came off as indeterminate. "The truth is," Bush reflected later, "at that moment, I had no idea what our options were." While he signed an executive order freezing Iraqi and Kuwaiti assets, he later recorded in his diary, "There is little the U.S. can do in a situation like this . . . This is radical Suddam [*sic*] Hussein moving . . ."

The same day, he flew to a scheduled joint-speaking engagement with Margaret Thatcher in Aspen, Colorado, providing an opportunity for the two to confer about what Bush publicly called Saddam's "naked aggression." Of immediate concern to both leaders were Saddam's larger ambitions. With the access to the Gulf that Iraq now had by seizing Kuwait, what would stop Saddam from invading Saudi Arabia and commanding 45 percent of the world's oil reserve? Thatcher's position on the situation was emphatic. "If Iraq wins, no small state is safe," she held. "They won't stop here. They see a chance to take a major share of its oil. It's got to be stopped. We must do everything possible." But the prime minister had far less to offer militarily and far less to lose than the president.

By Sunday, August 5, Bush was at Camp David where he brought together his core advisers—Dan Quayle, Baker, Scowcroft, Dick Cheney, John Sununu, and Colin Powell—to discuss military options. Powell had also brought in Norman Schwarzkopf, military commander of the Persian Gulf region. Most agreed that military action was needed. Baker argued that the Soviet Union would oppose aggression against Iraq, a client state to which the USSR sold weapons, while Cheney warned that the "American people might have a short tolerance for war."

As Bush considered the views of his team, his resolve hardened. Saudi Arabia was a vital American interest and had to be protected, he believed, and American military forces could protect Saudi Arabia from Iraqi invasion while liberating Kuwait from Iraqi forces. "My first objective is to keep Saddam out of Saudi Arabia," he told the group. "Our second is to protect the Saudis against retaliation when we shut down Iraq's export capability. We have a problem if Saddam does not invade Saudi Arabia but holds on to Kuwait."

When Bush boarded Marine One to return to the White House, Doro, who accompanied her father, noticed a change in his demeanor; there was a "smoldering intensity to him," she recalled. A small group of reporters met the president when he landed on the White House's South Lawn. He was going to keep his options open, he told them. Then, before setting back to work in the West Wing, he said calmly, "This will not stand, this will not stand. This aggression against Kuwait."

The same evening he reflected in his diary on the previous two days as "the most hectic 48 hours" since taking office. "[T]he enormity of Iraq is upon me now," he said.

gust 5, Bush was at Camp David where he brought
e advisers—Dan Quayle, Baker, Scowcroft, Dick
nunu, and Colin Powell—to discuss military op-
d also brought in Norman Schwarzkopf, military
e Persian Gulf region. Most agreed that military ac-
Baker argued that the Soviet Union would oppose
t Iraq, a client state to which the USSR sold weap-
ey warned that the "American people might have a
or war."

dered the views of his team, his resolve hardened.
s a vital American interest and had to be protected,
American military forces could protect Saudi Arabia
on while liberating Kuwait from Iraqi forces. "My
to keep Saddam out of Saudi Arabia," he told the
nd is to protect the Saudis against retaliation when
aq's export capability. We have a problem if Saddam
Saudi Arabia but holds on to Kuwait."

oarded Marine One to return to the White House,
ompanied her father, noticed a change in his de-
as a "smoldering intensity to him," she recalled. A
eporters met the president when he landed on the
outh Lawn. He was going to keep his options open,
en, before setting back to work in the West Wing, he
is will not stand, this will not stand. This aggression

ning he reflected in his diary on the previous two
st hectic 48 hours" since taking office. "[T]he enor-
pon me now," he said.

gress on a taxpayer-funded bailout of $100 million to alleviate the
crisis, a further drag on the sluggish economy.

The S&L failure hit uncomfortably close to home. Among the
eighteen Colorado S&Ls that required the federal bailout was the
Denver-based Silverado Banking, Savings and Loan Association,
on whose board of directors Neil Bush had served, a position he
had taken in 1985 at age thirty. Hardly an expert on banking or
finance—his only experience was a summer clerical job at a Dallas bank—Neil had no illusions about why he had been tapped. "I
would be naïve to think that the Bush name didn't have something
to do with it," he said in a 1990 interview. But as the savings and
loan crisis played out in 1990, he would also see the downside of
his last name, becoming the subject of a congressional investigation
charged with looking into the matter. Mirroring the intensive press
coverage, a *Washington Post* headline read, "Neil Bush Stands in the
Shattered Ruins of Denver Thrift."

Neil took the allegations of wrongdoing hard, obsessively monitoring the coverage and worrying what it would mean to his future
and to his father's. His plight weighed just as heavily on his parents.
Barbara wrote in her diary, "His whole problem is that he is our
son." The president blamed himself for his son's pain to the point of
contemplating not running for reelection in 1992. As Bush dictated
to his diary on July 11, 1990, "[I'm] . . . worried about Neil . . .
wondering in my heart of hearts, given what's happened to Neil,
whether I really want to do this after I serve this term."

In late July, Bush retreated to the tranquility of Walker's Point,
where the family joined him for the usual spate of manic summer
activities—fishing, boating, tennis, and rounds of "aerobic golf" pursued with such alacrity that they constituted a higher form of exercise. George W. was there with Laura and the girls and recognized
that after a difficult six months, his father's "spirits needed a lift."

The elder Bush shared his sentiments about Neil with George W., his closest adviser among his children, during a fishing outing. "I'm thinking of not running again," he told George W.

His son was taken aback. "Why, Dad?" he asked.

"Because of what Neil is going through," Bush replied.

"I know it's tough," George W. said, "but you've got work to do, and the country needs you."

The world, too, as it happened. Just a few days into Bush's Maine respite, he was called back to Washington. Ominous reports in the Middle East, his advisers told him, required his attention.

At 8:20 p.m. on Wednesday, August 1, Brent Scowcroft tracked down Bush in the basement of the White House as Bush was getting a deep-heat treatment for a sore shoulder. "Mr. President," Scowcroft told his friend, "it looks very bad. Iraq may be about to invade Kuwait." It confirmed the president's "worst fears" over the ambitions of Iraq's truculent dictator, Saddam Hussein; 120,000 Iraqi troops along with armored units had amassed on Iraq's southeastern border poised to invade its oil-rich neighbor. Seizing Kuwait, which Saddam had accused of stealing oil in the disputed Rumaila oil fields, would extinguish the debt Iraq had incurred from Kuwait to help fund its long-standing war with Iran, while providing unfettered access to the Persian Gulf. Moreover, it would double Iraq's oil supply, giving Saddam control of one-fifth of the world's oil reserves and making him a more significant figure on the world stage.

There was little doubt that Saddam was capable of large-scale military aggression. Saddam had been Iraq's dictator since seizing power in 1979, leading a totalitarian regime that was notoriously brutal. Political enemies and other antagonists were routinely shot in the name of patriotism as "traitors." In 1988, he had murdered thousands of his own citizens—75 percent of whom were reportedly woman and children—when he ordered poisonous gas to be dropped on the northeastern city of Halabja, to crush a mounting

resistance movement by Ku
World Report featured a m
headline "The Most Dange

By 10:00 p.m. the same
he and Bush had suspecte
border. Kuwait was under s

The following day, Augu
its way around the world. T
had taken the nation's oil fie
ing and wounding hundred
sheik into exile. The magnit
hit the president. "We are
when asked by a White Hou
But he added paradoxically,
tions even if we agreed upon
convey that he was keeping
indeterminate. "The truth is
I had no idea what our optic
order freezing Iraqi and Ku
ary, "There is little the U.S.
radical Suddam [*sic*] Hussein

The same day, he flew to
with Margaret Thatcher in
tunity for the two to confer
dam's "naked aggression." C
were Saddam's larger ambiti
Iraq now had by seizing Kuw
vading Saudi Arabia and com
reserve? Thatcher's position o
wins, no small state is safe,"
see a chance to take a major sh
must do everything possible."
offer militarily and far less to

By Sunday, A
together his co
Cheney, John S
tions. Powell h
commander of
tion was needed
aggression agai
ons, while Che
short tolerance

As Bush co
Saudi Arabia w
he believed, and
from Iraqi inva
first objective i
group. "Our se
we shut down
does not invade

When Bush
Doro, who ac
meanor; there
small group of
White House's
he told them. T
said calmly, "I
against Kuwai

The same e
days as "the m
mity of Iraq is

31
"YOU CAN'T GIVE IN"

THE "VISION THING," BY HIS own admission, eluded George H. W. Bush, at least manifestly. Oratory was not a strong suit, nor would it ever be. His ability to publicly articulate his aspirations was limited and generally uninspired, just as his syntax was often garbled and his manner awkward, sometimes goofy. Bush was never one to grandstand and speechify, anyway, but was more apt to reveal himself through small, quiet gestures that made a deeper impression. While it meant that he was an unexceptional communicator to the masses, it's also what made him a great statesman, which would come to bear in the months ahead in what would become Bush's finest hour as president.

But Bush did have a vision. The wheels in his mind turned as the Cold War wound down with the slow, irrevocable crumbling of the Soviet empire and the emergence of the U.S. as the world's lone superpower. America could lead the world, he believed, through the building of a coalition of the disparate powers throughout the globe. The Soviet Union and economic strongholds like Germany and Japan could partner with the U.S. toward mutual interests and in staving off tyranny that might threaten regional stability and economic prosperity—and he saw the mobilization of nations around the liberation of Kuwait as a chance to forge that coalition. On September 11, 1990, eleven years to the day before the attacks on American soil that would define his son's presidency, Bush addressed Americans with a view toward a "new world order," which he called "an historic period of cooperation . . . stronger in pursuit of justice, and more secure in the quest for peace . . . in which nations of the world, East and West, North and South, can prosper and live in harmony." He also stated in no uncertain terms America's goal in Kuwait: "Iraq must withdraw from Kuwait completely, immediately and without condition."

Throughout the balance of 1990, Bush used personal diplomacy to manage an unprecedented coalition of nations. Over his time in public life, especially in the presidency, he had cultivated the relationships with foreign leaders. The letters, notes, and photos he had sent and the phone calls he had made for no reason but to check in were small deposits that added up to a reservoir of trust with people who now came to his side. He and his team enlisted the Soviet Union in the cause. Gorbachev agreed to the use of military force in adherence to UN Resolution 678, passed by the UN Security Council in November, calling for member states to use "all means necessary" to compel Saddam to withdraw from Kuwait by January 15, 1991. Just as remarkable, King Fahd relented, allowing U.S. forces into Saudi Arabia, where they would remain even after the Gulf War.

By early November, 400,000 American troops were stationed in the Saudi kingdom. The U.S. was joined by forces from twenty-six other countries, later twenty-eight, forming a defensive front called "Desert Shield," comprising the greatest military alliance since the Second World War. Despite the condemnation of the world's strongest nations and the buildup of troops ready for battle against the 140,000 Iraqi troops now in Kuwait, Saddam defiantly held firm.

The forty-first president was roundly praised for his leadership around the crisis. "Bush, unlike Ronald Reagan, was no lone cowboy singlehandedly dispensing rough justice but a sheriff rounding up a posse of law-abiding nations," *Time* magazine wrote at year's end. "All along he has retained tight control of virtually every detail of U.S. action, revealing as little as possible about his plans to the American people and to Congress." By that time, Bush was determined to drive Saddam out of Kuwait with or without congressional support. If Congress failed to pass a resolution authorizing military force, Bush was prepared to be a "lone cowboy" after all. In November, he told his diary, "It is only the United States that can do what needs to be done. I still hope against hope that Saddam will get the message, but if he doesn't, we've got to take this action, and if it works in a few days, and he gives up, or is killed or gets out, Congress will say, 'Attaboy, we did it; wonderful job; wasn't it great we stayed together.' If it drags out and there are high casualties, I will be history; but no problem—sometimes you've got to do what you've got to do."

By Christmastime, Bush's resolve around the cause deepened, becoming a moral imperative. An Amnesty International report made a profound impression on him, revealing the atrocities inflicted on the Kuwaiti population by the Iraqi invaders. After an Oval Office plea for peace from the Right Honorable Reverend Edmond Browning, the presiding bishop of the Episcopal Church of the United States, Bush handed him the report. Given the human rights violations cited, Bush asked, "[W]hat do we do about peace? How do we handle it when people are being raped?" The same month he expressed his determination to Bob Gates, his deputy national security adviser. "If I don't get the votes" for war, he told Gates in an Oval Office meeting, "I'm going to do it anyway. And if I get impeached, so be it." As he explained in a letter to George W. and the other Bush children on New Year's Eve, just after they left Camp David where the family had spent Christmas together, "Principle must be adhered to—Saddam cannot profit in any way at all from his aggression and from brutalizing the people of Kuwait—and sometimes in life you have to act as you think best—you can't compromise, you can't give in . . ."

George W. saw his father's stance on Kuwait as the clear path of righteousness, one George H. W. felt viscerally and for which he was willing to act unilaterally, putting his presidency on the line, much as George W. saw his own position in Iraq a little over a decade later. It was an important similarity between the two that he saw as little appreciated. When historian Jon Meacham asked to interview him for his 2015 biography of George H. W. Bush, *Destiny and Power*, George W. initially declined. Meacham, he believed, would write a "pro-41, anti-43" book that would keep Meacham in good standing with the Manhattan media and literati set. He changed his mind when Meacham allowed him to read the manuscript, which highlighted his father's resolve around military action against Saddam despite the threat of impeachment.

Meacham's was a departure from the accepted thesis, and George W. thought aligned him more closely with his father. "It's conventional wisdom today that George W. Bush shoved aside doubt and debate in pursuit of war with Iraq in 2002," Meacham wrote in *Politico* after his book was published. "In a commonly accepted narrative, observers have long held that George H. W. Bush would have been more measured, less driven by gut and gut alone. It's become clear to me that George H. W. Bush was emotionally and morally attached to the idea of the Gulf War in the way that his son was to be about Iraq a decade later—even to the point of risking impeachment absent the approval of Congress."

It didn't come to that. On January 9, just under a week before the UN's deadline for Iraq's withdrawal, Bush got authorization to use military force. On January 16, at 12:30 a.m. Baghdad time, Operation Desert Storm began with dual bombing offensives from the air as U.S. and British aircraft pummeled targets in the Iraqi capital and against Iraqi troops in Kuwait. The White House announced the inception of the war at 7:00 p.m. eastern time, while the world watched it unfold on CNN, whose comprehensive coverage bested its major network rivals for the first time.

Over the next six weeks, U.S.-led coalition forces drove through Iraqi troops in Kuwait like a knife through butter. Saddam's only significant response was the launch of Scud missiles targeting Saudi

Arabia and Israel, which had remained out of the war at the urging of the U.S. to prevent the risk of alienating Arab allies. When ground forces were deployed in Kuwait on Sunday, February 24, Iraqi forces quickly surrendered, were killed, or limped impotently across the border. By Thursday the 28th, it was all over. Saddam agreed to honor a peace agreement drawn up by the UN, after which Bush ordered a cease-fire. The war had come off with a minimal sacrifice of blood and treasure. American troops had prevailed with few losses; 125 soldiers died in the conflict, and another twenty-one were missing in action. The bulk of the war's $61 billion cost was incurred by the Arab states, principally Saudi Arabia, the United Arab Emirates, and Kuwait, to whom Saddam posed the most immediate threat.

As American forces drove Iraqi troops out of the country, the question in the media became, Why not continue into Baghdad and take out Saddam? But Bush's objective was clear-cut: to drive Iraqi forces from Kuwait. Regime change in Iraq was never part of the plan or the often-impassioned discussions around the war within the Bush White House. "I was in those meetings, in all those rooms," Dick Cheney stated, "and I don't recall a single person among that group that the President looked to for advice ever suggesting that we go in and take down Saddam Hussein." There was also a general belief, promulgated by U.S. intelligence, that Saddam wouldn't survive politically, especially after his humiliation in Kuwait. Bush sustained criticism for refraining from hunting down Saddam. But as *Time* magazine's longtime president-watcher Hugh Sidey wrote later, "I had spent much of my time in Washington writing about wars of one kind or another that never seemed to end—the cold war, and wars on Korea, Central America and Vietnam. Bush had achieved his objective in Kuwait brilliantly. He refused to enlarge and prolong the war, an act that took more courage than launching Desert Storm."

Bush's victory in the Gulf came on the heels of another diplomatic coup: Within a year after the fall of the Berlin Wall, Bush and his team helped engineer the reunification of Germany, a feat that was expected to take at least several years. Despite resistance to the idea by Margaret Thatcher and French president François Mitterrand, Bush staunchly supported German integration, becoming the first Western leader to do so. He viewed it as "the final chapter in the transformation of Europe." In early October, a united Germany was reborn as a member of NATO, an act Bush quietly managed to balance, as he had with his muted reaction to the Berlin Wall, without isolating or compromising Gorbachev as the Soviet Union and Eastern Bloc nations remained in a tenuous state of transition.

On February 27, 1991, as he addressed a joint session of Congress, George Herbert Walker Bush seemed invincible. "Kuwait is liberated," he began. "Iraq's army is defeated." Throughout his speech, the House chamber erupted in a series of thunderous ovations for the hero of the Gulf War, the man who drew a "line in the sand," and led a fragile coalition of foreign powers toward a clear victory over tyranny. Any fears of the operation becoming "another Vietnam" were put to rest. American pride skyrocketed; so did Bush's approval ratings, which stood at 89 percent, the highest for a president since Harry Truman's after V-J Day.

But Bush would find that popularity, like military glory, was fleeting. After the parades were over and the confetti was swept up, the country went back to worrying about domestic issues at a time when the economy was sagging and crime was on the rise. At home, Americans increasingly came to believe Bush's vision was lacking.

32

COMING OF AGE

THINGS WERE GOING WELL FOR George W. Bush in 1991. He and Laura had settled comfortably into Dallas, where the twins, who turned ten in November, attended public school; the Texas Rangers, at 85 wins and 77 losses, had their best season in five years; and total attendance in Arlington was up by 45 percent since Bush became co-managing general partner of the organization. He had set his own course and was "living life to the fullest." "It was an accomplishment to have put together that baseball team and to have friends [invest], and we were having a wonderful time," he recalled. Life was good.

The year had not been as kind to his father back in Washington. *Time* magazine's last issue of 1990, its renowned "Man of the Year" issue, offered a political portent. On its cover were dual photos of the forty-first president morphed grotesquely together: one of him looking up, strong and statesman-like; the other of him looking down as though groping for an answer. The "Man of the Year" became the "Men of the Year: The Two George Bushes." Bush "seemed to be almost two Presidents last year," the story read, "turning to the world two faces that were not just different but also had few features in common. One was a foreign policy profile that was a study in resoluteness and mastery, the other a domestic visage just as strongly marked by wavering and confusion."

Dana Carvey, whose manic portrayal of Bush became a staple of *Saturday Night Live*, once described his impression of the president: "You start out with Mr. Rogers," he said. "You add a little

John Wayne . . . you put them together [and] you get George Herbert Walker Bush." The disparity—the strong John Wayne versus the meek Mr. Rogers—reflected Bush's bifurcated political image, which would become his central dilemma.

In early July, his nomination of Clarence Thomas to fill the Supreme Court vacancy of Thurgood Marshall, the first African American to ascend to the high court, began to go awry when Anita Hill, a former coworker, accused Thomas of sexual harassment. While Thomas, also African American, called the ensuing grilling he received by the Senate Judiciary Committee a "high-tech lynching," many in the court of public opinion, especially women, saw him as guilty of the charges while questioning his qualifications for the court. Bush stood by his nominee, who won his appointment by a narrow vote, but the controversy around it amounted to a net negative for the president.

Bush's Supreme Court appointment the previous year was problematic for different reasons. When William Brennan retired from the high court in 1990, Bush appointed New Hampshire Supreme Court judge David Souter in his place. Brennan had been a disappointment to Dwight Eisenhower, who believed he would be a reliable conservative vote on the court. Instead Brennan became a stalwart liberal who sat on the bench for thirty-four years. Bush befell the same fate with Souter. Instead of Souter's appointment being a "home run for conservatism," as promised by John Sununu, he soon drifted leftward, adding his vote to the court's moderate and liberal wing, another blow to Bush's conservative credentials.

Battling Democratic majorities in both houses of Congress, Bush had racked up some important legislative achievements, most notably the signing of the Clean Air Act, providing market incentives to compel companies to lower toxic emissions that cause acid rain, and the Americans with Disabilities Act, a comprehensive civil rights law protecting disabled Americans from discrimination, including those afflicted with AIDS. But Bush was seen as being out of touch with the concerns of everyday Americans. The country felt as stagnant as its economy, which was mired on the brink of recession, and the man at the helm, however well intentioned, seemed incapable of moving it forward.

A number of factors worked against Bush. The political mood in America was shifting. The Republican Party had had a lock on the White House for over a decade, the longest a party had held the presidency since Harry Truman left office in 1953, after a collective twenty years of Democratic rule under Truman and his predecessor, Franklin Roosevelt. There was also a generational shift afoot. Three decades earlier, John F. Kennedy became the first of those who had fought on the front lines of World War II to win the White House, proclaiming, "the torch has been passed to a new generation of Americans." For thirty years, it had remained with them; the age of every president since had been within nine years of Kennedy. Now, the baby boomer generation was gaining prominence, entering their thirties and forties, and becoming more active in the political process, with an appetite to see one of their own in charge.

It didn't help that Bush *seemed* older in 1991. Normally indefatigable, the president complained of a lack of energy throughout the spring. He had inexplicably slowed down and lost fifteen pounds. The culprit, it was quickly discovered, was Graves' disease, a rare disorder causing an overactive thyroid for which he was treated with radiation and hormone-replacement therapy. George W. was struck by his father's appearance when he and Laura flew to Washington for the state dinner his parents were hosting for the Queen of England in May. He had never seen his father look *old* before.

By the summer, George W.'s worry spread to his father's reelection chances. "He was beginning to fight for his survival," he said of his dad, "[and] his reelection was not looking good at that stage—at all." Just as his father had slowed down, so had the energy within the Bush camp, which had suffered from the loss of Lee Atwater to brain cancer earlier in the year, leaving the president's reelection effort rudderless. George W. expressed his concern to his father. He was taken aback when his father asked him to come to Washington after the Rangers' season wrapped up to canvass administration officials and offer his assessment on the functioning of the White House and the reelection campaign. "It was a big honor, a coming-of-age moment in a sense," George W. recalled. "When he called me to take a look at it as a loyal son, I was intrigued and amazed that he would ask me."

On the last day of October, Halloween, Bush sent a letter out to cabinet officials and White House staff. "I have asked [my] son George to very quietly make some soundings for me on 1992," it read. "I'd appreciate it if you'd visit with him on your innermost thoughts about how to best structure the campaign. My plan is to wait—defer final campaign structural decisions until after the first of the year at least; but there seems to be a fair amount of churning around out there."

In the late fall, George W. made several investigative trips to the White House. "I felt that whenever he came, he came as a sponge," Andy Card, the then–deputy chief of staff, said. "He was absorbing. He would come in and sit in my office and just kind of ask questions . . . collecting information . . . then he would allow his father to squeeze the sponge to get the information out of him." George W. realized quickly that much of the palpable anxiety within the White House revolved around Card's boss, Chief of Staff John Sununu, whom the *New York Times* called "a symbol of the increasing disarray in Mr. Bush's domestic policy team." Sununu, dubbed "King John" by members of the staff behind his back, was seen internally as a barrier to the president, from whom they felt disconnected. Additionally, he had been accused of using government planes and cars for personal use. "There was a lot of angst at the White House," George W. recalled, "and a lot of it had to do with John. And so it became apparent to me that a change would be helpful."

George W. offered his perspective to his father during a private dinner in the White House residence in December. "You're isolated," he told him during their first course. The White House, he contended, needed restructuring—Sununu should go. Bush, always a "good listener" by his son's view, took it all in without reacting. By dessert, he offered his response. "Who's going to tell John Sununu?" he asked.

"Why don't you talk to him," George W. suggested.

Never good at being the bad guy, Bush resisted. "I'd rather it be someone else," he said.

George W. offered some possibilities, Jim Baker or Bob Mosbacher, say. Then, despite misgivings about a president's son delivering a message on behalf of his father, he offered, "Dad, if no one else can do it, I can talk to Sununu anytime, if you'd like."

His father paused as he considered the suggestion. "Fine," he said.

It made sense. Barbara explained, "George can't do ugly things, so if something came up, George W. would feel perfectly free [to do it]—like his mother. [George W.] gets credit for doing stuff that [he] didn't like very much but that needed doing." A "myth" sprang up that the resulting meeting between George W. and Sununu produced Sununu's resignation, but George W. claimed, "that's not true." But he did go "talk to him about the concerns and suggested that he have a heart-to-heart with Dad about his future in the White House" and "over time they had a meeting of the minds." It was the "over time" part that wasn't part of the plan. As Andy Card explained, "George W. was given the role" of asking for Sununu's resignation, "and he did deliver the message, but it didn't take." Sununu remained in place, waging arguments to the president on why he should stay on. It fell to Card to deliver a more direct message to Sununu, an awkward situation given that Sununu was Card's boss. This time it took. On December 3, Sununu resigned, explaining to the press that he feared he would be "a drag" on Bush's reelection campaign.

The episode informed George W., who resolved to deliver tough messages to staffers either directly or through a "close confidant"— not a member of the family. During his presidency, when he later had concerns about his own presidential chief of staff, Andy Card, it was George W. who talked to him about the need for a change. And when he felt it was time to go for his secretary of the treasury, Paul O'Neill, and secretary of defense, Donald Rumsfeld, it was his vice president, Dick Cheney, who let them know.

But George W. was grateful to have had the chance to help put his father's administration back on track, and it "maybe created a new dynamic" between the two of them. It had been one thing to help his father get elected in 1988, but it was a far greater compliment to have been tapped to offer his view on his father's administration and to help put things right. He had come to the task having achieved his own success in Texas, and he left with the ultimate sign that his father trusted his judgment.

On December 4, the day after Sununu announced his resignation, George W. wrote his father.

Dear Dad,

This past week has been very tough on you, for the right reasons. You care about your friends and worry about people's feelings. Yet, as usual, you handled things just right.

For me, the past week was one of the highlights of my life. I loved the chance to help you—to be part of history. I'm sorry there is so much press, speculation, and leaks. But nothing could change the experience of hearing you talk about the historical significance of my small gesture or saying "thanks, son for a job well [sic]."

You are a great man, a great President, and a great Dad.

All of your boys are ready to go to war in '92.

Love,

George

The same month, George H. W. Bush had his own feeling of being "caught up in real history." As the Bush family celebrated Christmas Day 1991 at Camp David, the last of the Cold War breezes blew with another resignation. In Moscow, Mikhail Gorbachev relinquished power as president after six momentous years, and with a sweep of his pen, signed a decree dismantling the Soviet empire.

Two hours prior to his resignation, Gorbachev reached Bush by telephone at Camp David, where he was celebrating Christmas morning with the extended Bush family. In the last exchange between presidents of the United States and Soviet Union, the two men talked as friends. "At this special time of year," Bush told Gorbachev, "we salute you and thank you for what you have done for world peace. Thank you very much."

"Thank you, George," Gorbachev said. "I was glad to hear all of this today. I am saying goodbye and shaking your hand."

Later in the day, Bush recorded in his diary, "There was something very moving about this phone call—a real historic note . . . It was something important, some enormous turning point." The world was changing. The Soviet Union, for nearly half a century a menace posing a threat to human freedom across the globe, was no more.

33

"DEFEAT WITH DIGNITY"

CHANGE REMAINED IN THE AIR in 1992. As the year opened, the war for the presidency had already begun in earnest, with challenges to Bush not only from the Democrats but also, almost predictably, from the right wing of Bush's own party. Bush was vulnerable to a threat from the party's right wing, now galvanized by Newt Gingrich's conservative rebellion on Capitol Hill. It came from Pat Buchanan, a conservative commentator and former White House aide to Richard Nixon and Ronald Reagan, who believed Bush was anathema to conservative values. Outspoken and pugnacious, Buchanan spat nativist indignity along with a charge to "Make America First Again," which fast became a refrain in his insurgent, cash-strapped campaign.

In the first major contest for the GOP, the New Hampshire primary on February 9, Bush garnered 53 percent of his party's vote while Buchanan took 37 percent. The Granite State, which had rescued Bush from the political abyss four years earlier, now sent Bush a jarring message that an insurgent conservative rival could yield returns well into the double digits. Just five days earlier, Bush had taken a trip to Florida that included a tour of a grocery store where he marveled at the electronic scanner that had been common for a decade, raising hackles in the press that Bush was an exalted Washington insider oblivious to the everyday existence of most Americans. The story was misleading. Bush's wonder was around updated scanner technology that could decipher even damaged codes, not the scanner itself. But the episode underscored the out-of-touch image of Bush that followed him onto the campaign trail. Always humble, sometimes to his detriment, Bush made a statement recognizing that Buchanan had "reaped the seeds of discontent with the pace of the New Hampshire economy." He understood the voters' "message of dissatisfaction."

The Democrats understood it, too, and smelled opportunity. By the time of the New Hampshire primary, the Democratic field of candidates had narrowed. Party front-runners Mario Cuomo, the popular governor of New York, and Bill Bradley, the New Jersey senator and former NBA great, opted out of the race, leaving the nomination wide open to candidates of lesser national name recognition. Former Massachusetts senator Paul Tsongas, former California governor Jerry Brown, and Bill Clinton, the young Arkansas governor of protean talents, emerged as the most viable.

Tsongas leveraged regional advantage into a win in New Hampshire with 35 percent of the vote, but it was Clinton, the second-place contender, who came out as the Democrat to beat. With 26 percent of the vote, well ahead of the rest of the Democrat pack, he was well placed for the next round of primaries in the South where his Arkansas roots would play nicely. Looking ahead and taunting the incumbent president, he told reporters, "In November, we will win a great victory against Pat Buchanan."

It had been a rocky road for the silver-tongued Clinton, who at forty-six, was a month younger than George W. After throwing his hat in the ring the previous October, Clinton was soon beset with character issues. He had been charged with smoking pot in college, evading the Vietnam draft, and serial womanizing, including a longtime affair with Little Rock lounge singer Gennifer Flowers. Revelations of philandering had sunk the candidacy of Democratic presidential hopeful Gary Hart just four years earlier, but Clinton continued to advance his candidacy despite his outsize character flaws, declaring himself the "Comeback Kid" after the returns in

New Hampshire were in. In the end, none of it seemed to matter, anyway. James Carville, Clinton's campaign strategist, summed up what was really on the minds of voters by scratching out the words "The economy, stupid!" at Clinton's campaign headquarters in Little Rock. Later coined as "It's the economy, stupid!" the admonition may as well have served as Clinton's campaign slogan.

Moreover, Clinton connected with people, showing an ability to find commonalities in contrast to Bush who came off as remote. As Bush was asking for "just a splash" of coffee at a New Hampshire diner and insisting incongruously that pork rinds were a favorite snack, Clinton was wolfing down Big Macs and wooing voters as though he was one of their own. In many ways, he was. The first child of a single mother who had lost her traveling salesman husband in a car crash three months before she gave birth to their son, Clinton's "I feel your pain" sensitivity tapped into the Oprah Winfrey baby boomer–oriented cultural zeitgeist. Even his foibles made him more relatable. With Southern wind at his back after a decisive victory in South Carolina's primary in March, the Democratic nomination was well within Clinton's grasp.

Bush's campaign lacked the same momentum and energy. After the New Hampshire primary, Bush called George W. to get his take. His son told him what was already plain: The right wing of the party, which had propelled Reagan's victories in '80 and '84, was in serious jeopardy. New Hampshire represented Buchanan's high-water mark—he would go on to garner just under a third of the vote in the next few GOP primaries before teetering out when Bush's nomination was inevitable—but conservatives remained wary of Bush.

George W. chose not to go to Washington to aid in his father's campaign as he had in '88. His responsibilities with the Rangers made it difficult to get away, and as he explained to the *New York Times* in April, he offered his father an "added dimension" by remaining in Texas. "Do I bring a different perspective than what he usually hears?" he asked. "You bet. Does he do everything I suggest?

No way. But he can always be totally sure that my agenda is his agenda."

The first son showed the value his father placed on his perspective in May. After momentum on Capitol Hill started building for a balanced budget amendment (BBA) that the administration had proposed earlier, George W. sent a memo to his father and high-level campaign staffers urging the White House to take ownership of the issue as a means of appealing to conservatives and overcoming the perception of domestic impotence. "The American people want governmental reform," he wrote. "Likewise, Bush supporters want the President to get a victory. They call it 'leadership.' I call it getting in the news with a conservative idea that gets enacted. BBA is the opportunity . . . Let us not let Congress take credit for our idea." The business school graduate also outlined an "action plan" around the issue, including holding an evening press conference and doing an interview with conservative radio personality Rush Limbaugh. Bush forwarded the memo to his chief of staff, Sam Skinner, with a handwritten note: "Sam—'W' has a good point here. If *he* doesn't know we're out front on this, who will?—GB."

Among the things George W. saw clearly from his perch in Dallas, adding to his concern for his father's political future, was the momentum around the Independent candidacy of Ross Perot, the diminutive Dallas billionaire and founder of Electronic Data Systems. George W. watched the movement for Perot building from his ninth-floor Dallas office, which looked out onto Perot's campaign headquarters. Alarmed, he called his father at the White House. "I want to tell you what's going on outside my window," he said. "We've got a problem here. There are like ten cars deep with people stopping to get bumper stickers. You can't write this guy off." It was, as Bush wrote later, "like watching the disintegration of a political base in slow motion."

The anticipation of a Perot candidacy had begun in February when Perot indicated that he may be willing to get into the race if volunteers saw to it that his name was placed on every state ballot. During an appearance on CNN's *Larry King Live*, he repeated the pledge, and a buzz around Perot's entry into the race began to grow. Perot was gaining traction by appealing mostly to disaffected white voters who were drawn to his folksy, antiestablishment "it's time to take out the trash and clean out the barn" message and the prospect of him applying his business savvy to fix a stagnant economy—despite a lack of specificity on how he might do it. By June, when it appeared that he was in the race, a Gallup poll had Perot yielding 39 percent of the vote.

Ross Perot and George H. W. Bush had been friends for most of three decades, two powerful Texans who would naturally fall into the same lofty business and social orbits. While the relationship hit a snag when Bush turned down Perot's offer in 1977 to lead his oil business, it "turned sour" altogether a decade later. "I think he was driven by a personal dislike, a personal resentment of me, you might say," he said. Dismissing his son's warning—and those of GOP leaders—Bush believed that Perot would be "seen as a weirdo, and we shouldn't be concerned with him." When Perot pulled out of the race in mid-July later claiming that Bush forces were out to sabotage not only his campaign but also, bizarrely, his daughter's wedding, it appeared that Bush was not far off the mark. George W.'s concern, however, didn't dissipate with Perot's retreat. "He'll be back," he told friends.

Perot's puzzling exit came as the Democratic Party was in the midst of its national convention in New York's Madison Square Garden where it made Clinton's nomination official. The first baby boomer to become the standard-bearer for a major party, Clinton rounded out his ticket with another of his generation, Al Gore, who at forty-four was two years Clinton's junior. The Clinton-Gore message of change, reflected in their relative youth, resounded in the line of a Fleetwood Mac song that repeated as a soundtrack throughout their campaign: "Don't stop thinking about tomorrow."

Benefiting from the Perot withdrawal, Clinton-Gore left Manhattan with a convention bounce of 24 percent—the biggest since modern political polling began—as Bush's disapproval rating rose to its highest level at 63 percent. The only incumbent president to emerge victorious amid such dire circumstances was Harry Truman, whose long-shot come-from-behind reelection win against Republican challenger Thomas Dewey in 1948 would become the touchstone for political underdogs. Perhaps groping for inspiration, Bush lugged *Truman*, David McCullough's thousand-page biography of the thirty-third president, on the campaign trail.

Among the things that George W. had suggested to his father the previous fall, when offering his assessment of the reelection effort, was the reassignment of Jim Baker from secretary of state to chief of staff, where he could oversee Bush's campaign, just as he had for Bush in '80 and '88. Bush was reluctant to make the change. Baker was ensconced at State at a consequential time. But in July, as the Democratic ticket surged ahead in the polls, Bush made the ask of Baker during a fishing outing at Baker's Wyoming ranch.

"You know, I think I'm going to need you at the campaign," Bush said. "You won't need to come till after Labor Day, that's when the campaigning will really begin."

"Nuh-uh," Baker replied. "We got to go for sure before that."

Baker "didn't want to leave" his post for the gritty business of running a campaign. But as he said, "if he was going to ask, there was never any hesitation. I mean, hell, I owed my public service career to him." He incurred criticism from the Bush family for not going to Bush's aid in advance of Bush's request. "I think [W.] thought . . . I didn't go over soon enough. But I went when I was asked. I thought it would be a little arrogant of me to go to my buddy and say, 'Hey your campaign is messed up. You need me.'"

When the Republicans gathered in Houston for their own convention in mid-August, Baker was firmly in place on the campaign. In '88, he had helped to engineer Bush's win after battling an eighteen-point deficit in the polls at the time of the convention. Things were different this time. The proceedings seemed defensive, including the party platform, which contained a repudiation of Bush's own tax increase. George W. was struck when he saw senior campaign workers scrambling to finish his father's acceptance speech at the last minute. *We're reacting*, he thought, *not leading*.

Pat Buchanan offered an appeal to the party's base to stand behind their candidate, but it went much further. His fiery speech decried a "cultural war" that was occurring across America for the nation's soul. Implicitly, though, he was also battling for the soul of the Republican Party, which had conservatives and evangelicals who identified with moral and social issues increasingly at odds with establishment and country club Republicans. "The agenda Clinton and Gore would impose on America—abortion on demand, a litmus test for the Supreme Court, homosexual rights, discrimination against religious schools, women in combat—that's change, all right," Buchanan said indignantly, "but it's not the kind of change America wants." While it fired up an important part of the Republican base, it did not, as George W. later put it, "convey a kinder, gentler Republican party."

Bush appeased the base—and showed his inherent loyalty—by sticking with his vice president, Dan Quayle, as his running mate, despite the recommendation of George W., Baker, and others to make a change. George W. liked Quayle's conservatism. When the vice president raised controversy by denouncing the popular TV sitcom *Murphy Brown* for featuring a plot in which its eponymous protagonist has a child as a single, unmarried parent, George W. supported Quayle's stance. In his May memo to his father on an economic matter, he added a handwritten P.S.: "Quayle was right. I don't want my daughters to think it's *cool* to have a child out of wedlock." But he also thought Quayle had become a drag on his father's presidency and provided a recommendation for his replacement: Dick Cheney. Though Quayle had dodged the bullet, the Bush-Quayle ticket, while narrowing the gap against Clinton-Gore, still trailed by ten points.

In late August, Hurricane Andrew blew viciously across the Atlantic and swept into the Southern states resulting in sixty-five deaths and over $26 billion in damages, as Bush interrupted his campaign schedule to manage a response. His campaign woes were compounded when Ross Perot jumped back into the race on the first day of October. Three televised debates between the three candidates did little to change the views of voters. Clinton continued to beguile; Perot to intrigue. Bush's performance was less inspired. In the second debate, a town hall, Bush glanced at his watch, reinforcing the perception of indifference.

Bush sprinted through the final weeks of the campaign. By the end, he was too exhausted to sleep. After a final campaign appearance in Akron, Ohio, on November 2, he and Barbara boarded Air Force One bound for Houston, where they would vote and watch the returns. George W. joined his parents for the trip. The mood was grim. Only Mary Matalin, Bush's deputy campaign manager, remained optimistic. "We're going to win, we're going to win," she repeated. Eventually, George W. took her aside. "We're gonna lose," he said. "We all need to be strong. We all need to be realistic. If you love my father, what you don't want him to do, and I know he will do, is try to make you feel better because he lost." The protector, George W. didn't want his father's disappointment compounded by knowing he had hurt those close to him. "It was George W. Bush in the final hours that gave us the strength to accept defeat with dignity," Matalin said. "He was the most mature, and must have been hurting worse than anybody else."

As Air Force One lumbered toward Houston, the Oak Ridge Boys, a favorite country band of Bush's, which had joined him for the last campaign swing, sang "Amazing Grace." Bush looked at George W.

as they sang. Both had tears in their eyes. "Boy," George H. W. Bush said, "would my father have loved to have been here hearing these guys," he said. The same evening, Bush the father thought of his sons. "I've given it my best shot," he reflected in his diary. "I've run the extra course, and George and Jeb say, 'Dad, you've run a great campaign—there's nothing more you can do.' And I'm grateful to them. They are the ones who have done so much work, taken so much flak on behalf of their father. They are the ones, all of them, who have lifted me and given me strength."

Change was delivered the following day, Election Day. As exit polls came in, it was George W. who delivered his parents the news in their suite at the Houstonian Hotel. "It looks like you're going to lose," he told his father. Bill Clinton would be the next president. When all the results were counted, Clinton captured 43 percent of the vote, while Bush and Perot trailed with 37 and 19 percent respectively. Clinton's take in the Electoral College was bigger, 370 votes to Bush's 168.

The smile Bush would flash stoically in the weeks that followed belied the regrets he harbored. He thought of a Kenyan long-distance runner in the Summer Olympics earlier in the year who had run through an injury, crossing the finish line forty-five minutes behind the race's victor. When asked why he had continued in pain knowing that he had no chance of medaling, he replied, "My country didn't send me here to start the race. They sent me here to finish it." Bush lamented that he hadn't finished the race and that he had let down his family, friends, and party. He had another regret, too, a bigger one for which he could only blame himself. As president he had worked hard, done his best, and cared about the American people. Yet he was convinced that they just didn't know his heartbeat.

34

"FINISH STRONG"

A LITTLE AFTER MIDNIGHT ON November 4, the day after the election, Bush coached himself through the balance of his administration. "Be strong, be kind, be generous of spirit, be understanding and let people know how grateful you are," he dictated in his diary. "Don't get even. Comfort the ones I've hurt and let down. Say your prayers and ask for God's understanding and strength . . . do what's right and finish strong." He would put a brave face on and be gracious, just as his mother would have expected.

Dorothy Walker Bush was much on his mind in his last days in the White House. On November 19, at age ninety-one, she died peacefully in Greenwich, another doleful chapter in a year Bush was eager to put behind him. Earlier in the day, he and Doro went to her hospital bedside in Greenwich to see her in her last hours. Both of them sobbed as they looked at her tiny figure, struggling to breathe, her tattered Bible at her side stuffed with notes her second son, the future president of the United States, had written her from Andover. She had been slipping mentally in the past couple years, and it had been painful to see her diminished. Once the undisputed leader among the expanded Bush families and the most vivacious mother among his friends—the one who won the mothers' race at Greenwich Country Day School, leaving others well in her wake, and the captain of the mothers' baseball team—she had become, in Bush's words, a "tired old lady." Now she was gone.

In many ways, it was harder for George W. Bush to accept the election defeat than it was for his father. Nearly three decades earlier, he was hurt and angry when William Sloane Coffin, a one-time friend of his father at Yale, told him that his dad, his hero, had been beat in his 1964 Senate race by a "better man." But how could the American people not see that it was the better man who had been beat on November 3? How could the American people have thrown George Bush over for Bill Clinton? Sure, the economy mattered, but what about character? George W. was the family's most astute politician, the one his father most relied on for his political view. Still, it was difficult for George W. to separate the political reality that had been so apparent in 1992 from his own disappointment. That was the way between the Bush father and son, to feel pain for the other more acutely than they themselves did.

Almost immediately after the election, George W. began training for the Houston Tenneco Marathon, to be held a few days after Clinton's inauguration, when his parents would be back home. It was a chance to exercise discipline and focus on something else—and to purge all the bile that had accumulated in the last four years: the media's scrutiny and knocks against the old man; the obsequiousness over Clinton; Neil getting dragged through the mud on Silverado. The eighteen-mile training runs leading up to the race would give him a chance to let it all go and move forward. Maybe it would help his parents to move on, too. (When the race came off, they would be there to cheer him on at mile 19, where his irrepressible mother yelled, "Hurry up, George, there are fat people ahead of you!")

On January 20, 1993, Bill and Hillary Clinton arrived at the White House late for the traditional coffee between the incoming and outgoing first couples. The Bushes received them cordially, exchanging pleasantries and talking convivially before taking the 1.8 mile ride up Pennsylvania Avenue, where Clinton would go through the rituals of his inauguration. Well afterward, George W. tried to "psychobabble" his father by asking, "What did it feel like to welcome Clinton? I hear you were very gracious." His father replied, "I didn't have a choice." When Clinton returned to the White House later in the day as its chief resident, he found a letter Bush had left him on the Oval Office's Resolute Desk.

Dear Bill,

When I walked into this office just now, I felt the same sense of wonder and respect that I felt four years ago. I know you will feel that, too.

I wish you great happiness here. I never felt the loneliness some Presidents have described.

There will be very tough times, made even more difficult by criticism you may not think is fair. I'm not a very good one to give advice, but just don't let the critics discourage you or push you off course.

You will be president when you read this note. I wish you well. I wish your family well. Your success is our country's success. I am rooting hard for you.

Good luck—

George

Earlier in the month, Tip O'Neill, who had been a tough opponent on Capitol Hill, told Bush, "Don't worry. You'll leave this place with a lot of people loving you. You're a good man. You've been a good president." The American people seemed to agree. Bush's approval rating, which had sunk to 29 percent the previous summer, stood at 56 percent. "I've done what I could," Bush said before departing Washington. "I intend to do everything I can to honor the office of the presidency . . . I feel good that I handed over the office so that Iraq is no problem for President Clinton . . . I wish him well." The Bush children could be forgiven for not being quite as cordial.

At the close of an interview with the *New York Times*, Jeb sarcastically jibed at the new president. "Remember," he said, alluding to the chorus of Clinton's campaign theme song, "Don't stop thinking about tomorrow."

As for what tomorrow would hold for George H. W. Bush, it was anyone's guess. Even his family didn't know what would come next. When George W. was asked the following day what his father would do in his post-presidency, he replied, "I've asked him myself and he said, 'I don't know.' I just don't think he's worked it out yet."

On January 21, the coffee machine began percolating at around 5:00 a.m. in the Bushes' two-story colonial brick town house in the Tanglewood section of Houston, a rental they would occupy for seven months as construction was completed on their modest new home in the West Oaks subdivision not far away. Informally dressed in a checkered shirt, windbreaker, and running shoes, and clutching a briefcase, Bush left the house at 7:30 a.m., well before rush hour, en route to his new office in the pink granite building at 10000 Memorial Drive. Lying in wait to capture the former president, a lone press photographer accompanied Bush on the elevator up to his ninth-floor, eight-office suite. Bush explained that he could go no farther than his office door. "I've been in public life for over twenty years," he said. "Now, I just want a little time for myself."

PART V

"His Turn Now in the Family"

35

"THE L-WORD"

THE FORTY-FIRST PRESIDENT WASN'T FORGOTTEN after he left Washington. Seven hundred letters or more flooded into his office every day. When he went out to restaurants, the other diners applauded. Abundant speaking offers poured in with handsome honoraria attached, temptations Bush would yield to on occasion calling them "white collar crime." The brave face he had worn during the seventy-seven days between the election and his departure from the White House remained in place in his new life. "We have much to be grateful for," he wrote in early February to a columnist who had written a flattering piece on his presidency—one of many to come from others through the years. "And once in a while someone says something very pleasant and nice and reassuring; and it all seems like the whole journey was worth every single minute of it."

But though he didn't wear it on his sleeve, the reelection loss pained Bush more than people knew. George W. saw it in quieter moments with him. Some of it came with the territory. "All presidents go through a period of withdrawal, I don't care what they say," said George W., who would go through his own withdrawal sixteen years later. "It's just a natural phenomenon. It's a deflation." Some of it was compounded by the circumstances. "I don't think depressed is the right word, but definitely—kind of a downer. It was a painful year for Dad, really painful to be rejected," George W. added. The repudiation of the American people, the inability to complete the mission he had started; it was hard to accept. "Barbara is way out ahead of me," Bush wrote of his wife.

Indeed, Barbara Bush was quicker in moving on; she had spent her life looking new circumstances in the eye—however daunting—and unflinchingly adapting. This was no different. "On January 20, we woke up [at the White House] and we had a household staff of 93," she recalled. "The very next morning we woke up and it was just George, me, and two dogs—and that's not all that bad." The former first lady readapted quickly to suburban life, taking the wheel of a car for the first time in a dozen years after she and her husband bought her a new Mercury Sable. Among other things, she did her own grocery shopping. When the curious asked, "Aren't you Barbara Bush?" she had a stock reply: "Oh, she's much older than I am."

Both got on with their lives. Every morning, the Bushes awoke to a new day, sharing a light breakfast while reading the *New York Times* and the two local Houston papers before walking the dogs, then moving on to busy schedules that had marked most of their lives. "Along the way," Bush wrote to a friend, "we count our blessings."

Shortly into their time after the White House, the Bushes accepted an invitation from the Kuwaiti royal family to visit Kuwait. It had been over three years since the Gulf War's end, and Bush had yet to see the Middle Eastern country his leadership had liberated. On April 13, the Bushes left Houston with a party that included Jeb and Columba, Neil, Laura, Jim Baker, and John Sununu. They boarded a chartered blue-and-white Kuwait Airways jetliner with amenities that rivaled Air Force One, and set off for a three-day journey.

A hero's welcome awaited Bush as the plane touched down in the country's capital, Kuwait City. Thousands of schoolchildren, accorded a special holiday for the festivities, lined the streets waving

American flags, holding balloons, placards, and flowers, and cheering their country's champion. Among the gifts Bush received from the Kuwaiti government were an antique door (symbolizing a Kuwaiti proverb, "If a man gives you a key to his home, you are friends. But if he gives you a door, you are family") inscribed with the names of the fallen U.S. soldiers in Operation Desert Storm, and the Mubarak the Great medal, Kuwait's highest civilian honor.

While the Kuwaiti head of state hadn't forgotten Bush's leadership role in driving Iraqi forces out of the country just three years earlier, neither had Iraq's. The day before Bush's arrival, sixteen Iraqis were taken into custody at the Kuwait-Iraq border after a plot to assassinate Bush was uncovered. Officials traced materials found in their vehicles to the Iraqi government. Their mission had been to drive a vehicle filled with explosives near the parade route where Bush's motorcade would be traveling, and to detonate the bombs with a remote control, or if the remote failed, with a timing device. The backup plan would be a suicide mission in which one of the leaders of the operation would strap on a belt laden with explosives and detonate it while approaching the former president. Mounting evidence in the weeks that followed pointed to a chief culprit: Saddam Hussein.

On June 27, President Clinton gave the order for a "firm and commensurate response." U.S. Navy ships in the Persian Gulf launched twenty-three Tomahawk missiles toward the headquarters of the Iraqi Intelligence Service, which had directed the assassination attempt against Bush. In a televised Oval Office address explaining the attacks, Clinton said, "The attempt at revenge by a tyrant against the leader of the world coalition that defeated him in the war is particularly loathsome and cowardly." Asked why Saddam himself hadn't been targeted, Secretary of Defense Les Aspin replied, "It's very difficult to capture a single individual. Dropping bombs on the hope that you're going to get a single individual is a very, very demanding task." When Bush was reached for comment in Kennebunkport, where the Bushes would annually spend the late spring, summer, and early fall months, his response was brief. "I'm not in the interview business anymore," he said. "But thank you very much for calling."

Bush was delighted to be out of the "interview business," taking pleasure in turning down the bulk of the media's frequent requests. He didn't need the spotlight and had no grand plans for an activist post-presidency. In recent years, he had seen Richard Nixon rise like a phoenix from the ashes of Watergate to become a kind of self-appointed secretary of state and eventually a respected elder statesman, someone whose foreign-policy judgment Bush himself had come to value as president—just as Ronald Reagan had and Bill Clinton would. He had also seen Jimmy Carter overcome the humiliation of his reelection defeat in '80 by banging nails for Habitat for Humanity and launching the Carter Center, a nonprofit organization attached to his presidential library devoted to peacemaking and human rights. Soon, Carter, a future Nobel Peace Prize laureate, became what many considered America's finest former president.

Bush, on the other hand, had given up on "saving the world" and saw no need for rehabilitation, content instead to be a "point of light for the community." Eventually, his efforts on behalf of various philanthropies throughout his post-presidency would yield over $670 million. Among other things, the former president busied himself as a board trustee for Houston's MD Anderson Cancer Center, eventually becoming chairman, and overseeing the planning of the $83 million George Bush Presidential Library and Museum and the George Bush School of Government and Public Service on the edge of the campus of Texas A&M University in College Station, an hour-and-a-half drive from Houston.

Likewise, Bush dismissed talk about his legacy—the "L-word," he called it when the subject arose. While Barbara spent much of her first year out of the White House on an IBM laptop laboring away on *Barbara Bush: A Memoir*, the autobiography she had been contracted to write for Scribner, he opted not to do so. The first ex-president since Woodrow Wilson to decline to write a memoir—and the sizable paycheck that went with it—Bush had no interest in making a case for the historical record. He was content instead to leave his political past behind him. On his last day in the White House, as he bade farewell to his staff, he avoided self-aggrandizement. "Maybe they'll say we did some good things," he said. But he would leave it up to historians to make the call. As he put it later, "Let history be the judge without my pushing and pulling."

What he couldn't have known was that history was not done with him. Bush's proudest moments were rooted in family. Ultimately, nothing mattered more to him. As he settled into his post-presidency, his two eldest sons began focusing on their own political ambitions. His days of "saving the world" may have been behind him, but the Bush patriarch's legacy—the "L-word"—was still growing.

36

JOY AND HEARTACHE

"WHEN ALL THOSE PEOPLE IN Austin say, 'He ain't never done anything,' well, this is it," boasted George W. Bush as he toured a Texas journalist around the Ballpark at Arlington, the $191 million, 49,166-seat future home of the Texas Rangers. Though it imbued an old-time baseball feel, the stadium gleamed with state-of-the-art modernity, featuring all the latter-day bells and whistles: 120 luxury suites, four stories of offices within the ballpark, a baseball museum, a children's learning center, and a youth baseball park. It infused civic pride in the locals of north Texas, especially the 300,000 residents of Arlington—and in George W. Bush, who had done as much as anyone to bring the project to fruition. And it came at a good time in Bush's career, as he set his sights on the governor's mansion in Austin, the state's capital.

Bush ran the effort to get the stadium built as though it were a political campaign. Behind the scenes, he relentlessly lobbied Arlington officials, plying all the persuasion of a Lyndon Johnson backroom deal while making public appearances to lobby support among the local citizenry. A $130,000 public relations effort aimed effectively at appealing to Arlington voters to approve the half-percent sales tax increase to raise most of the funds. In January 1991, voters passed the referendum by a two-to-one margin, providing 70 percent of the stadium's cost, which added up to a cool $135 million. "I have to make a fairly big splash in the pool for people to recognize me," he said to a *Newsday* reporter upon the sale of the Rangers to his buyers' group in 1989, alluding to the high bar his father had set for him. The Ballpark at Arlington was his big splash.

Since buying the team, Bush had taken it on himself to embark on a statewide PR tour for the Rangers, pressing the flesh as booster in chief for the team on the "mashed-potato" circuit. "Ladies and gentlemen, I know you wished the most famous Bush could be here tonight," he would josh during his father's presidency in what amounted to a stump speech, "but Mom was busy." Now, the new ballpark made George W. Bush a famous Bush in his own right—at least in Texas. "For those who mock George W. Bush as a daddy's boy who slouched through life reaping the dividends of his name," wrote the *New York Times* in 2000, as Bush sought to become the forty-third president, "the lovely Texas Rangers baseball stadium here [in Arlington] is a useful rebuttal . . . This stadium is the house that Bush built."

The deal was not without controversy. Especially when Bush ran for president, the irony of the fiscal conservative perpetrating "corporate welfare" by soliciting a government subsidy for a private enterprise and securing the land it sat on under eminent domain, something he opposed as Texas governor, was not lost on his political opponents. But there was no denying that ballpark's success. By 1996, two years after the ballpark's opening, the Rangers would yield the highest revenue in Major League Baseball, bringing in $25.5 million according to *Financial World* magazine. Moreover, the ballpark would make Bush wealthy. When the team sold in 1998 for $250 million, Bush had leveraged his $606,302 initial investment into $14.9 million. Less than a decade earlier, in the oil business, Bush had hoped to hit the "big one." Now he had done it, not in the oil patch of West Texas but in the corporate world of Major League Baseball. "[W]ith the Rangers, he really hit his stride," said Joe O'Neill, who had known him for four decades. "It took some hard times and big jobs to bring out the bigness in him."

It was Jeb, not George W., who first revealed his gubernatorial ambitions to the Bush family. In the fall of 1993, he talked about throwing his hat in the ring in Florida, where the incumbent Democratic governor, Lawton Chiles, was vulnerable to a challenge. Jeb's announcement came as no surprise to the Bushes. Jeb had been plotting for some time to launch himself into politics, having taken prudent steps toward high political office as though he had meticulously followed the Bush family playbook. He had achieved financial security in the South Florida real estate business, amassing a net worth of $2.25 million; established a solid Rolodex of valuable political contacts throughout the state; and served as chairman of the Miami-Dade County Republican Party in 1986 before his two-year appointment as state commerce secretary by then–Florida governor Bob Martinez. Among the Bush children, Jeb was a good bet to pursue the family business. As a child, while George W. dreamed of being Willie Mays, Jeb's childhood ambition was to be president. It made sense that George W. was running the Texas Rangers and Jeb was running for the Florida governorship.

George W.'s intention to capture the Texas statehouse came as a greater surprise to his parents. When he told them he was "fixin'" to take on Texas's incumbent Democratic governor, Ann Richards, his mother rejoined, "You can't win." Two decades later, when George W. related the anecdote, well known in Bush lore, he added with a laugh, "True story."

In the larger narrative of Bush political dynasty, a common conception is that it was Jeb, not George W., who was his father's political heir apparent, the family's choice to carry the political torch—maybe even back to the White House. Jeb, the story goes, was the serious one; George W. was the screw-up, or at least the cutup, not as well suited to the task. The buzz that George W. was the lesser of the two gained traction in the late 1990s as momentum built for George W. to run for president. "[It's inaccurate] to suggest that Dad or Mom thought I would be president and George wouldn't," Jeb said in 2014, as he was contemplating his own bid for the presidency. "I don't know where [that speculation] comes from . . . It could have been one of those wild things my mother says sometimes. For a relatively long period of time, I was always used as the basis to disparage my brother. A role I did not expect and to this day it just kind of pisses me off."

George W., however, could understand why the perceptions formed early on in their adult lives. "I was always viewed as kind of the frivolous guy," he said later. "You know, I was single and moved around quite a bit, so to speak," he said. "And Jeb on the other hand was viewed as the serious dude. He got married very early, fine student, and by nature we're very different." But the narrative disregards the fact that George W. had made his political aspirations known since his midtwenties when he considered running for the Texas State Legislature. He had thrown himself in the arena, making a respectable run in the '78 congressional race, and had openly chewed on a run for governor in 1990, just four years earlier. In between, he had proven himself to be a valuable political asset to his father, helping in his campaigns behind the scenes, as a surrogate on the hustings, and offering his counsel on his White House and the electorate. And though it had taken him significantly longer than Jeb, he had achieved his own business success. It was logical, maybe inevitable, that Jeb would get into politics, but it was for George W., too.

While it's true that Barbara harbored doubts that George W. could win, her hesitation was less rooted in her lack of confidence in her son's political ability than it was trepidation over his opponent, Ann Richards. Boasting a 63 percent approval rating, she was riding high in Texas and ostensibly poised to coast to reelection. George H. W. Bush, who often played the optimist to his wife's pragmatist, had greater faith that his son could prevail. But overall there was concern that Richards might be invincible. "We weren't sure if W. could beat Ann Richards," Jim Baker said. "We—the family and those close to them—thought Jeb could beat Lawton Chiles." Regardless, it didn't much matter to George W. who sought neither their approval nor counsel. "[I]t wasn't like I wanted [their] advice," he explained. "I'd made up my mind." His response to his mother's concern that he couldn't beat the popular incumbent was simple. "Yeah, I can," he said.

His conviction wasn't based on just bluster or maternal defiance. The timing for his run, he believed, was actually *good*. In essence, his father's loss in '92 paved the way for him and Jeb to pursue their own ambitions largely unencumbered by their father's political baggage. "If [Dad] had won in '92," George W. maintained, "it would have required an unbelievable sense of bravado to run because I'd have been [running for governor] in '94 defending him with two years to go in his presidency. The election would have been, 'Well, your father said this, and your father did that.'" Furthermore, he believed that his opponent, as popular as she was, was vulnerable to a challenge because of his father's successor. "The environment was conducive to a challenger because of Clinton," he said. "People were starting to get buyer's remorse—especially in Texas."

Jeb worried that two Bush brothers concurrently vying for governor's mansions would turn "into a *People* magazine story." But he conceded that he had "no control" over his brother. It became clear when George W. was asked what he would do at the next meeting of the National Governors Association if both brothers were attendees. "Jeb's my little brother," he quipped. "He's done what I've told him to all his life."

"Former President's Shadow Looms Over 2 Bush Sons in Governors' Races," read a headline from the Associated Press in late October 1993, a little over a year before Election Day, one of many stories generated as their races began. Along with it, speculation arose that the two Bush sons were rivals locked in a competition to become the family's next-generation standard-bearer. Like the narrative of him as the "Prodigal Son," George W. dismissed the rumors of a competition between them as "a total myth." "First of all," he explained, "when you're running . . . you're absorbed in your own race and you don't have time to analyze somebody else's race, but of course I wanted him to win . . . I don't think you're a better man for hoping somebody else fails."

While it's impossible to fully plumb the psychological undercurrents that stream from within a successful, hypercompetitive family, it's doubtful either of the brothers wanted to win at the other's expense. The Bush children jibed and jabbed at each other—along with winning, it was the family currency—but it would have been uncharacteristic for one family member to wish another ill. Though they kept their distance as they immersed themselves in their campaigns, they rooted for each other's success while striving for their own victories. "George and I didn't talk much. We'd talk every once in a while but we were on two separate paths," Jeb explained. "I wished my brother would win and he wished I would win . . . It's a mistake to say that being competitive means you're rivals—and that's the distinction."

If their father's shadow loomed over them, the candidates didn't let it show. "I'm not running because I'm George Bush's son; I'm running because I'm George P., and Noelle, and Jeb's father," Jeb said repeatedly as a refrain. Recognizing a good line, George W. unabashedly co-opted it, substituting Jeb's children's names for those of Barbara and Jenna. But he didn't take his brother's lead on using his father on the campaign trail. Jeb and his father appeared together regularly at Florida rallies, an experience that proved emotional for both of them. George W., on the other hand, while tapping his mother for campaign appearances, pointedly relegated the old man to the sidelines. In a rare reference to him, he said, "I can take care of my own self, but I hope he raises me a whole lot of money."

He did. The former president's drawing power at fundraisers and influence with high-level donors helped bolster the war chests for both of his sons. Over three decades earlier, Prescott Bush, after forgoing a reelection bid in the U.S. Senate at the advice of his doctor, lamented, "Once you've had the exposure to politics that I had . . . it gets in your blood, and then when you get out, nothing else satisfies that in your blood." While it was true for George H. W. Bush, too, his own sons—his flesh and blood—had brought him back into the game. The blueness of his first year out of the White House began to wash away as he looked toward their future. "I surmise Jeb's and my race helped him to recover," George W. observed. It provided him a sense of purpose that had been lacking in his post-presidency. But he also realized that while the Bush brand name would be valuable in gaining them political advantage in the early stages, it—he—could also prove to be a liability to them, as well. Moreover, he knew the pain of losing and privately worried about the prospect of their defeat—especially George W.'s, whose race came with greater odds against him.

There was an added paternal dimension to George W.'s race. Richards's speech at the 1988 Democratic National Convention, in which she derided George H. W. Bush as being "born with a silver foot in his mouth," had propelled her onto the national stage, setting her up nicely to become the Democrat's gubernatorial candidate two years later.

Few had forgotten it, let alone George W. Bush. Yet as he told reporters, "I bear no ill will personally toward the current Governor," while adding that the job required more than just "funny sound bites." The sixty-one-year-old Richards, on the other hand, thought that it was Bush, the callow businessman with the big last name, who wasn't serious. "This is not a joke," she said in late October at their only debate. "We're talking about who is going to run the state of Texas. You have to have some experience in the public sector before you get the chief executive's job." Deriding him as "Shrub" and, of course, "Junior," and on at least one occasion, "a jerk," she seemed resentful that she was being challenged by the former first son at all.

It didn't rattle George W., who kept his irascibility in check, running a campaign that was far more focused and disciplined than many expected. Defining himself as a "conservative" and his opponent as a "liberal" who was "constrained by the current way of doing things in Austin," he hammered away at her as being soft on crime and weak on education, while aligning her with Clinton. "He's too liberal," Bush declared of his father's successor from the stump. "And in 1996, you can be sure that Ann Richards will do everything to elect Bill Clinton, and George W. Bush will do the opposite."

Jeb ran a similar campaign in Florida, touting his conservatism while blasting away at Chiles's tax-and-spend liberalism as crime rose and schools foundered. The sixty-four-year-old Chiles, who prided himself as a true Floridian who spoke "cracker," played the country card, painting his opponent as an elite carpetbagger and political neophyte who couldn't understand the average Floridian. In a debate in early November, he predicted a come-from-behind advance, warning, "let me tell you about the old liberal—the old he-coon walks just before the light of day." Jeb was puzzled by the colloquialism, a reference to a rural Florida legend about the shrewdest member of a pack of raccoons. Like his brother in his 1978 congressional race, Jeb had been out-country'ed. For the balance of the campaign, Chiles referred to himself as the he-coon, bolstering his campaign in rural areas and narrowing the gap with his rival.

By early November, the races in Florida and Texas were both dead heats. Even then, Barbara had concerns about George W.'s chances. After campaigning with him in the final days, she told George, "You know, he thinks he's going to win."

"He is," George replied.

He was right. On November 8, when the returns came in, George W. won comfortably, taking 53 percent of the vote to 46 percent for Richards, who became the fifth Texas governor out of the last six to fail to win reelection. Jeb realized a different fate. Chiles, the "he-coon," walked away with 51 percent of the electorate, edging out Jeb by two percentage points.

For George and Barbara, watching the returns from their Houston home, the high of George W.'s victory was nearly overshadowed by the pain they felt over Jeb's defeat. George W. was in a suite at the Sheraton Crescent Hotel near the capitol when his father phoned to congratulate him on his victory. Karl Rove recalled the governor-elect taking the call in the bathroom of his room, the only quiet place in the suite. While Rove could hear only one side of the conversation, it was "clear that his dad was in agony that Jeb [was] losing." "But Dad," George W. said, "*I* won."

When asked by reporters the same evening what he was feeling, the former president said, "The joy is in Texas; my heart is in Florida." Yet there was great pride in his elder son's victory, and maybe a little vindication, too. "I must say," he told *Time*'s Hugh Sidey years later, "I felt a certain sense of joy when [George W.] took [Ann Richards] down. We showed her what she could do with that silver foot, where she could stick that now."

Just over two months later, the whole family gathered in Austin on the seventeenth day of January, a Tuesday, to witness the inauguration of George W. Bush as Texas's forty-sixth governor. Billy Graham, who had helped Bush forge a different path nearly a decade earlier, offered an invocation for the new governor while recognizing the "moral and spiritual example his mother and father set for all." Then, Bush laid his left hand flat on the thick Bible of the Republic of Texas's first president, Sam Houston, held by Laura, raised his right hand, and recited the oath of office. Later, Governor Bush would occasionally look at a photograph he kept of that moment. He's wearing a dark suit and gray tie, just as his father had at his presidential inauguration in 1989. His parents are there, his father wiping a tear from his eye, and all his siblings, basking in the moment. Jeb clearly shares their pride, but there's something else in him, too—a sadness. It was a "tough moment" for him, Bush recognized, and given his fraternal love, tough for him to look at. Had the situation been reversed, his own sadness would have been profound.

As it was, the torch, and all the implications and expectations that went with it, was being passed to him.

Inevitable analogies to the Kennedys arose, though not ones the Bushes invited. On Inauguration Day, Bush's campaign adviser, Karl Rove, crowed, "On this day, George W. Bush clearly comes into his own, and no one can dispute that. We really do have a generational shift passage here. Sort of like Joe Kennedy handing off to young John."

Earlier in the morning, Barbara handed the governor-elect an envelope containing a letter from his father along with his "most treasured possession," the gold cufflinks Prescott Bush had presented to him in Corpus Christi, Texas, after he had earned his Navy wings on June 9, 1943. George couldn't bring himself to give them to George W. personally—the moment would have been too fraught with emotion.

"At first I didn't think about the continuity, the grandfather part," he said later. "The main thing I thought was that it was from my dad. He was saying that he was proud of me. But later I reread the letter and thought about it. It ended with, 'Now it's your turn.'"

37

"CHART YOUR OWN COURSE"

GEORGE W. BUSH NEVER LACKED confidence. Since his bike-riding, baseball-playing Midland youth, he had shown himself to be a natural leader, brimming with self-assurance. It could come off as frat-boy cockiness, even arrogance—the smirks, the nicknames, the swagger—and at times shrouded the discipline, focus, and hard work that went into nearly everything he did. All those traits were on display as Bush plunged into his new duties in the Texas state-house, and despite having never held public office, Bush settled into his state's highest elected office as comfortably as his father—with a résumé of public service as long as his forearm—had taken to the Oval Office. As he did, he quickly began to win over critics.

Being comfortable in his skin was part of Bush's charm. Paul Burka, a hard-bitten veteran journalist for *Texas Monthly* and former legislature staffer, wrote after Bush had left the statehouse, "I have never seen anyone that good at the game of politics. It was impossible to be around the guy and not like him. He filled a room. He was always himself. He said what he thought. He had the ability to let his guard down without losing the dignity of 'I am your governor.'" After Bush's first hundred days in office, Burka judged that the former co–managing partner of the Texas Rangers was "batting a thousand" as governor.

Bush's success came in large measure due to his courtship of the state's Democratic lieutenant governor, the hard-drinking, chain-smoking, big-as-Texas Bob Bullock, who controlled the state's legislative agenda and reigned supreme over the Democratic-dominated House and Senate. As Bush wrote, "Bullock was not of my party or my generation; he was a crafty master of the political process not inclined to think much of a rookie like me." In fact, Bullock was the kind of authentic folksy Texas politician who might have nipped at Bush's Achilles' heel by caricaturizing him as a carpetbagging Ivy Leaguer. Instead the lieutenant governor took a shine to Bush. "I am genuinely fond of him," he said of the governor two years into the latter's tenure. "He's a fine person. He says that when he makes a campaign promise, he's given his word. I admire that a lot."

The two worked together to enact much of Bush's gubernatorial agenda, taking aim with focused precision at four key issues Bush had built his campaign around: improving public education, enacting tougher juvenile and criminal laws, enacting fairer tort laws, and implementing welfare reform. The former, an overhaul of public schools, was prompted by Bush's concern over the high rate of illiteracy among Texas youth. Working with education experts and lawmakers, Bush granted schools greater autonomy while holding them accountable to meet standards and achieve results. The changes, a template for the No Child Left Behind education reform he would implement as president, saw appreciable boosts in student test scores, including those of minorities, within his term in office. In 1994, the year before Bush took office, 58 percent of Texas's third graders passed the state's standardized academic test, a number that would increase to 76 percent by 1998. During the same period, employment grew by 15 percent while juvenile crime and welfare payouts fell by 30 and 47 percent respectively. In addition, Bush would put money back in the wallets of Texans by helping to steer through a $1-billion tax cut.

Though Governor Bush did not invite comparisons to former president Bush, his father was never far from his mind. He used the old oak desk that George H. W. Bush had used as a U.S. congressman—or so he thought. When his father saw the desk, which George W. had affixed with a brass plaque that read "Representative George Bush," the elder Bush broke the news that the desk had never been in Washington; he had picked it up in a Midland sidewalk sale for a hundred dollars as an up-and-comer in the oil business. Despite abhorring interruptions as he pursued his gubernatorial duties, George W. put everything on hold when aides told him his father was on the line. "When somebody says, 'The president is calling,' we know which one it is," said his legislative director. The calls were not to proffer unsolicited advice or live vicariously through him or reflect in his political glory, but just to check in—"how are you doing, how's the family?"– type conversations. In fact, Bush seemed genuinely glad to be "out of it," by which he meant the day-to-day political issues—and content to see George W. inherit the mantle.

The elder Bush sought to downplay his rarefied status as a former president as the email domain for his office—FLFW, an acronym for "former leader of the free world"—suggested. It didn't mean, however, that he was leading a quiet life. After leaving the White House, Dwight Eisenhower remarked that retirement for an ex-president was "just a word in the dictionary." So it was for Bush. His was a crowded existence that included campaign appearances and fundraisers, lucrative speeches, dedications, award ceremonies, board meetings, and philanthropic activities, as well as a frenetic leisure schedule. In 1996 alone, Bush spent 143 nights on the road.

Additionally, Bush added author to his spate of activities, writing, along with Brent Scowcroft, *A World Transformed*, published in 1998, about the dramatic world events that transpired during his presidential watch. A year later, he offered *All the Best, George Bush*, a compendium of his profusion of letters, notes, and diary entries culled from fifty-seven years of adult life, providing as intimate a reflection of Bush as might have come from a memoir.

He relished doing whatever he wanted, unconstrained by the weight of public office. After addressing the International Parachute Association in Houston and relating his terrifying World War II plane crash during which he aborted his plane and pulled his chute too early, he resolved, at age seventy-two, to make one more jump. When he informed his children of his plans, George W. responded as though his father was having a midlife crisis. "Just don't tell anybody about your girlfriend," he joked. In March of 1997, the former commander in chief journeyed to the Arizona desert, where Bush donned a white Elvis jumpsuit—"The King would have approved," he maintained—and along with members of the army's Golden Knights, plunged from a plane at 120 miles per hour with a "twinge of fear," pulling the chute with a jolt at five thousand feet before landing safely on terra firma. Instead of getting the jump out of his system, Bush became an adrenaline junkie of sorts, making several more jumps in the years to come, including those on his eightieth, eighty-fifth, and ninetieth birthdays, the last of which his son, the forty-third president, was there to greet him as he touched down.

On November 6, 1997, he went through a rite of passage accorded to all modern ex-presidents as he and Barbara christened the George Bush Presidential Library and Museum, which housed the forty million documents and papers of his administration, joined by their successors, Bill and Hillary Clinton. The president was nearly a year into his second term in office after winning reelection in a rout against two other former Bush opponents, Republican nominee Bob Dole and Independent candidate Ross Perot, with Clinton yielding 49 percent of the vote versus 41 and 8 percent for Dole and Perot respectively. Also on hand at the unveiling of the Bush Library were Jimmy and Rosalynn Carter, Gerald and Betty Ford, Nancy Reagan, Lady Bird Johnson, and members of the Bush family. Jeb played master of ceremonies, introducing the day's speakers including his brother, the governor of Texas. "He stood always ready to serve, as war hero, loving husband, world leader, dedicated father, and invincible optimist," said George W. of his dad while adding a subtle dig at Clinton, "and a president who brought dignity and character and honor to the White House." When it was his turn to speak, the forty-first president apologized to his late mother for violating her "no bragging rule" given the museum's depiction of his presidency, then waxed sentimental about his days in the White House. "Now that my political days are over," he said in closing, "I can honestly say that the most rewarding titles I've held are the only three I have left—a husband, a father, and a granddad."

Bush the father would see his second son climb back into the political ring in the following weeks. Jeb had spent the previous several years shoring up his marriage, which had suffered from the strain of the first campaign, converting to Columba's Catholicism, and reconnecting with his three children. Now, at forty-four, he was ready for another go at the Florida governorship. Losing one's first political race was virtually a Bush tradition—Prescott and George had lost their Senate bids in 1950 and 1964 respectively, and George W. his congressional campaign in 1978. Another Bush tradition was winning in the next contest having taken to heart the lessons from the first. As he squared off against his opponent, Florida's lieutenant governor Buddy MacKay, Jeb retooled his strategy, tamping down his natural conservatism, which had come off as strident four years earlier, and stressing compassion and inclusiveness. The husband of a Mexican immigrant and the father of Hispanic children played up his own family's diversity as a means of relating to Latinos, African Americans, and liberal Democrats.

Jeb's venture came as George W. prepared for reelection in Texas. In early December 1997, he announced for a second term at Sam Houston Elementary School, his grade school in Midland, a contest that would have him matched up against Garry Mauro, the state's Democratic land commissioner. He did so with his father's '92 defeat in mind, "knowing the perils of incumbency," as well as the fickleness of the Texas electorate. "The only reason to look back," he reminded voters who might be wary of him resting on his laurels, "was to determine who is best to lead us forward."

As his boys were in the thick of their campaigns, George got wind through Barbara that they were uncomfortable with the way the media was diminishing him; their political attributes were being touted at his expense. *They* had vision. *They* spoke in complete sentences. On the first day of August 1998, three months before Texans and Floridians went to the polls, Bush wrote his sons a letter giving them his blessing to distance themselves from him as they sought to establish themselves as leaders on their own. As it related to the family business of politics, the missive—a Bush family manifesto, of sorts—was as revealing of the Bush patriarch's relationship with his sons as any he would write.

Dear George and Jeb,

. . . Your mother tells me that both of you have mentioned to her your concerns about some of the political stories—the ones that seem to put me down and make me seem irrelevant—that contrast you favorably to a father who had no vision and who was but a place holder in the broader scheme of things.

I have been reluctant to pass along advice. Both of you are charting your own course, spelling out what direction you want to take your State, in George's case running on a record of accomplishment.

But the advice is this. Do not worry when you see the stories that compare you favorably to a Dad for whom English was a second language and for whom the word destiny meant nothing.

First, I am content with how historians will judge my administrations—even the economy. I hope and think they will say we helped change the world in a positive sense . . .

It is inevitable that the new breed journalists will have to find a hook in stories, will have to write not only on your plans and your dreams but will have to compare those with what, in their view, I failed to accomplish.

That can be hurtful to a family that loves each other. That

can hurt you boys who have been wonderful to me, you two of whom I am so very proud. But the advice is don't worry about it. At some point, you may want to say, "Well, I don't agree with my Dad on that point" or "Frankly, I think Dad was wrong on that." Do it. Chart your own course, not just on the issues but on defining yourselves. No one will ever question your love of family—your devotion to your parents. We have all lived long enough and lived in a way that demonstrates our closeness; so do not worry when the comparison might be hurtful to your Dad for nothing can ever be written that will drive a wedge between us—nothing at all . . .

And that's not just the journalists. There is the Washington Establishment. The far right will continue to accuse me of "Betraying the Reagan Revolution"—something that Ronald Reagan would never do. Then they feed the press giving them the anti-Bush quote of the day. I saw one the other day "No New Bushes" an obvious reference to no new taxes . . .

Nothing that crowd can ever say or those journalists can ever write will diminish my pride in you both, so worry not. These comparisons are inevitable and they will inevitably be hurtful to all of us, but not hurtful enough to divide, not hurtful enough to really mean anything. So when the next one surfaces just say "Dad understands. He is at my side. He understands that I would never say much less do anything to hurt any member of our family."

So read my lips—no more worrying. Go on out there and, as they say in the oil fields, "Show 'em a clean one."

This from your very proud and devoted,

Dad

On November 3, the eve of the election, Bush wrote his friend Hugh Sidey. Nervous as always before the polls came in, he expressed concern about the possibility of Jeb's loss while exuding unabashed pride over George W., who had been in Houston the previous day for a rally.

Should Jeb lose in Florida I will be heartbroken—not because I want to be the former President with two Governor sons. No, heartbroken because I know how hurt Jeb will be . . .

[George W.] is good, this boy of ours. He's uptight at times, feisty at other times—but who wouldn't be after months of grueling campaigning.

He includes people. He has no sharp edges on issues. He is no ideologue, no divider. He brings people together and he knows how to get things done. He [has] principles to which he adheres but he knows how to give a little to get a lot. He doesn't hog the credit. He's low on ego, high on drive.

All the talk about his wild youth is pure nuts. His character will pass muster with flying colors.

Bush ended the letter hopefully.

Six years ago I was president of the United States of America. Tonight, maybe, the father of two governors. How great it is!

But then tomorrow a whole new life begins.

The following day, he got his wish, becoming the father of the governors of the second and fourth largest states in the union. Jeb captured 55 percent of the vote versus MacKay's 45 percent. Along with his win came a Republican majority in the legislature, marking the first time the GOP controlled Florida's executive and legislative branches in the twentieth century. George W. won bigger in Texas, taking a landslide 69 percent of the vote—including just under half of the Hispanic vote—and bucking the prevailing Texas trend of rejecting the incumbent that the popular Ann Richards had seen four years earlier. He had achieved statewide election in Texas twice, something that had eluded his father as many times in his failed runs for the U.S. Senate.

If "a whole new life" had begun for George H. W. Bush in the wake of his sons' victories, the change was far greater for George W. Bush. In spite of—or rather because of—his impressive win, Bush's second-term agenda as governor in 1999 was less on the minds of those in the media and in political circles than the inevitable question: Would he run for president in 2000? "I hardly ever think about it, except every day when people ask me about it," he claimed. Even before his reelection win, as speculation intensified around a possible presidential run commensurate with his rising national poll numbers, Bush said he felt like "a cork in a raging river." But he was quick to add, "I'm interested or I would have said no." The main consideration was whether he and Laura wanted to "make a lifestyle change," especially as they considered their sixteen-year-old twins. "The girls matter a lot," he said. "If the answer is to go, they'll be good soldiers. But they won't like it. I wouldn't have liked it [at their age, either]."

Given his last name, the word *dynasty* became a staple in coverage of him. He spurned it, just as his father shunned *legacy* as the "L-word." "In a dynasty, you don't have to earn anything," he told a reporter with a hint of irritation, just after his reelection. "In a democracy, you have to earn it." As for an attempt to *earn* the presidency, that was a decision that George W. Bush knew he had to make—and soon.

38

"NO TURNING BACK"

THE MOMENT MAY HAVE BEEN staged, but it made it no less poignant. On Sunday, June 13, 1999, a day after his seventy-fifth birthday, George H. W. Bush stood at a microphone on his lawn at Walker's Point with his eldest son, the governor of Texas, at his side. Against a backdrop of the churning Atlantic pounding the rocky Kennebunkport coastline, and rambling homes on Ocean Avenue in the distance, he addressed the press corps that had amassed for the occasion, including twenty-eight television crews from as far away as Europe and Japan. "I had my turn," the Bush patriarch said in a blue windbreaker fixed with the presidential seal as he gestured toward his son. "I got some things right. Maybe I messed some things up. But it's his turn now in the family."

The day before, George W. Bush had made it official. At a stop in Amana, Iowa, with an entourage in tow that suggested power in motion, he said what most had expected since he formed an exploratory committee for his party's presidential nomination three months earlier, in March. "I'm coming here today to tell you this," he said standing on bales of hay that had been arranged for the occasion, a red pitchfork stabbed in the straw behind him. "I'm running for president of the United States. There's no turning back. And I intend to be the next president of the United States." At least three other GOP candidates were stumping in the Hawkeye State the same day, but it was Bush who commanded the media's attention as the race's immediate front-runner, having already racked up key endorsements and $36 million in financial commitments. Campaigning on a platform similar to that in his first run for governor—education reform and lower taxes—he used his stop to firmly distinguish himself from other prospects as a "compassionate conservative." "Is compassion beneath us?" he asked. "Is mercy beneath us? Should our party be led by someone who boasts of a hard heart? I know Republicans are generous of heart. I am proud to be a compassionate conservative. And on this ground, I will make my stand."

Now on Walker's Point, the Bush family's most hallowed ground, George H. W. Bush was making a stand of his own, relegating his political career to the past and showing total support and deference to his son. "He doesn't need advice from me," he said of his son who stood in shirtsleeves beside him, and a soft breeze blew in from the sea. "I'll be there to support him if he just needs help and someone to love him. But we're not, I'm not, in the advice mission business." The elder Bush's humility and self-imposed subordination set the tone for his involvement in his son's political life going forward, and would change little after his son became president. Before the press conference was over, George W. slipped inside the family home for a nap as his father fielded additional questions; there would be plenty of opportunities for the candidate to face them himself in the days ahead.

George W.'s decision to seek the White House hadn't been an easy one. It wasn't his destiny to become president. In fact, at least ostensibly, he had no burning desire for the job. He despised the culture of Washington, as well as the national media and the agonizing scrutiny that went with it, all of which he had seen up close when he lived in Washington in 1988, working on his father's first presidential campaign. Lord knows he, Laura, and the girls had no desire to leave Texas. It wasn't about avenging his father's loss eight years earlier—Why would his father need to be vindicated? He had served honorably and done his best. History, he was confident, would reflect that.

If there was a consideration by virtue of his last name, it was that it was instilled in him early by the examples of his father and grandfather that public service was the most worthy course one could take in American life. He honored the Bush family by serving, and there was no higher office to which he could aspire than the presidency. As a natural politician and competitor, there was no greater test, and he knew that timing was everything in politics. With no incumbent president in the race due to the constraints of the Twenty-Second Amendment to the Constitution preventing a third White House term, a solid record in Texas, and stratospheric approval numbers, his time was now.

Still, assuming he captured the nomination of his party, his path to the White House would be no cakewalk. Twelve years before, his father had blazed a trail from the vice president's residence to the White House aided by a purring economy, bullishness about the future, and the lofty popularity of his boss, the president. Al Gore, who declared his candidacy for the presidency three days after George W., had the same advantages. Despite the whiff of scandal that had trailed Bill Clinton discernably throughout his presidency, including his impeachment over his affair with twentysomething White House intern Monica Lewinsky, his approval ratings, which dwelled between the high-fifties and mid-sixties were more than respectable. With Gore as his number two, Clinton had brought the longest period of sustained economic growth in American history, lower crime rates, and optimism. Times were good.

Though Gore was as wooden as Clinton was charismatic, he had brought a wonkish sincerity and competence to the White House. Like Bush, Gore had spent much of his life in the shadow of his notable paternal namesake. Al Gore Sr. was an establishment Democrat from Tennessee who had served in the House and Senate for a collective thirty-two years. The future vice president had grown up inside the Washington Beltway with an eye toward the White House, following his father to the House and Senate, where he spent eight years in each chamber before being tapped as Clinton's running mate in 1992, at age forty-four. Still, Bush saw a weakness. As he later told a journalist, "I know who I am and Gore doesn't know who he is."

While Bush entered the race as his party's presumptive nominee, other challengers entered the fray as primary season approached in 2000. The previous year, *Fortune* magazine wrote that the Republican Party "is an unofficial disaster area, badly split between moderates and right-wing ideologues, and led by a bunch of men who range from untested to unknown." It boded well for Bush, whose message of compassionate conservatism resonated with both the moderate and right-wing camps. His last name gave him instant name recognition—and the rub-off effect of his father's honor and integrity, which stood in contrast to the incumbent president. Bush's most serious competition came from Arizona senator John McCain, a moderate, whose underdog status and unvarnished "straight talk" appealed to the press who followed him.

Bush took an early lead in the Republican field with a win in the Iowa caucus on January 24, 2000, but McCain gave him chase a week later, on February 1, taking the New Hampshire primary with 49 percent of the vote to Bush's 30 percent. The South Carolina primary, on February 19, became a must-win for both camps and proved as ugly as it was decisive. McCain kicked off his campaign in the state controversially with a speech at Bob Jones University, which banned interracial dating. "We are the party of Abraham Lincoln, not Bob Jones," McCain contended, but it didn't seem to matter. A racially tinged smear campaign against McCain, which included at least one incident involving a Bush supporter emailing fellow Republicans alleging that McCain "chose to sire children without marriage," including one nonwhite child, proved effective in impeding McCain. The Bush campaign denied any involvement in the slander, but when Bush put his arm around McCain after a debate and told him his team had nothing to do with it, McCain growled, "Don't give me that shit. And get your hands off me." Bush took the primary with 53 percent of the vote, topping McCain's 42 percent. It was enough to give Bush the advantage going into Super Tuesday on March 7, where Bush cemented his nomination by capturing the lion's share of contests in sixteen states. Tapped of money and momentum, McCain withdrew from the race shortly thereafter.

Gore, meanwhile, overcame a challenge from New Jersey senator Bill Bradley, mounting clear victories in the Iowa caucus and New Hampshire primary that carried him to the nomination with relative ease.

Ross Perot refrained from a third-party challenge, but political activist and consumer advocate Ralph Nader stepped in to fill the third-party void, running as the nominee of the fledgling Green Party. At the urging of its most prominent member, Minnesota governor Jesse Ventura, Donald Trump flirted with the idea of running for the nomination of the Reform Party, appearing on CNN's *Larry King Live* in October to announce an exploratory committee. Trump saw the necessity of his own candidacy, even if his rationale was based more on dystopia than reality. "All of the polls are saying that I win or do very well in this election," he told a reporter hyperbolically in the latter part of 1999. "I'm not happy with what's happening in the country—nobody else is either. The spirit, the whole—it's gone, it's just terrible." By mid-February, as polls showed him in single digits in a matchup against Bush and Gore, he announced on NBC's *Today* show that he wouldn't go forward. Trump's rhetoric didn't fit the zeitgeist. But his time would come later in the millennium when a perfect storm of the times and political circumstance fit his unchanging message.

George H. W. Bush stayed mostly on the sidelines as his son pursued the presidency. Still, his protective instincts and pride sometimes got in the way. When Clinton characterized George W.'s turn for the presidency as "How bad can I be? I've been governor of Texas. My daddy was president. I've owned a baseball team," the "daddy" struck back. "I'm going to wait a month and if he continues that, then I'm going to tell the nation what I think of him as a human being and as a person," Bush said in an interview with MSNBC. Just prior to the New Hampshire primary, he realized he had inadvertently diminished his son when he gushed praise for "this boy, this son of ours." Thereafter, he did his best to rein himself in.

George W., in turn, made clear that his would not be his father's White House. From the earliest days of his pursuit of the job, he would begin most meetings with potential political benefactors by stating, "This is not going to be George H. W. Bush, Part Two. It's going to be George W. Bush, Part One." His senior campaign strategist, Karl Rove, with George H. W. Bush's 1992 loss in mind, had a ready answer when he was asked how George W.'s White House would differ from his father's. "Well," Rove would say, "he's more ideological, more conservative. He's just much more interested in domestic policy than his father."

But behind the scenes the Bush machine hummed with George H. W. Bush playing a vital role. When Dan Quayle, contemplating a presidential run himself, called his former boss to suggest that the Bush and Quayle camps go easy on each other, Bush dispensed with sentimentality. "We'll do what we have to do," he told his vice president. They did. Prior to George W. announcing his candidacy, Bush had called GOP heavyweights to ask if they would refrain from endorsing candidates until George W. had made his decision to run. Once his son was in the race, Bush thumbed through his Rolodex and worked the phones, reaching out to his network of party leaders, state operatives, big donors, and Bush family loyalists to enlist them in the cause.

"My dad," George W. once explained, "plays a big role in my life as a shadow government." But he offered his own unsought advice to his son rarely and sparingly. When he felt strongly about something, he seeded his views through those on the campaign. "Dad said that same thing this morning," the candidate might say when someone offered their opinion or counsel. It helped that a lot of the people that George W. brought into the campaign were former Bush White House staffers. Brent Scowcroft and Condoleezza Rice schooled George W. on foreign policy; Andy Card aided in running the campaign; and Dick Cheney vetted a field of potential running mates, just as he had done for Gerald Ford in 1976.

Over time, Cheney's ascent from running mate recruiter to actual running mate took on Machiavellian undertones. Did Cheney manipulate the process so that he could land the job himself? Cheney, the elder Bush's secretary of defense and the CEO of Dallas-based Halliburton, an oil and gas conglomerate, took the task of helping to choose the younger Bush's vice presidential nominee only after initially rejecting the offer to be vetted for the job himself. During the primaries, Cheney agreed to a meeting with Bush's campaign manager, Joe Allbaugh, who asked if he was interested in being considered for Bush's running mate. Cheney declined, claiming it would be "bad optics" given his current profile in the oil business and the past involvement with the Bush father and son. There were also, he told Allbaugh, his three heart attacks to consider, and the fact that his home state was Wyoming, a Republican stronghold with only three electoral votes. "If you need me on the ticket to carry Wyoming, you have bigger problems," he told Allbaugh. But when Governor Bush himself called to ask if Cheney would lead the search for his running mate he "jumped at the chance" to lend his support. It was a "definable [task] and it would be over by convention time," Cheney reasoned.

Cheney spent the next several months whittling down candidates, before meeting with Bush at his ranch near Crawford, Texas, just outside Waco, to "wrap things up" shortly before the Fourth of

July weekend. After having a lunch with Laura, Cheney and Bush stepped out onto the porch. "Look, you're the solution to my problem," the governor told Cheney, asking him to reconsider the notion of being his running mate.

Cheney told him he would consider it. Several days later, he met in Austin with Bush and members of his team to further explore the notion. "This is a bad idea," Rove contended. Among other things, he worried that it would look as though Bush was "falling back" on his father's administration. Cheney didn't think it would work either, as he recounted his "warts" including a "misspent youth" during which he had been arrested for drunk driving and failed out of Yale—twice. He also pointed out that his ideological outlook was further right than his White House tenure with moderate Gerald Ford might have suggested.

"Look," Cheney said as a warning, "I'm conservative."

"We know that," Bush said.

"No, I'm *really* conservative," Cheney said.

Two days later, during an early morning workout on a treadmill in his Dallas home, Cheney got a call from Bush. "Look," Bush said. "You're my guy. I want you to run with me as my vice president." Why did Cheney agree when he initially rejected the position? "He said that's what he wanted—and he said it enough times in the marching orders when we were evaluating people—that he wasn't looking for a conventional vice president," Cheney said. "What he wanted was somebody who could be part of his team. He persuaded me that it would be a consequential post, that I would have walk-in rights, that I could get involved in anything I wanted to get involved in, though he wanted me to especially get involved in national security matters." Bush's choice was not about politics but about Cheney's experience. The fifty-nine-year-old Cheney—with a résumé that included stints as Ford's chief of staff, minority whip in Congress, and his father's secretary of defense—was a heavyweight, especially in foreign policy where Bush's knowledge and experience were limited, and Bush was confident enough to pick someone whom he could draw on for guidance. To those who accused Cheney of scheming to get the post, Cheney said if he wanted it in the first place, he would have agreed to be vetted for the position by Joe Allbaugh months earlier.

Cheney came with the imprimatur of George H. W. Bush. "He was very supportive of the decision," said George W., who recalled his father saying, "You'll never have to worry about Cheney going behind your back." Plus there was the fact that Cheney had no aspirations of running for the presidency himself. In early 2001, George would write to his friend Hugh Sidey, "George W. Bush is lucky to have Dick by his side." His view would change over time as Cheney's conservatism, which had been topped in his own administration by the more moderate voices of Brent Scowcroft, Colin Powell, and Jim Baker, came to bear in his son's war on terror. But as the Bush-Cheney ticket rode into the Republican National Convention in Philadelphia in early August, running neck and neck with Al Gore and his running mate, Connecticut senator Joe Lieberman, optimism abounded.

39
GREAT EXPECTATIONS

"**PLEASE STOW YOUR EXPECTATIONS SECURELY** in your overhead bins, as they may shift during the trip and can fall and hurt someone—especially me," George W. Bush had said just before his plane, christened "Great Expectations," took off for Iowa where he would announce his candidacy. "We know you have a choice in candidates when you fly and we appreciate you choosing Great Expectations." It was vintage Bush: flippant, loose, self-deprecating—everything Al Gore was not. It was also characteristic for Bush to manage expectations downward, which would serve as an effective tactic against his whip-smart, overachieving, often smug opponent.

After sailing through the smooth conventions the parties staged with red-white-and-blue flair—the Democrats in Los Angeles, the Republicans in Philadelphia—the standard-bearers went into the fall to a series of three televised debates with Gore showing a 1 percent lead in the polls. Citing recent media coverage, the Bush team readily conceded that the glib Gore was an expert debater; sure, Gore would do well in the debates. What they didn't reveal was that they had been prepping Bush for the debates since April. "Keeping quiet was a way to keep expectations low for Governor Bush," said campaign spokesperson Karen Hughes. "In debates, you run against expectations almost as much as you run against your opponent." Sometimes, underestimation was self-fulfilling prophecy. After Bush had won the South Carolina primary against McCain, he said that his opponent had "misunderestimated" him, one of many malapropisms throughout his political career that caused people to lowball him.

Gore was prepping, too, but as facile as he was on the issues, his advisers became concerned. "Right away I picked up a problem about Gore during debate prep," said Paul Begala, a campaign strategist who played the role of Bush in mock debates. He saw in Gore "a raw, unbridled contempt he had for Bush . . . He would sometimes sigh when I was talking, or frown, or roll his eyes . . . it all communicated that Gore thought Bush was an idiot. 'You don't deserve to be on stage with me,' was Gore's basic attitude."

When the first debate came off at the University of Massachusetts in Boston on October 3, with a viewership of over forty-six million, nothing much had changed. "For all of Gore's good answers," Begala recalled, "his feelings for Bush were his fatal flaw. You can't afford to look across the stage with that kind of contempt at someone who millions of people had nominated as their standard-bearer, and not have it bleed over to the audience and have them see you negatively." Bush, on the other hand, came off as composed, humble, likable, and more knowledgeable than expected. "The initial encounter between Vice President Al Gore and Gov. George W. Bush produced a highly unusual impact," the New York Times wrote, "if the polls are accurate, Mr. Gore 'won' the debate, in the view of voters, but Mr. Bush picked up more ground in the race for the White House."

The results were devastating for Gore. By late October after the last debate, the polls showed Bush with a lead as wide as ten points. If the election had been held then, Bush believed he would have won by five or six points. But the race took another turn in the week before Election Day when it was revealed that Bush had been arrested for driving while under the influence in Kennebunkport in the summer of 1976. "I've often said that years ago, I made some mistakes. I occasionally drank too much," Bush told the press. Regardless, the fact that the incident had not been disclosed—a conscious decision by the Bush campaign—tarnished Bush, who had campaigned with the promise of restoring "honor and dignity to the White House," a pointed departure from scandal-ridden Clinton. The race went back to being a dead heat.

If the campaign had seesawed in its last weeks, so would November 7, election night. As rain fell on Austin, George and Barbara, Jeb, Laura's mother, Jenna, and a handful of others all gathered with George W. and Laura in the small upstairs rooms at the governor's mansion to watch the results.

At 7:49 p.m. eastern time, NBC called the swing state of Florida for Gore, despite the fact that the polls weren't yet closed in the largely Republican Florida panhandle, which stretched west into the central time zone. Jeb hugged his brother and said he was sorry he let him down, then feverishly worked the phones to get a read on what had happened in his home state. With Florida in Gore's win column, and other states following, the networks proclaimed Gore the election's victor just before 10:00 p.m. eastern time. As they did, both the Bush and Gore camps began getting reports that Florida, which still had returns coming in, was too close to call. The networks retracted their earlier verdict; the election was still in the balance as the Florida results continued to be tabulated. Around 2:00 a.m., with the Florida vote more definitive, Fox reported that Bush was the winner. Other networks followed.

Gore dutifully called Bush from Nashville to congratulate him on his win before moving on to Nashville's War Memorial Plaza to make a concession speech to supporters. Just prior to moving to the podium, Gore's aides informed him that Florida was once again in the undecided column, with results that could be contested. Gore called Bush again.

"Circumstances have changed since I first called you," the vice president informed the governor.

Bush was incredulous. "Are you saying what I think you're saying? Let me make sure I understand. You're calling back to retract your concession," he said.

"You don't have to be snippy about it," Gore said.

When Bush added that his "little brother" had told him that Florida was decided in his favor, Gore responded, "Let me explain something, your brother is not the ultimate authority on this."

Afterward, the Bushes hunkered back down in their small living room to continue to watch the ongoing coverage, exhausted by the hour and turn of events. Jeb recalled his father being "totally cool, calm, and collected," even taking a nap on the couch at one point in the early morning hours. "It didn't seem to phase him," he said. Laura Bush alleviated the stress by washing dishes; Barbara Bush immersed herself in a needlepoint project while listening to a Sandra Brown novel through earphones. As early morning hours passed, the networks pulled back again on naming a victor. Eventually Laura Bush slipped her arm around her husband. "Bushie," she asked her husband, invoking the nickname they had for each other, "would you rather win or go to bed?"

"Go to bed," he replied. Election night was over; the election was not.

The following morning a little after 8:00 a.m., Jim Baker was on his way to work for a meeting at Baker Botts in Houston when he got a call from Don Evans. The Bush camp had gotten wind of the fact that Warren Christopher, secretary of state during Clinton's first term, would be heading up Gore's efforts to challenge the election results in Florida. The selection of Christopher indicated that the Gore campaign was looking to a high statesman, once the nation's top diplomat, as their gray eminence. The Bush campaign hoped to match them in kind. "We're kicking your name around to maybe help us in Florida," Evans told Baker. "Looks like Gore is going to pick Warren Christopher. We want to know if *our* former secretary of state [will] do it."

"I'd be happy to if the governor wants me to," Baker replied.

Twenty minutes later, Baker received another call from Evans. Yes, Evans confirmed, "the governor wants you to do it." By 2:00 p.m., Baker was on a private plane bound for Tallahassee. Aside from Thanksgiving, when he returned to Houston for the day, he would spend the next seven weeks in the Florida capital.

When the dust began settling in Florida and all the votes were tallied, Bush had edged out Gore by less than about two thousand votes out of the 5.8 million cast, triggering a state law requiring that all votes tabulated by machines be recounted. After the recount, Bush maintained his win, though it had shrunk to a margin of just under three hundred. Baker immediately claimed victory. A recount had been completed, he said; Bush won. The Gore campaign, though, challenged the results based on voting irregularities. The Florida Supreme Court, dominated by Democratic-appointed justices, rendered two verdicts ordering manual recounts, and ambiguous ballots, many with nebulously punched "hanging chads," were scrutinized to infer voter intent.

Gore was front and center during the challenge, zealously appearing in public to show that he was actively engaged in preparing for office while orchestrating legal strategy behind the scenes. Bush, by contrast, maintained a low profile, more fatalistic about it all. "I'll be at the ranch," he told his aides. "Let me know what happens." He had done his best to win the election; now it was out of his hands. For much of the time, he and Laura retreated to their Crawford ranch, nearly sixteen hundred acres of blackland prairie, which fell in the farthest reaches of Texas Hill Country. "The only way to stay sane was to live in the moment," Laura reflected, "and that's how George [W.] became the brush-clearing Zen master of Crawford."

Just two days after the election, on November 9, George and Barbara joined the Carters, the Fords, and Lady Bird Johnson at the White House as the Clintons hosted a dinner in the State Dining Room to celebrate the mansion's two hundredth anniversary. An air of bipartisanship and presidential decorum prevailed, but the elephant in the room was whether the next president would be the vice president of the occasion's host or the son of two of his guests. The same day, George H. W. Bush told reporters, "It's a nervous time in my life. Whatever happens my pride and Barbara's pride, knows no bounds, and moreover, our democracy will go on."

Nervous he was, calling Baker for updates at least daily, sometimes two or three times. He also called Jeb, who was working diligently in Florida to sort out the mess. "I wasn't getting advice from my Dad, I was giving the advice, and that was, 'Dad, chill out,'" he recalled. "My Dad was a living wreck. It was horrible. He was so into this, and to have a son . . . I don't think there would be a way to describe the powerful emotions of pride, anger, love, and all the emotions to the max. And it just stretched out—ebbing and flowing."

The ebb and flow lasted thirty-six interminable days. On December 12, the matter was put to rest by the United States Supreme Court. In the case that would become known as *Bush v. Gore*, the former prevailed by a ruling of five to four. The court overruled the Florida Supreme Court's earlier verdicts by accepting the certification of Florida's Republican secretary of state declaring Bush the winner of Florida's electoral votes. With Florida in Bush's win column, despite losing the national popular vote by over half a million votes, Bush earned 271 of the 270 electoral votes necessary to swing the election. Jim Baker's cool, canny leadership through the crisis was instrumental in the outcome.

When the high court's decision was released on the morning of the 12th, Bush was in his pajamas reading in bed at the governor's mansion. Karl Rove called to deliver the news. "Congratulations,

Mr. President," Rove greeted him. Bush, who didn't have his television on, was unaware of the development. He turned on his set to get a definitive announcement, but couldn't find one. "I'm calling Baker," he said, abruptly hanging up. He then reached Baker who confirmed the verdict.

It was settled. George W. Bush would be the forty-third president.

PART VI

43

40

"MR. PRESIDENT"

JUST AS IT HAD RAINED in Austin on election night, a cold drizzle
fell on Washington as George W. Bush went through the rites of
his inauguration on January 20, 2001. Some 300,000 stood in the
muddy grass on the Mall under a pewter sky to watch the new president take office thirty-nine days after the Supreme Court rendered
its judgment making it so. His left hand on the same King James
family Bible his father had used for his own inauguration a dozen
years to the day before, Bush, in a black overcoat and blue striped tie,
recited the oath of office before kissing his wife and twin nineteen-
year-old daughters, now college freshmen at Yale and the University
of Texas. Bush's parents watched just steps away from him, a tear
falling down his father's cheek. While the sting of his 1992 defeat
had largely receded, remnants of it still lingered within George H.
W. Bush even after his sons had successfully launched their political
careers. Now, eight years after yielding the office to Clinton, who
in turn had given it over to George's first son, the page had finally
turned.

After a divisive and breathtakingly close election, the new president sounded a message of unity, using the word *civil* or *civility* six
times during his fourteen-minute inaugural address. "Civility is not
a tactic," he said just after noon. "It is the determined choice of trust
over cynicism, of community over chaos. And this commitment, if
we keep it, is a way to shared accomplishment." To those abroad posing a threat to America, Bush had a far different message, one that

portended the central crisis of his presidency. "The enemies of liberty and our country should make no mistake," he warned. "America remains engaged in the world, by history and by choice, shaping the balance of power that favors freedom. We will defend our allies and our interests."

Later in the day, as the rain-soaked inaugural parade wound down by late afternoon, the new president and first lady repaired to the White House, where they would take up residence for the next eight years, turbulent beyond their imaginations on that day. Earlier, George H. W. Bush had retreated from the parade-reviewing stand to the Queen's Bedroom in the White House to escape the rawness of the weather. He was warming in a hot bath when he was interrupted by a knock on the door. "Mr. President," an attendant said, "President Bush would like you to meet him downstairs to walk over to the Oval Office." The elder Bush quickly toweled dry, threw on a suit and tie, and took the residence elevator to the ground floor to find that the president had already gone over to the Oval Office. The old man followed after him.

Andy Card, Bush's chief of staff who had served as deputy chief of staff in the Bush 41 administration, was checking on the progress of the office's hasty redecoration when George W. Bush arrived. "It was cold and dark outside, but it was warm and bright in that room," Card recalled. "You could smell the paint drying; they had just painted it. And I'm just standing there, and George W. Bush comes in and he doesn't say anything. And he walks in kind of puffed up." Bush sat down in the black leather chair behind the desk, joking with Card about a vibrating feature in it that plugged into the wall. Several minutes later, the elder Bush arrived thrilled "again to walk in the terrace door" to his old office. "Mr. President," he said, greeting his son. "Mr. President," his son responded in kind, as he rose and embraced him, both men standing on the newly installed ivory rug adorned with the presidential seal. The forty-first president cried for a second time that day; the forty-third president cried a little, too.

Despite Bush's promise that his administration would not be "George H. W. Bush, Part Two," he brought back more Bush 41 alumni than just Cheney and Andy Card. Colin Powell served as secretary of state; Condoleezza Rice, as national security adviser; and Paul Wolfowitz and Richard Armitage, as deputy secretary of defense and deputy secretary of state respectively. Bush 43 also brought in members of his own team: Don Evans was appointed as secretary of commerce; Karen Hughes oversaw communications strategy; and Karl Rove was his chief political strategist.

His appointment of Donald Rumsfeld as his secretary of defense, however, sent a signal that this was not his father's White House. Rumsfeld had crossed George H. W. Bush during the Ford administration, preventing him from being considered as Ford's pick for running mate in 1976, and it didn't escape the attention of Bush 41 loyalists. "Dad didn't tell me directly [about his disapproval of Rumsfeld], but I think at times Mother reflected his point of view," Bush 43 said, "and I could hear the buzz amongst the old Bush [41] hands: 'I can't believe [it], Rumsfeld this, Rumsfeld that!'" As with other gossip in the Bush 41 camp, 43 disregarded it. "Their history didn't bother me, because I was looking for competency," he contended. Specifically, Bush wanted to reorganize the Department of Defense and, after FedEx CEO Fred Smith declined the post, concluded that Rumsfeld was the best man for the job.

In addition to his confidence in Rumsfeld's ability as an administrator, Bush had other reasons for choosing Rumsfeld. According to Card, when 41 was making staffing choices he named John Sununu as his chief of staff to show Jim Baker, "You're not the copresident." Card saw the appointment of Rumsfeld in "the same way." Appointing Rumsfeld would show the Bush 41 team—maybe his father, too—that 43 was his own man. As 43 was quick to point out, it was something he had done throughout his life, forming relationships with those whom his father or those around him regarded as enemies. They included former Texas governor John Connally—who had consistently tried to block his father in Texas politics and within the Nixon administration, where Connally had served as secretary of the treasury—and Ross Perot Jr., the son of Ross Perot, his father's nemesis in the 1992 election. Politics, Bush 43 understood, often meant strange bedfellows, and he was confident that his father, who he knew would never question his loyalty, knew that, too.

While many of his predecessors, including Bill Clinton, had done their best to stay in the headlines and lead national television news coverage, Bush's approach was to plod along quietly and competently. In the first placid month of his administration, Bush turned his attention toward domestic matters. The federal funding of embryonic stem cell research became an area of singular focus for Bush, testing the definitions of his compassion and conservatism. Scientists argued that government-funded embryonic stem cell research was needed to cure myriad diseases including diabetes, Parkinson's, and Alzheimer's. Among those who advocated it was Nancy Reagan, whose husband, now in his ninety-first year, was in the last stages of Alzheimer's disease, unable to recognize anyone but her. On the other side of the issue were evangelicals and other conservatives who equated stem cell research with abortion, considering embryonic stem cells a form of life that warranted protection. In a rare televised address in early August, Bush announced a compromise, banning the harvesting of new embryonic stem cells for government-funded research but allowing research to be done using stem cell lines that had already been destroyed.

Tax cuts were another early hallmark of the Bush 43 administration. As in Texas, Bush advocated tax cuts under the Keynesian economic doctrine, adhered to by Kennedy and Reagan before him, that prescribed them as a short-term means of stimulating economic growth and curbing recession. In May, Congress passed a tax cut to the tune of $1.3 trillion. The measure, proposed as a temporary fix, would become a lasting policy fixture as the federal deficit, like it had under Reagan, distended throughout Bush's White House tenure. It was a clear signal that Bush was not going to fall prey to his father's "no new taxes" political blunder. He would also steer away from other mistakes his father had made, giving less power to his chief of staff than his father had given his (prior to the ouster of John Sununu) and appointing proven conservatives to the U.S. Supreme Court—in his case, Chief Justice John Roberts and Associate Justice Sam Alito—unlike his father's choice of David Souter.

Internationally, it wasn't what Bush did in his first months as president that garnered the most attention but what he didn't do. In March, he refused to sign the Kyoto Protocol, an international treaty to reduce greenhouse gas as a means of combating global warming. Bush's rejection of the treaty, based on the "incomplete state of scientific knowledge" on climate change, was seen as his bent toward isolationism. But as future events of his presidency would reveal, Bush was no isolationist. America, as Bush foretold in his inaugural address, would remain engaged in the world.

41

"SHOCK AND AWE"

ON THE MORNING OF TUESDAY, September 11, 2001, as he took a predawn jog around the Colony Beach and Tennis Resort outside Sarasota, Florida, President George W. Bush almost certainly would have been content with going down in history as the "education president." He had held out the prospect in the heated presidential campaign the previous year, touting the education "miracle" he had seen to in his six years in the Texas statehouse and denouncing the "soft bigotry of low expectations" for poor and minority children. Early in his administration Bush had cultivated a relationship with Ted Kennedy, the stalwart liberal senator from Massachusetts and aging emblem of the Kennedy dynasty. "Let's show them Washington can still get things done," he challenged Kennedy. At the top of the list of the things on which he sought bipartisan support was education reform. Bush's appearance later that morning before a first-grade class at Sarasota's Emma E. Booker Elementary School was meant to align him with the issue and serve as a backdrop for his legislative plan, No Child Left Behind. But the education president he would not be.

At 8:54 a.m., as he entered the school, Bush was told by Karl Rove that one of the World Trade Center towers in lower Manhattan, the north tower, had been struck by an aircraft at 8:46 a.m. Initially, it looked like an accident caused by pilot error, maybe a heart attack. Before Bush entered the classroom, he received a call from Condoleezza Rice indicating that the plane was a commercial aircraft. Nothing more was known. At 9:06 a.m., as Bush listened to a teacher conduct a reading exercise with the book *The Pet Goat* to a class of largely African American students, Andy Card quietly approached him. "A second plane hit the second tower," he whispered in the president's ear. "America is under attack." Bush sat there for five minutes taking in the news, looking slightly dazed, his lips tightening. He consciously remained in place to project calm and not alarm the students—a decision that would draw controversy later as Bush was criticized for not reacting more urgently. He knew in that instant that he had become a "wartime president," something he "never anticipated" and "never wanted." Then and there George W. Bush's life, like that of so many others, had changed. So, too, had his presidency.

After commending the students on their reading skills, the president and his entourage went to a holding room in the school where they watched the crashes on television, as Bush, sitting on a child's desk chair, pulled a Sharpie from his pocket and began writing out a statement on a yellow writing pad. Minutes later, he entered the school's gymnasium crowded with reporters, school administrators, teachers, and students. Many were unaware of what had transpired. "Today we've had a national tragedy," Bush said in his first words to the nation after the attacks, as Americans gripped by horror and uncertainty stationed themselves in front of televisions. "Two airplanes have crashed into the World Trade Center in an apparent terrorist attack on our country." He added, "Terrorism against our nation will not stand," the latter three words echoing those his father had used as president after Saddam Hussein had invaded Kuwait, but misreading the written text, which read, "Terrorism against our nation will not succeed." "Dad's words must have been buried in my subconscious, waiting to surface during another moment of crisis," Bush wrote later. Bush, though, sounded less resolute than his father had seemed twelve years earlier.

Minutes later, the president and his long motorcade bounded

down U.S. Route 41 at 85 miles per hour toward Air Force One, flanked on all sides by police cars. During the short ride, he called Rice, who gave him the latest development: A plane crashed into the Pentagon at 9:37 a.m.—a third commercial airline hijacked by terrorists and used as a guided missile. "Is Rumsfeld alive?" Bush asked. He was, Rice told him. The first plane could have been an accident, Bush thought. The second plane was definitely an attack. The third was a declaration of war.

By 9:54 a.m., the president and his party were airborne. The question became, Where would they go? When it was learned that Air Force One was a possible target for the terrorists, the plane, which bolted upward to forty-five thousand feet, far higher than normal cruising altitude, abruptly banked a turn and sprinted a thousand miles toward Louisiana's Barksdale Air Force Base at 630 miles per hour. F-16 fighter jets, not even a stone's throw away, hugged the aircraft on either side, providing a protective flank. Bush talked to Cheney, who had been hurtled by the Secret Service to the Presidential Emergency Operations Center, a bunker deep under the White House's East Wing. Bush would make decisions from Air Force One; Cheney would carry them out from the bunker. Bush gave him clear direction on the rules of engagement: An effort should be made to contact suspicious planes to get them to land peacefully. If not, the military had Bush's authority to shoot them down. Bush gave the order somberly, knowing that innocent passengers on board those aircraft may be at risk. As subsequent reports made their way to Bush, some later proven to be false, he learned that a plane had crashed near Shanksville, Pennsylvania, in the western part of the state. It wasn't known whether it was shot down by the military under Bush's orders or if it went down on its own. *Am I responsible for the deaths of those passengers?* Bush wondered.

Andy Card was taken with how Bush "immediately started thinking about the greater burdens." "It wasn't tunnel vision," Card said. "[Bush had] terrific peripheral vision about what was happening."

Among the first calls Bush made was to Russian president Vladimir Putin. Bush wanted to make clear to Putin not to make the mistake of Russia exploiting the situation to go to war with America given its vulnerability after the attacks. Card explained that Bush wanted to "make sure [Putin] knows" not to "use [9/11] as an excuse to go to war" with the U.S. "[H]e was just saying, there is a big consequence here if people miscalculate," said Card.

As Air Force One sped westward, news reports spotlighted the rapidly evolving developments on the East Coast. None of them was good. At 9:59 a.m. the south tower collapsed, rolling downward with colossal might and disintegrating into a mushroom cloud of gray ash, which rose up in its place. Eleven minutes later, at 10:10 a.m., a western wing of the Pentagon fell. At 10:28 a.m., the north tower descended with the same apocalyptic force as the south tower. In less than two hours, the 110-story Twin Towers, symbols of American possibility and prosperity, exclamation points at the end of Manhattan's glittering skyline, were both down in heaping ruins, as the Pentagon, a fortress housing defenders of American security, was in flames. Untold thousands were dead. As images of the crippled towers falling played over and over, Bush felt like almost every other American on that morning: shocked, angry, and overcome with feelings of helplessness, a fierce will for justice, and a desire to hear the voices of those he loved.

Air Force One touched down at Barksdale a little before noon eastern time. Bush made another statement to the American people. He was firmer than he had been in Florida. "Make no mistake, the United States will hunt down and punish those responsible for these cowardly acts," he said, but as David Frum, who had written the statement, observed, he "looked and sounded like the hunted, not the hunter." Aides who were with Bush on 9/11 saw the president's steely resolve, his calm, and his ease with making big decisions. It was less evident over the airwaves.

After his remarks, upon conferring with Cheney, Bush was soon airborne again. The communications system at Barksdale was deemed insufficient, and Cheney recommended that Bush go to Offutt Air Force Base in Nebraska, which housed the U.S. Strategic Command. Bush resisted. As in Florida, his instincts told him he needed to return to Washington. "We don't need some tinhorn terrorist to scare us off," he said. "The American people want to know where their dang president is." But the situation was too unstable. At 9:45 a.m., when the White House was thought to be a target, staffers were ordered to evacuate the complex, as agents shouted instructions to women to take off their heels and run. Chaos reigned in Washington, with little reliable information filtering through the fog of war. Better that the commander in chief stay in a secure environment until things stabilized.

At Offutt, where he arrived just before 3:00 p.m. eastern time, Bush was rushed to the base's command center. There he held a videoconference with Cheney, Rumsfeld, CIA director George Tenet, and FBI director Robert Mueller. Tenet told the president what they all had suspected: al Qaeda, the militant Islamist terrorist group founded by Osama bin Laden during the presidency of Bush's father, was the likely perpetrator. "We're at war against terror," Bush told the group. "From this day forward this is the new priority of our administration."

Afterward, Bush issued an order: He was going back to Washington. Air Force One rode tailwinds over twelve hundred miles of a changed America. By 6:55 p.m., Marine One, hovering as low to the ground as possible for security reasons, far lower than the Washington Monument as it approached the White House, deposited him on South Lawn. A few minutes later, he was reunited with Laura underground in the White House bunker.

Like her husband, Laura Bush had begun her day campaigning for No Child Left Behind. She learned from a Secret Service agent on a limousine ride to Capitol Hill, where she would meet with the Senate committee on education to discuss early childhood development, that the first tower had been hit, then the second. By the time she met with Ted Kennedy after 9:00 a.m. in his office on the third floor of the Russell Senate Office Building, the world knew that the U.S. was under attack. A small television in the corner of Kennedy's office broadcast the developments, but the senator disregarded them. Instead, Kennedy, whose life had been embossed by tragedy including the assassinations of two of his brothers, offered the first lady a tour of his office, the walls of which were a collage of Camelot glory. "We never talked about the terrorist attacks," Laura recalled. "He just kept up small talk. I don't know if he thought I would fall apart if we started talking about it, or if it was just a . . . psychological mechanism because he had had so many terrible shocks in his own life—if it was just the way he could deal with it."

Bush's parents had been with Laura at the White House on the evening of September 10, while their son was in Florida. The following morning, as Laura was on Capitol Hill, they made their way back to Maine by way of Saint Paul, Minnesota, where they both were slated to give speeches. When their private plane, along with all commercial aircraft, was ordered to land at the nearest airport, they found themselves in Milwaukee where they checked into a suburban motel in which the bed was made up with brown sheets. There they glued themselves to the television. When asked later what his reaction to the attacks was that day, George H. W. Bush replied, "Shock and awe," the phrase that was used during his presidency when he ordered the raining of bombs on Baghdad in an overwhelming show of force after Iraq's invasion of Kuwait. "It was traumatic for me and Barbara," he said, "but no more so than for anyone else, except the president was our son." The Bushes were relieved when George W. reached them from Air Force One several hours after the towers were struck. They spoke briefly. When he asked what they were doing in Wisconsin, Barbara replied, "Son, you grounded our plane." His father suggested, "the sooner [you] get back to Washington, the better," counsel with which the president agreed.

The Bushes would discover in the days ahead how close the attacks had come to their family. The elder Bushes' youngest son, Marvin, had been in a Manhattan subway car under Wall Street on his way to a meeting when the attacks occurred. After the train was evacuated and orders were given to passengers to walk uptown or across the Brooklyn Bridge as soon as possible, he trekked seventy blocks to his midtown hotel with hordes of others, like war refugees, through smoke, debris, and blind confusion. Nine-eleven did not discriminate.

A little over ninety minutes after arriving back at the White House on the evening of 9/11, Bush was behind the Resolute Desk in the Oval Office, where he briefly addressed the nation at 8:30 p.m. "The search is underway for those who are behind these evil acts," he assured Americans. "I've directed the full resources for our intelligence and law enforcement communities to find those responsible and bring them to justice." Then he added, "We will make no distinction between the terrorists who committed these acts and those who harbored them." He hadn't discussed the declaration or its implications with any of his core foreign-policy team—Cheney, Rice, Rumsfeld, or Powell—but on September 11, 2001, after the bloodiest attack ever perpetrated on American soil, the forty-third president had just sowed the seeds of the Bush Doctrine.

42

MOURNING IN AMERICA

THE TOLL WOULD BECOME FULLY realized in the weeks and months ahead: 2,606 dead in Manhattan; 125 at the Pentagon; 246 on board the four crashed aircraft, excluding the nineteen al Qaeda terrorists; $55 billion in physical damage; $133 billion in economic impact. The nation reeled; fear, uncertainty, and anger pervaded. *"Nous sommes tous Américains"*—"We Are All Americans"—read the headline in the French newspaper *Le Monde* on September 12, a sentiment shared by much of the world.

While George W. Bush had projected calm on 9/11, the American people didn't see the steely, resolute leader his aides saw throughout the day. His rhetoric hadn't stirred or soothed the public, and his absence in Washington in the immediate hours after the attack, while necessary as a security precaution, held the image of retreat. But on Friday, September 14, the day the White House called for national mourning, Bush hit his leadership stride. In the morning, he presided over a national prayer service at Washington National Cathedral. The packed congregation included members of Congress, the cabinet, Al Gore, and all the former presidents except for Ronald Reagan, who was ailing from Alzheimer's disease in Los Angeles. The Fords, Carters, elder Bushes, Clintons, and the president and first lady sat in two pews in the front of the church. Protocol dictated that the presidents and first ladies be seated by rank, with the most recent former president, Clinton, sitting closest to the president, and others seated in descending order. But when the elder

Bushes asked if it could be broken for them to sit next to George W. and Laura, the Clintons were glad to oblige.

A Jewish rabbi and Muslim cleric offered words of prayer, and Billy Graham gave a short sermon in which he claimed that the attacks left Americans "more united than ever." The president then made his way to the lectern to deliver to the congregation, the nation, and the world the most important speech of his life. "Just three days removed from these events, Americans do not yet have the distance of history, but our responsibility to history is already clear: To answer these attacks and rid the world of evil," he said, slowly rising to rhetorical heights he had not seen in his political life.

War has been waged against us by stealth and deceit and murder. This nation is peaceful, but fierce when stirred to anger. This conflict was begun on the timing and terms of others. It will end in a way, and at an hour of our choosing . . .

In every generation the world has produced enemies of human freedom. They have attacked America because we are freedom's home and defender. And the commitment of our fathers is now the calling of our time.

George W. Bush's own father watched his son, his pride in him never greater than at that moment. When the president returned to his pew, the Bush patriarch reached across Laura, who sat between them, and patted his son's arm. George W. then clutched his father's hand. While "his turn" in the family had come long before, it may have been then that George W. Bush fully came into his own.

The memory of the moment would be the most indelible each would hold of the other during the younger Bush's eight years in the presidency. "Feeling his arm on mine, reaching across Laura, it was very emotional," 43 said. "And I was determined not to be emotional at the pulpit. I needed to make it through, it was very important." Forty-one knew it, too. Prior to the speech, he told Condoleezza

Rice that he was concerned that 43 might let his emotions get the better of him; 41 was sure that if he himself were called on to deliver it, he would have descended into tears. "The president really came through," Rice told her old boss afterward. "Yeah, he's strong," 41 said, relieved.

He remained strong the rest of the day, the most poignant and affecting of his presidency. Later in the morning, he and a group of White House aides that included Karl Rove, Karen Hughes, and Andy Card left for New York to tour Ground Zero, the ten square city blocks that had been destroyed in the attacks. Air Force One was met at New Jersey's McGuire Air Force Base by New York mayor Rudolph Giuliani and New York governor George Pataki, who joined the presidential entourage as they boarded helicopters and flew toward Manhattan. Thirty-seven members of Congress followed them. The dense acrid air met them as they approached Ground Zero. "[Y]ou could smell it," recalled Karen Hughes. "It permeated the air. We had seen it on TV, but somehow being there . . . you really saw the horror." Smoke still rose steadily from the site of the attacks, which looked like a war zone, a gaping open wound at the end of Manhattan Island. After landing, they were driven in armed SUVs toward Ground Zero, which was shrouded in despair.

As they approached the site of the attacks, they could detect a hum. Rove thought it was coming from the trucks. He soon realized it was the crowd outside as the motorcade drew nearer to the site of the attacks. "It was like you were hit with a wall of sound; they were chanting, 'USA! USA!' and waving flags," he said. The party was struck by the raw emotion. Bush had not planned to speak during the trip, having already given his remarks at the National Cathedral earlier in the day. "They want to hear their president," a member of Bush's advance team, Nina Bishop, implored Rove. "He needs to speak." Rove agreed. He scrambled to find a bullhorn, securing one from a Con Edison worker, and handed it to Bush.

Among the smoking ruins and heroic firefighters who stood

ghostly gray from ash and numb with exhaustion, Bush, in a tan jacket and open-collared shirt, bullhorn in hand, met the moment. He stood tall on a mound of rubble that had been a gleaming red fire engine just three days earlier, his arm around Bob Beckwith, a sixty-nine-year-old firefighter who had come out of retirement to help find survivors after the attacks. Bush began speaking. When someone from the crowd standing above him cried, "I can't hear you!" Bush paused, looked up, and placed the bullhorn to his mouth. "I can hear you," he declared in response. "The world can hear you! And the people who knocked down these buildings will hear from *all* of us soon!" After fist-pumping chants of "USA! USA!" he continued. "The nation sends its love and compassion to everybody who's here. Thank you for your hard work. Thank you for making the nation proud, and may God Bless America." The "bullhorn moment," as it would be called, was the public moment, Bush's most iconic—the one that hit the TV news, the front pages, and the magazine covers.

The private moment, more emotional and ultimately as uplifting, was to come as Bush and his party traveled north to the Jacob K. Javits Convention Center on Manhattan's West Side for their next stop. One of the center's parking lots had been pipe-and-draped as an impromptu gathering place for some four hundred loved ones of first responders—firefighters and police—who were missing or dead after the attacks. They all awaited a visit from their president. The room was quiet, still, and thick with sadness when Bush arrived. Karen Hughes called the gathering, "the most intense experience of my life." Finding the mood unbearable, she stayed for twenty minutes or so then left, as Bush went from family to family. "When I walked back in about a half hour later, the entire mood had changed," she recalled. "He had managed to get people to talk about their loved ones." Bush worked through the room, putting his arms around the grieving family members and encouraging them to share memories of those they had lost. Laughter mixed with tears. The visit, which was supposed to last under an hour, stretched to well over two. Just before Bush left, he was approached by the mother of a missing Port Authority officer, Arlene Howard, who pressed her son's police badge into his hand. Bush kept it in his suit pocket throughout his presidency. Like the small American flag his father placed on his Oval Office desk, the one given to him by an army ranger wounded during the Panama invasion, the badge was a symbol of the responsibility 43 bore, and as he said later, "a reminder of all that was lost."

When the presidential motorcade left Manhattan, New Yorkers stood along the streets ten and twelve deep in a show of unity, ubiquitously cheering "USA! USA!" and singing, "God Bless America." Bush had come to Manhattan to offer comfort to those in need. He left inspired by their strength and resilience. The terrorists didn't win, he realized; they were never going to defeat America. Yet those good and decent people had been unmercifully attacked. Now, it was time to do something about it.

43

PRELUDE TO WAR

AFTER THE VISIT, BUSH WAS spent. As Air Force One headed south from New York toward Washington, aides saw him depleted as never before. But there was work to be done. The following day, he was at Camp David where he spent the weekend huddling with the group that would become known as his war cabinet: Cheney, Rumsfeld, Powell, Rice, and Card. Others from the administration were also on hand. Their charge was to plot immediate retaliatory measures and safeguard against future attacks.

Afghanistan, the primitive central Asian nation on Iran's eastern border, was a clear target for military action. Al Qaeda, Osama bin Laden's Islamic terrorist operation, was located in the country's remote southern mountains near the Pakistan border, where thousands of operatives, including the nineteen 9/11 hijackers, were trained to carry out their missions. Bin Laden, the seventeenth son of the fifty-two children of a billionaire construction magnate with close ties to the Saudi Arabian royal family, financially supported Afghanistan's militant Taliban government in exchange for the country's refuge and protection. Several military options against Afghanistan were presented to Bush. He chose the most intensive: a full-scale military invasion of Afghanistan to depose the Taliban government. In 1998, after al Qaeda had bombed U.S. embassies in Kenya and Tanzania, Clinton ordered the deployment of cruise missiles to take out bin Laden at the camps, narrowly missing him. It had only emboldened bin Laden, who saw the U.S. as a "paper tiger" devoid of the will to launch a sustained military effort. "This time we would put boots on the ground," Bush decided, "and keep them there until the Taliban and al Qaeda were driven out and a free society could emerge."

It would be a formidable task. Afghanistan's history of fending off larger military powers, including the Soviet Union in its failed invasion of the country from 1979 to 1989, made it a graveyard of empires. But it was clear that the country was harboring the terrorist operation behind the 9/11 attacks, and U.S. allies were poised to offer support around a U.S.-led military operation.

Iraq was a less obvious military target. While providing no evidence, Paul Wolfowitz, Rumsfeld's deputy secretary of defense, asserted that there was a 10 to 50 percent chance that Saddam Hussein had a hand in the 9/11 attacks. Saddam had continually flaunted UN resolutions that had been in place since his defeat in the Gulf War, and Iraq, Wolfowitz believed, was of greater strategic significance than Afghanistan. Powell, Rice, and Card resisted the notion. Powell, who had seen the importance of maintaining an international coalition in the Gulf War, believed, "No one will understand or support us doing anything but going after those who attacked us." Wolfowitz continued to make his case. "[W]e really have to think broader than [Afghanistan] now," he told Bush. "We've got to make sure we go ahead and get Saddam out at the same time—it's a perfect opportunity." Bush, growing impatient, put an end to the discussion. "How many times do I have to tell you we are not going after Iraq at this minute?" he snapped at Wolfowitz. "We're going to wait to go after the people we know did this to us. Do you understand me?" Iraq would be put aside so that they could focus on Afghanistan. For now.

Back in Washington on Monday, September 17, Bush took care to show Americans that while al Qaeda was the enemy, Islam was not, promulgating a message of peace and unity during a visit to

Washington's Islamic Center. The president met with members of the Muslim clergy, proclaiming, "Islam is peace."

> These terrorists don't represent peace. They represent evil and war.
>   This is a great country. It's a great country because we share the same values of respect and dignity and human worth. And it is my honor to be meeting with leaders who feel just the same way I do. They're outraged, they're sad. They love America just as much as I do.

As for the terrorists, Bush wanted justice. Asked the same day by reporters what outcome he was seeking in the wake of the attacks, he replied, "There's an old poster out west as I recall, that says 'Wanted Dead or Alive.'" While he was knocked for the hyperbolic cowboy rhetoric, his unshrinking swagger, a natural part of his demeanor since his Texas childhood, played like Churchillian resolve in the days after the attacks, showing a president intent on keeping the country safe.

Bush's focus was to make good on that implicit charge. He appointed Pennsylvania governor Tom Ridge as his official on Homeland Security, later to be a cabinet post. When airports reopened the weekend after the attacks, intensive new security measures, already standard in many countries around the world but draconian relative to previous U.S. security checks, were put into place to become permanent realities in commercial air travel. Plain-clothed air marshals traveled on flights, blending in with other passengers. When airlines saw demand plunge in the wake of the crashes, slowing to a trickle, 43's father offered to fly commercial to instill confidence in a wary public that it was safe to fly, granting an interview to NBC's Tom Brokaw before hopping on a Continental flight from Boston to Houston.

On October 6, Bush signed the USA Patriot Act, enhancing the ability of U.S. intelligence apparatus to probe personal and financial records in the interest of deterring future attacks. Bush explained the law as a means of enhancing "the penalties that will fall on terrorists or anyone who helps them," though it would draw controversy and legal challenges throughout his tenure for its alleged infringement of individual rights.

On September 20, two days after signing a joint resolution of Congress allowing the use of force against the culprits of 9/11, Bush appeared in front of a joint session of Congress, delivering what *Time* magazine called "the finest, strongest, clearest, several-times-chill-giving speech of his life." Bush spoke to the nation for just under thirty-five minutes. Dick Cheney, who as vice president would have sat behind the president to the right of the speaker of the House of Representatives, was instead at an "undisclosed location," a sign of the trepidation that pervaded Washington less than a week after the attacks. Bush was undaunted. He began his remarks by demanding the Taliban's immediate cooperation in delivering the leaders of al Qaeda to American authorities. "They will hand over the terrorists, or they will share their fate," he warned.

He then prepared the nation for the task at hand. Immediate military victory, as his father had seen in Kuwait, would not come quickly, he indicated. It would be a far more sustained effort, and it would be directed at any nation that harbored terrorists, beyond al Qaeda. "Our war with terror begins with al Qaeda, but does not end there," he said. Just as every nation on the globe was seen in his father's political era through the lens of the Cold War, defined as either allied with the communist nations of the Soviet Union and People's Republic of China or with the U.S., every nation would now be seen by America as either being on the side of the U.S. or on the side of the terrorists.

Americans should not expect one battle, but a lengthy campaign, unlike any other we have seen. It may include dramatic strikes, visible on television, and covert operations, secret even in success . . .

From this day forward, any nation that continues to harbor
or support terrorism will be regarded as a hostile regime . . .
We will take defensive measures against terrorism to protect
Americans.

Every nation, in every region, now has a decision to make.
Either you are with us, or you are with the terrorists.

Bush's speech was interrupted thirty-one times by thunderous ovation in the House chamber. Afterward, his approval rating
soared to 90 percent, making it the highest in presidential history,
edging out his father's 89 percent approval following his Gulf War
triumph. It would be short-lived. Like his father, Bush would see the
ephemeral nature of public approbation.

44

A BROADER BATTLE

GOING TO WAR WAS NOT a decision George W. Bush took lightly. During his father's presidency, he had witnessed firsthand the toll war could take on its commander in chief. "I was able to see personal moments," he recalled, including his father's bearing after making the decision to send troops to Panama, his first major military action. "I could see how stiff his neck was when he slept in his office at Camp David, worried about the troops he'd put into combat. When he put troops into Panama or Iraq, I'd seen his concern, I'd seen his empathy for the wounded and the troops who'd lost their lives."

Bush was no less empathetic and heedful to war's costs than his father had been—nor was he any less steadfast. When the Taliban refused to give up the terrorists and release foreign aid workers the government held captive, he knew that war with Afghanistan was inevitable, and he wanted to get on with it. In late September and early October, when the Pentagon was slow in getting him a comprehensive war plan, he grew impatient. "I want a plan tomorrow," he demanded of Condoleezza Rice during a meeting with the CIA. "Call Don [Rumsfeld] and make sure I have one." Bush soon had his plan, Operation Enduring Freedom, conceived by Tommy Franks, commander of the U.S. Central Command, with its objective of toppling the oppressive Taliban regime and replacing it with a democracy and free society.

The war began on October 7, when American and British cruise missiles and long-range bombers pounded targets in Afghanistan including the country's capital, Kabul, Taliban military bases, and al Qaeda training camps. The air campaign was followed by boots on the ground. Overcoming concerns in October among the media punditry that the effort was becoming a "quagmire," an instinctive worry given the lingering stigma of the Vietnam War, U.S. Special Forces took the northern city of Mazar-i-Sharif in early November. Other northern cities fell in quick succession before the troops worked their way south liberating Kabul on November 15 and Kandahar on December 7. With them came the collapse of the Taliban government, which retreated, many fleeing into the copious mountains along the country's porous eastern border with Pakistan, with members of al Qaeda, though many leaders were hunted down, providing useful intelligence on the organization. Among those who eluded capture, however, was Osama bin Laden; the whereabouts of the world's most wanted man remained unknown. Still, the 2001 calendar year ended hopefully as Afghanistan was liberated in a decisive U.S.-led military victory, and Hamid Karzai, a military commander and former Afghan government official who had led an anti-Taliban movement, was installed as Afghanistan's interim president with the support of the White House.

When Bush announced in October that the U.S. had gone to war, he was careful to note that the military's initial target was Afghanistan, but added, "the battle is broader." In his State of the Union address on January 29, 2002, just over a year after taking office, he spelled out the greater security concerns of his administration, hinting at the battles to come and making the most definitive declaration of his administration. In what would become known as the Bush Doctrine, Bush warned of an "axis of evil," comprising rogue states like North Korea, Iran, and Iraq, and those like them connected to terrorism, proclaiming:

States like these, and their terrorist allies, constitute an axis of evil, arming to threaten the peace of the world. By seeking weapons of mass destruction, these regimes pose a grave and growing danger. They could provide these arms to terrorists, giving them the means to match their hatred. They could attack our allies or attempt to blackmail the United States. In any of these cases, the price of indifference would be catastrophic.

The doctrine held that the U.S. would strike any nation, preemptively and unilaterally if necessary, considered a threat to national security, while committing to fostering democracy and human freedom abroad. Espoused by the neoconservatives in his administration—Cheney, Paul Wolfowitz, and Elliott Abrams, chief among them, dubbed the "neocons"—the policy marked a departure from the U.S. Cold War foreign-policy paradigm, which emphasized containment and the avoidance of war. It was not, in other words, his father's foreign policy. The neocons believed that foreign policy had suffered in the hands of Bush 41 and Clinton, and looked for a chance to demonstrate U.S. might along with its idealism. Critics quickly took aim at the administration, deriding the neocon set and Bush's bellicose rhetoric. They blasted the notion of waging war without provocation and the naïve idea that democracy could be exported like fast food and Coca-Cola, creating a more peaceful world. Conveniently, the European goodwill America had engendered after 9/11 began to slip away and would continue to fray throughout the balance of Bush's presidency.

Bush summarily dismissed the critics and their views. With 9/11, the world had changed and U.S. foreign policy needed to change with it, he held. "Behind my presidency was a firm set of beliefs," he said after leaving office, "all of which must have rattled the nerves of people who felt there is a different way to conduct foreign policy, which is: manage to tyranny." Those who expressed doubt, he said, "weren't in the classroom" in Florida when he received the news of the attacks on September 11, or on Air Force One when he "made the decision to shoot down airplanes" considered to be a threat, or privy to meetings when he was "developing a strategy necessary to deal with an ideological conflict."

In fact, Bush was "amused" by the glib term "neocon," which he considered a "meaningless labeling," all part of "stereotyping" meant to "demean" his administration's point of view. During the latter years of his father's presidency, he had seen how the press had portrayed the old man as an out-of-touch aristocrat who didn't understand the economy, which to his mind amounted to "propaganda." He refused to let those who stereotyped him affect his policy.

The tenets of the Bush Doctrine would be memorialized in a document written by Rice titled "National Security Strategy: September 2002," which stated that the U.S. "will not hesitate to act alone, if necessary, to exercise our right of self-defense preemptively . . ." These precepts would soon be manifest in Bush's actions toward Iraq. In his State of the Union speech, he condemned Iraq for continuing "to flaunt its hostility toward America and to support terror," and its plot to "develop anthrax, and nerve gas, and nuclear weapons for over a decade." He went on to cite Saddam's use of poisonous gas to murder Kurdish citizens in the northern part of Iraq who had risen up against his regime, and his ousting of UN inspectors charged with ensuring that Iraq was not developing weapons of mass destruction (WMD). While, in Bush's words, "the thrill of liberation" in Afghanistan would soon "[give] way to the daunting task of helping the Afghan people rebuild," it was Iraq that would consume much of Bush's second year in office; the U.S. increasingly rattled sabers in the face of an insolent Saddam, who, it would be known later, didn't believe the U.S. would make good on its threats to depose him. Soon Iraq would become the overriding trial and burden of Bush's administration.

Saddam had been among the few who condemned the U.S. action in Afghanistan. After convening an emergency meeting with his deputies when bombing in the war began in October 2001, he issued a statement that read, "True believers cannot but condemn this act, not because it has been committed by America against a Muslim people but because it is an aggression perpetrated outside international law." Regardless, as 2002 neared its end, it became increasingly plain that Iraq was next.

45

41 AND 43

ON DECEMBER 23, 1990, PRESIDENT George H. W. Bush slept restlessly. The extended Bush family had gathered at Camp David for the holidays as they did every year of his presidency, boosting his spirits, but this year especially he had a lot on his mind. The ground war in Kuwait looked increasingly likely as Saddam Hussein continued to ignore the warnings of the U.S. and allied nations. It preoccupied Bush, running continuously through his mind. On Christmas Eve morning, he awakened with the remnants of a dream in his head: He was driving into a hotel near a golf course. Across a fence was another golf course, a lesser one. He heard his father was there and went looking for him, finding him in a hotel room just as he remembered him, "big, strong, highly respected." The two men embraced. "I miss you very much," the son told his father.

A dozen years later, during Christmastime of 2002, the extended Bush family once again found themselves at Camp David, as President George W. Bush, was faced with the possibility of a war with Saddam Hussein, just as his father had been. But while 41 had unconsciously yearned for his father in 1990, 43 had his own father to lean on—and he was right there. As 43 labored to find a diplomatic solution to his standoff with Saddam, who repeatedly breeched UN sanctions, he briefed his father on the situation and solicited his view. "You know how tough war is, son," 41 told him, "and you've got to try everything you can to avoid war. But if the man won't comply, you don't have any other choice." The elder Bush's advice went no further. "[H]e didn't need to tell me, 'I hope you're concerned about the troops,'" 43 said of his father. "He knew me well enough to *know* that I'd be concerned."

It was unprecedented. Never before in American history had a president had a father and presidential peer whom he could draw on immediately for counsel. When John Quincy Adams took office in Washington in March 1825, twenty-four years after his father left office, the decrepit elder Adams, at eighty-nine, was in Quincy, Massachusetts, in his last fifteen months of life. John Quincy learned of his father's death days after his passing on July 4, 1826, arriving in Quincy's United First Parish Church just in time for his funeral. "It is among the rarest ingredients of happiness," he wrote a friend, "to have a father yet living till a son is far advanced in years." Yet, distance hadn't allowed the elder Adams to be an active resource for his son during his first year and a quarter in the presidency. But the younger Bush's father—*big, strong, highly respected*—was accessible to offer guidance. Regardlesss, by both Bushes' accounts, 41's succinct advice at Camp David was the only time 43 solicited his view on anything of consequence regarding Iraq—and it seemed to belie concerns 41 quietly harbored about a war.

By all inside accounts, George H. W. Bush was first and foremost a loving father during his son's White House years, refraining from imparting unsolicited advice even as he worried about 43's administration, especially later as the war in Iraq got mired in mission creep. "I would definitely not characterize 41 as counseling his son in a reproachful way," said Jim Baker. "If he were counseling him he would say, 'Are you really sure this is something you want to do?' Now, I know that 41 thought that some of the advice that 43 was getting in the foreign policy was not the right advice. That's no secret. But he had the view that, 'Look, we had our chance; now it's [his] turn.'" Forty-one also conceded that the world had changed since his administration. "He always said to me, 'Well, the world was different when I was there,'" Condoleezza Rice said. "People who try to say, 'Well, 41 would have been more circumspect,' or 'Jim Baker would have handled it differently'—with al Qaeda having blown up the World Trade Center, really?"

It was in large measure *because* 41 had been president that he didn't tender advice more readily. "George Bush knew what it meant to be briefed as president," 43 said. "If I had asked [him], 'What should I do on embryonic cell research?' his response would have been, 'Please send your briefers down here.' He also knew presidents don't need frivolous, shallow advice: 'Even though it may not be as informed as your aides', here's my opinion . . .'" It was a lesson that 43 himself had learned during his father's presidency. "At one point in time, I said something that was clearly an extra burden," he recalled, "and he's not a lasher, but you could just tell by his body language that what I said was clearly unnecessary. And I said to myself, 'Wow, I'm not going to do that again.' I just wanted to be part of an environment that [made] him relax." Forty-one strove to do the same. Mainly, he played the paternal role of comforter for his son, who explained it this way:

If you've been president, you can see the stresses of the job; if you've been around a president you can see the stresses of the job. And most of the conversations were between father and son. "Son, how are you doing?" "Aw, I'm doing fine, Dad." A loving father is one who understands it's important to comfort in times of stress. To provide love in an environment that frankly is not very loving at times. To be a listener. That's the crux of the role . . . [N]ever before have there been conversations like this between two people who've both been president, who love each other.

It gave me comfort to talk to someone who knew what I'm going through, to hear, 'I love you.' Because of all the people in the country, he knew the pressures of the office. Nobody knows what you're going through. They just don't know. The most likely person though, is somebody who's been through what you've been through.

Often, 41 tried to offer comfort through humor, sometimes bawdy, which had always been a Bush family staple. In his post-presidential office in Houston, he kept a sterling-silver-framed photograph of himself smiling broadly with George W. and Jeb, which George W. had inscribed, "To Dad, who taught us to laugh." During his son's presidency, 41 had jokes faxed to the White House to provide 43 some levity. Among them was a story about a man who faced sentencing after stealing a can of peaches. When the judge sentenced him to six days in jail, correlating to the number of peaches in the can, the thief's wife blurted out, "He stole a can of peas, too!"

Forty-three allowed that "few are going to believe" that his father's influence on his presidency wasn't deeper, adding, "It's so simple, it's going to be hard for people to grasp the truth." He conceded, "the big speculation" about his father's involvement was "about Iraq." Indeed, Iraq and Saddam, echoes of his father's presidency, cast Shakespearean overtones onto 43's presidency. He stared down the same enemy who had been his father's chief antagonist—the malevolent dictator who had been driven out of Kuwait by his father and who later plotted to kill his father; the one who spat defiance at the U.S. still. Yet the father had declined to overturn Saddam's regime to avert the risk of alienating member states in the UN coalition and creating "more instability in Iraq," which would have been "very bad for the neighborhood." And the son, it would soon become known in 2003, fatefully chose otherwise.

The world was unaware of the conversation that transpired at Camp David between the father and son presidents, but great speculation arose among the public and in the media as to 43's motivation behind the war. Was he trying to prove something to his father? Or avenge Saddam's attempted assassination of his father? In September 2002, 43 had played into the latter conjecture by stating

of Saddam, "There's no doubt his hatred is mainly directed at us . . .
After all, this is a guy that tried to kill my dad at one time." The
statement left many to surmise that 43's targeting of Saddam was a
vendetta. "Some Americans have wondered whether the president's
determination to take on Saddam is a personal obsession," wrote
ABC.com in 2003, "one born in the aftermath of the Gulf War his
father launched, when Saddam was left in power."

At the same time, the media speculated that the elder Bush was
sending his son a message to abstain from war with Iraq using his
friend and former national security adviser, Brent Scowcroft, as a
proxy. On August 15, 2002, Scowcroft, joining a growing chorus
of those opposed to the war, rendered his view in a stinging *Wall
Street Journal* op-ed headlined "Don't Attack Saddam." In the piece,
Scowcroft challenged 43's rationale for the war by asserting that it
would represent a diversion from the war on terror and that Iraq was
not linked to al Qaeda in any direct way. Unlike the Gulf War, he
contended, international opposition against the war would necessi-
tate "a virtual go-it-alone strategy" that risks "unleashing an Arma-
geddon in the Middle East" and would "seriously jeopardize, if not
destroy, the global counterterrorist campaign."

In a subsequent column in the *New York Times* titled "Junior Gets
a Spanking," Maureen Dowd wrote, "On everything from taxes to
Iraq, the son has tried to use his father's failures in the eyes of conservatives as a reverse playbook . . . Poppy bequeathed his son, a
foreign affairs neophyte, his own trusted Desert Storm team, with
Dick Cheney as a surrogate father."

Scowcroft said 41 "knew nothing about" the op-ed, pointedly
sending him a copy as a courtesy at the same time he submitted the
piece to the *Wall Street Journal*. But he "did seek [41's] permission
to go public with his misgivings," according to 41's chief of staff,
Jean Becker. "Forty-three would have loved if his dad had said, 'Put
a muzzle on it,'" she said. "But [41] felt, 'That isn't fair.'" Scowcroft,
he believed, had "earned the right" to express his opinion.

But while 41 had no hand in the content of Scowcroft's piece, Scowcroft was confident that he saw the war similarly. "I think I know 98 percent of what he thinks about foreign policy," he said. "I can guess his reaction to most things. Do I think it reflected his view? Yeah. Yeah." Baker said Scowcroft's frankness "gave 41 some heartburn," adding, "in retrospect, [Brent] was right about a lot of it, and I felt the same way, too, but I wasn't gonna go out there and say it. I didn't think it was my place."

"The question people will be asking is, 'Is this your opinion?'" Bush told his father in a phone conversation after the piece ran. "You don't need a PhD in political science to know the ramifications. He didn't do us any favors, Dad!"

"Brent's a friend," 41 countered.

"Some friend," 43 said.

To his staff, Bush raised the question: "Why did [Scowcroft] feel he needed to do [the op-ed] in the first place?" Scowcroft was, after all, the chairman of the President's Foreign Intelligence Advisory Board, an independent body formed for the express purpose of weighing in on the quality of intelligence that reached the president's desk. "He's in my administration, and he communicates to me through an op-ed piece!" 43 vented incredulously to Card. "Why didn't he call Condi or [Steve] Hadley?"

"It's an interesting incident reflecting Washington," 43 said later. "It's interesting that a former national security adviser to [my dad] would express his opinion, which of course delighted the chatterers. 'Even Brent Scowcroft can't believe what's taking place! [George W. Bush is] clearly captured by the Neocons!' I can hear it all."

In the end, Bush chalked up the incident to "how it works inside the Beltway." Scowcroft became something of a pariah in 43's White House, even though he had been a mentor to Rice and Hadley and had helped to school Bush on foreign policy during the campaign. But tellingly, Scowcroft's relationship remained intact with 41 and Barbara, who had pulled away from friends and aides in the past for perceived disloyalty. Forty-one, while largely circumspect in his own views on his son's administration, would continue to carefully consider those of Scowcroft and Baker, his closest friends and confidants. "They're very close to George," Barbara said in 2014, describing her husband and them as "like brothers."

Ten days after the publication of Scowcroft's piece, Baker wrote his own op-ed, which appeared in the *New York Times* on August 25. Less of a rebuke than Scowcroft's, Baker's commentary urged the president not to "go it alone" in Iraq, but to "reject the advice of those who counsel doing so" and secure UN authority as a means of occupying "the moral high ground." He warned, "Unless we do it in the right way, there will be costs to other American foreign policy interests, including our relationships with practically all other Arab countries (and even many of our customary allies in Europe and elsewhere) and perhaps even our top foreign policy priority, the war on terrorism."

Baker's piece drew less fire than that of Scowcroft, but while it didn't give 41 "heartburn," it did, along with Scowcroft's, give him pause. As Andy Card put it, eluding to Baker and Scowcroft, "I think people around 41 were disappointed" about the path that 43 was taking in Iraq, "which made 41 disappointed." But 41 reserved judgment. He had faith that his son "made the decisions he thought he should make given the information he had," Card said. Forty-one also remained largely silent in the media. "What I want to do," he said, "is support [him], period. And because of that [I don't] get into the depths of these issues as I might otherwise be inclined."

The elder Bush lamented the fact that the media tried to "read" his relationship with his son as "some sort of competition." "It wasn't. Ever," he said flatly. "Just love between a father and a son." Even in the most strained of times, the love between them was never in jeopardy. Still, 41 fretted privately about the course his son was charting in.Iraq. "I know that [41] was worried about the beating of drums for war, and worried about how Iraq would turn out," Baker said. "Now, how much of that did he communicate with 43? I'm not privy to that." There was another thing Baker observed in his longtime friend and former boss, too: "The one thing that stuck in 41's craw was when someone would ask, 'Why didn't you take care of Saddam Hussein when you had the chance?'"

The answer would come soon enough.

PART VII

"The Decider"

46

TO WHOM MUCH IS GIVEN . . .

IN A PREDAWN HOUR OF 9/11, before the world changed suddenly and irrevocably, George W. Bush started the new day with a portentous reading from the Old Testament's book of Psalms:

> *My heart is in anguish within me;*
> *The terrors of death assail me.*
> *Destructive forces are at work in the city;*
> *Threats and lives never leave its streets.*
> *Let death take my enemies by surprise;*
> *Let them go down alive to their grave.*
> *For evil finds lodging among them.*

As he wrestled with the weighty issues of his presidency, Bush found solace in the Bible, which offered him guidance and moral clarity. When *Washington Post* reporter Bob Woodward pressed him on his father's influence on his administration during an interview for his book *Plan of Attack*, Bush said he was "the wrong father to appeal to in terms of strength; there is a higher father I appeal to." Distancing himself from his father "was understandable as a purely political issue," Jim Baker said later. Surely, the elder Bush saw it that way, too; it was part of George W. charting his own course—George H. W. had, after all, assured his son, "No one will ever question your devotion to your parents."

But 43, in large measure, meant what he said. The greatest influence in his life prior to his spiritual awakening in 1986 had been his father, he said in 2016; afterward it was Jesus Christ. While he spoke with his dad by phone three or four times a month, usually by placing early morning calls himself from the White House to his father's Houston office, he began every day before sunrise with readings from the Bible, sometimes followed by Bible study. His father, too, had found comfort in spirituality in his own presidency. "I thank God for the faith that He's given me," 41 said. "And as I grow older I'm more aware of the spiritual elements of life, and I ask for God's help."

Forty-three, though, wore his faith more openly and comfortably, offering a political advantage. He engendered the trust of the evangelicals, which would become the most powerful single bloc of the American electorate, just as conservative hero Ronald Reagan had found favor with the Moral Majority, which regarded 41 warily as a closet moderate. Forty-three's worldview was also more like that of Reagan, who saw the world in black-and-white terms—good versus evil, freedom versus oppression—and the Wilsonian fostering of American-style liberty as America's responsibility to the world and a means of achieving peace. Reagan had bucked the status quo strategy of managing Cold War tensions solely through peaceful coexistence; he was determined to engineer the clear triumph of American ideals over totalitarianism and rhetorically upped the ante by labeling the Soviet Union "the evil empire." Likewise, 43 pursued an aggressive "Freedom Agenda" that included "axis of evil" rhetoric against those countries that posed the greatest threat to America and its ideals. Bush believed that the war in Iraq not only would result in the democratization of Iraq and its liberation from the tyranny of Saddam Hussein but also would have a transformational effect in the Middle East. "If you believe that America is a force for good, then you're willing to use U.S. influence to affect people's lives," he said after his presidency, "and that informed my foreign policy."

It was a biblical passage from the Gospel of Luke in the same spirit that most guided Bush in his presidency: *For everyone to whom much is given, of him much shall be required*, or as Bush quoted more commonly, *To whom much is given, much is required*. The United States was a blessed nation, he believed, and had a moral obligation to lend a hand to those in most desperate need. When the HIV/AIDS crisis ravaged much of the third world, principally sub-Saharan Africa, Bush made a five-year $15 billion commitment to the President's Emergency Plan for AIDS Relief (PEPFAR), allowing for the treatment of 2.1 million afflicted with the life-threatening disease while caring for another ten million, including four million indigent children. By 2016, twelve million were receiving anti-retroviral treatment of AIDS through PEPFAR. Bush's initiative represented the largest international health drive to combat a single disease. (Among those who lobbied Bush to commit to AIDS relief was U2 front man Bono. Before an Oval Office meeting with the rock star and philanthropist, Bush, who prided himself on his detachment from popular culture, was asked by his deputy chief of staff, Josh Bolten, if he knew who Bono was. "Of course, he's a rock star," Bush replied, before joking, "Used to be married to Cher, didn't he?") Bush's decision to support PEPFAR was reinforced by 9/11. "I clearly saw the ideological conflict we faced with our enemy," he said. "If you're a young child and you've lost your mother and dad to AIDS, and rich nations sit on the sidelines not caring, you will become a hopeless person. We face an enemy that can only recruit when they find hopeless people. So, my justification for the program was one of moral justification and national security justification."

He saw the war in Iraq in much the same way; it was not only a national security offensive but a moral imperative as well. America's national security, he came to believe along with Cheney, the neocons, and Rice, was at stake as intelligence reports—later proved erroneous—pointed to stockpiles of WMD Saddam had amassed that could be sold to terrorists. And just as his father's resolve to liberate Kuwait hardened after reading an Amnesty International report revealing the brutality of the Iraqi invaders, Bush's intent to depose Saddam was bolstered in reading human rights reports disclosing the methods of torture he routinely used on his own citizens to quash dissent.

It was an admonition from a devotional by Scottish minister Oswald Chambers with which Bush began his day on March 19, 2003.

> Living a life of faith means never knowing where you are being led. But it does mean loving and knowing the One who is leading . . . Faith is rooted in the knowledge of a Person, and one of the biggest traps we fall into is the belief that if we have faith, God will surely lead us to success in the world.

When Bush met with his war team in the Situation Room later the same morning—with generals, including Tommy Franks, participating virtually on video screens—he was satisfied that they had what they needed to accomplish their mission. He turned to Rumsfeld. "Mr. Secretary, for the peace of the world and the benefit of the Iraqi people," he said somberly, "I hereby give the order to execute Operation Iraqi Freedom. May God bless the troops." When the president left the room, White House photographer Eric Draper noticed his eyes were red and flush with tears.

Trailed by his springer spaniel, Spot, Bush adjourned to South Lawn and walked the grounds for a moment of reflection, just as his father had done after giving the final order to execute Desert Storm.

As the reality of the war weighed on him, Bush prayed for the troops. Then his thoughts moved from the "higher father" to his dad. He repaired upstairs to the Treaty Room and handwrote a letter.

Dear Dad,

At around 9:30 a.m., I gave the order to SecDef to execute the war plan for Operation Iraqi Freedom. In spite of the fact that I had decided a few months ago to use force, if need be, to liberate Iraq and rid the country of WMD, the decision was an emotional one . . .

I know I have taken the right action and do pray few will lose life. Iraq will be free, the world will be safer. The emotion of the moment has passed and now I wait on the covert action that is taking place.

I know what you went through.

Love,

George

A few hours later, he received a fax in return.

Dear George,

Your handwritten note, just received, touched my heart. You are doing the right thing. Your decision, just made, is the toughest decision you've had to make up until now. But you made it with strength and with compassion. It is right to worry about the loss of innocent life be it Iraqi or American. But you have done that which you had to do . . .

Remember Robin's words "I love you more than tongue can tell."

Well, I do.

Devotedly,

Dad

At 10:15 p.m. that evening, Bush spoke to the nation, delivering a trim four-minute address from the Oval Office. "My fellow Americans," he began, "at this hour, American and coalition forces are in the early stages of military operations to disarm Iraq, to free its people, and to defend the world from grave danger." As he continued, he said, "We have no ambition in Iraq except to remove a threat and restore control of that country to its people."

47

"REGARDS FROM PRESIDENT BUSH"

THE MANHUNT—ONE OF THE LARGEST in history—lasted just a few days shy of nine months. Acting on a tip that Saddam Hussein was in the small Iraqi town of Ad Dawr, fifteen miles from his hometown of Tikrit, members of the Raider Brigade of the Army's Fourth Infantry Division arrived at a farmhouse. There the outlaw labeled by the military as "High-Value Target Number One" was found bearded, mangy, and bedraggled, standing in a ventilated "spider hole" no deeper than eight feet and no wider than a whiskey barrel. Buried with him were two AK-47s, a pistol, and $750,000 American dollars, all in one hundred dollar bills. Once the dictator of twenty-five million people, with his pick of palaces throughout the country, his life had been reduced to that of a cockroach who scurried underground at the threat of danger.

"My name is Saddam Hussein," he said after being dragged out of the ground by the soldier who discovered him. "I am the president of Iraq and I want to negotiate."

"Regards from President Bush," the soldier replied.

Bush learned of Saddam's capture through a call from Rumsfeld. Myriad congratulatory calls followed, including one from Bush's father. Forty-one generally refrained from being "the father calling all the time," but this was an exception. The American collaring of the nemesis of both their presidencies was a shared triumph, but the call was short. "It was a touching moment," 43 said later in an interview with ABC's Diane Sawyer, "[but] I was busy . . . Look, I had phone calls stacked up, and I wanted—I didn't want to keep other foreign leaders waiting." Shortly afterward, in a nationally broadcast address from the Cabinet Room, Bush said, "The capture of this man was crucial to the rise of a free Iraq. It marks the end of the road for him and for all who bullied and killed in his name."

Baghdad had fallen the previous spring, on April 9, just three weeks after the U.S. and coalition forces had begun the war, sending Saddam into hiding. Footage of a massive statue of the despotic leader being brought down with a crane by U.S. troops to the cheers of a handful of liberated Iraqis showed that while Saddam had yet to be apprehended, Iraq had entered an auspicious new era. It was a moment of pride for Americans, including the president's father, who faxed his son the same day.

> You have borne the burden with no complaining, no posturing. You have led with conviction and determination; and now the whole world sees that more clearly . . .
>
> We will stay out of your way, but I am there beside you, my heart overflowing with happiness on this day of vindication.
>
> No doubt tough times ahead, but, henceforth, here and abroad, there will never be any doubts about our Commander in Chief, about his leadership, about our boy George.

It looked as though Rumsfeld's war plan for Operation Iraqi Freedom, based on greater speed and mobility and fewer troops than previous U.S. military operations, had succeeded with remarkable alacrity. The secretary of defense had consciously departed from Colin Powell's "Powell Doctrine," which had stood since the Gulf War advocating the exercise of military force only with overwhelming force. Forty-three's invasion of Iraq would eventually involve just 240,000 American troops, under half the number his father had dispatched in the Gulf War.

On May 1, shortly after Baghdad's liberation, Bush co-piloted a

U.S. Navy plane, dubbed "Navy One," which landed on the flight deck of the USS *Abraham Lincoln* off the coast of San Diego. The president emerged dramatically from the plane in a flight suit and was greeted by the ship's sailors, before reverting to a business suit and speaking to the country. "Major combat operations in Iraq have ended. In the Battle of Iraq, the United States and our allies have prevailed," he declared. As he continued, he said, "We have difficult work to do in Iraq . . . Our coalition will stay until our work is done." Trumping the latter part of Bush's message was a red-white-and-blue banner behind him blaring from the ship's bridge that read, "Mission Accomplished," suggesting a finality to American entrenchment in Iraq.

Democrats hoping for a PR blunder along the lines of Michael Dukakis's tank debacle in the 1988 presidential campaign against Bush's father were disappointed. Bush, who had already racked up his share of triumphal photo ops including his "bullhorn moment" at Ground Zero and the perfect strike he drilled to open game three of the 9/11-delayed 2001 World Series at New York's Yankee Stadium, added another to the list. Pundits gushed. *U.S. News & World Report* reported afterward, "The top gun cut a striking figure in his Top Gun duds, surrounded by admiring men and women in uniform—an image that's likely to turn up in Republican campaign advertising next year," while MSNBC's garrulous Chris Matthews pronounced Bush a "hero," effusing, "He won the war. He was an effective commander. Everybody recognizes that, I believe, except a few critics."

By the waning months of 2003, even before Saddam's capture, the critics would come home to roost. Rumsfeld's plan for waging the war in Iraq had worked, but there was no clear strategy in place for rebuilding the country. The tide in Iraq turned to a postwar quagmire unleashing insurgency and stirring up ancient tribal hatred between Sunni and Shiite Muslim sects that resulted in terrorism, violence, and political dispute that no one in the administration seemed to anticipate. Chaos swept the country. In October 2003, *Time* magazine published a cover story titled "Mission *Not* Accomplished," as the military operation in Iraq dragged on perniciously and inconclusively. A year later, with the U.S. death toll exceeding one thousand, *Time* delivered a follow-up feature headlined "Mission *Still* Not Accomplished," while *Newsweek* ran a story titled "It's Worse Than You Think." Car bombs, kidnappings, beheadings, and suicide bombs in Iraq became staples of foreign news coverage. By the middle of 2006, an average of 120 Iraqis would be killed daily.

The promise that U.S. troop involvement would diminish after Iraq's liberation was dashed due to the power vacuum created by the dismantling of the Iraqi Army. There was also the matter of weapons of mass destruction. Though Saddam Hussein had been found, WMD—the impetus for the war—had not, manifesting a glaring intelligence failure and eroding Bush's credibility. By 2004, only one in five Americans believed Bush was telling the entire truth when addressed with the subject of Iraq. Conversely, while Saddam had been toppled to prevent the threat of global terrorism, al Qaeda found Iraq fertile ground for the recruitment and training of terrorists.

The war's mounting price was also at issue. In September 2002, Bush's director of the National Economic Council, Lawrence Lindsey, guessed the cost of the war to be between $100 to $200 billion, an estimate Rumsfeld called "baloney," asserting that it would be more like $50 to $60 billion. In fact, five years after the war began, the *Washington Post* reported that its tally had topped $3 trillion, making it the second-most costly American war after World War II. Finally, there was the unexpectedly high cost in blood. In 2017, well after U.S. troops were pulled from Iraq under the order of Barack Obama, the Department of Defense reported that a total of 4,424 American soldiers died in Operation Iraqi Freedom while another 31,942 were wounded.

As the situation in Iraq worsened, the media once again saw

patriarchal undercurrents. The *New Yorker* wrote in the summer of 2004, "[I]n denying his real father's influence, George W. reaffirms its importance. The very decision to stage a celebration on the [USS] Abraham Lincoln flight deck can be read as a rebuke to his father, signaling that he forged ahead where his [father] had held his fire. It was also at odds with the elder Bush's self-effacing style. Today, as the Iraq adventure slogs on, that stunt is looking less like a moment of glory than like a moment of vainglory—and a mistake that George H. W. Bush, whatever his shortcomings, never would have made."

Indeed, among those who were disquieted about the situation in Iraq was George H. W. Bush. Though he eschewed expressing his concerns to his son, he conveyed anxiety privately about the influence of his former rival Donald Rumsfeld, and the neoconservative Elliott Abrams, whom he had pardoned in his own administration for misdeeds in the Iran-Contra affair, a decision, according to Bush insiders, he came to regret. At the same time, he worried about Colin Powell's diminished role as secretary of state, as Powell's more moderate voice on foreign affairs was drowned out by those of Cheney, Rumsfeld, and the neocons. "I know that [41] was concerned about the way Powell was treated early on because of the model we had when he was president," allowed Jim Baker, who believed "the most important thing for a secretary of state is to have a seamless relationship with his president."

Forty-three harbored his own concern that Powell was sounding off to the 41 camp about his marginalization and lack of presidential access in the first place. He also had his own view: It wasn't that Powell didn't have access to him but that he simply "didn't agree" with him. During the latter days of Bush's first term, Powell was increasingly at odds with Cheney and Rumsfeld as tension in the White House mounted. Rice and Hadley were put in the middle of the conflict, 43 contended, but "didn't know how to handle it." It was no great surprise when Powell stepped down as secretary of state after Bush's first term, replaced by Rice, who in turn was succeeded as national security adviser by Hadley.

The influence of Cheney on Bush's presidency took on a life of its own during Bush's first term. The media often depicted the vice president as a Machiavellian puppet master who pulled the strings on policy decisions, straying from the more moderate path he had tread in 41's administration and leading 43 down the garden path on Iraqi regime change. Cheney's conservative drift was a matter of some debate. Some chalked it up to his heart condition affecting his mind. Brent Scowcroft, who had known Cheney since the two worked together in the Ford administration thirty years earlier and who went on to work with him in 41's administration, said in 2006, "I don't know him anymore. He's not the same guy." Cheney, for his part, said in 2013, "I don't think I changed ideologically. What happened was 9/11 . . . that was a sobering moment. Now, was I different after 9/11 than before? Well, different in most regards as the fact that I think we faced an entirely different threat compared to what we were dealing with in Desert Storm." Lawrence Wilkerson, Powell's chief of staff, said in 2011 what many critics suspected. Despite being the only sitting vice president since the 1920s not to have expressed presidential ambitions, Cheney "wanted desperately to be president of the United States," asserted Wilkerson. "He knew the Texas governor was not steeped in anything but baseball, so he knew he was going to be president and I think he got his dream. He was president for all practical purposes for the first term of the Bush administration."

Well after his son had left office, 41 observed that "Cheney had his own empire and marched to his own drummer." If so, it wasn't something 41 addressed with his son during his administration. Any feelings 41 had about the matter were outweighed by his confidence in his son and his inherent optimism that everything would turn out all right. He "didn't worry" about Cheney's influence on 43's presidency, he said in 2013. "It's true," Barbara Bush confirmed in the same interview, "he didn't worry about that. He had great faith in George." Instead, 41 used whatever sway he had with his son to gently question Cheney's recommendations, not Cheney himself.

Neither did he take his concerns about Iraq to Cheney. "I never talked to him about it," Cheney reflected. "He never expressed views of it one way or the other. I've assumed that 41 and 43 talked about it, but I wasn't there. He was careful. He didn't come in and say, 'Dick, you need to do X or Y.' That just wasn't his style. And it would have been kind of infringing on 43's turf, and he wouldn't have done that." Tellingly, though, 41 said in a 2006 interview that he and Cheney "used to be close," while he remained more closely connected to other alumni in his administration who were then serving 43.

Barbara Bush was more vocal in her criticisms of Cheney, citing her belief that he had changed discernibly between her husband's administration and her son's due to the heart attacks he had suffered. "I do think he was different," she maintained, indicating that her view was largely influenced by Baker and Scowcroft. "I think his heart operation made a difference. I always liked him but I didn't like him so much for a while because I thought he hurt George. I wasn't that fond of him. I think he pushed things a little too far right."

The president was aware of his parents' wariness of the influence of Cheney and the neocons on him. "I'm confident they concerned Dad and Mother," he said, believing that they, in turn, were influenced by the "inside-the-Beltway chatterers" he grew to disdain. Forty-three was appalled by his mother's privately stated belief that he was "unduly influenced" by the neocons "clearly steering him to the right." "Surely, you've got more confidence in your son that I would make up my own mind," he told her on more than one occasion. "If you don't agree with it, it's one thing, but I'm plenty capable of making my own decisions."

Barbara recalled her son's admonishment. "Mom, when you're

criticizing someone in my administration, you're criticizing me," he had said. Afterward she kept her doubts to herself. "That sort of cut it off," she said.

Forty-three was incredulous that anyone—let alone his mother—would believe that he wasn't the one calling the shots of his presidency. "I hear the voices and I read the front page and I hear the speculation," an exasperated Bush said in mid-April 2006, as Washington buzzed that he should replace Rumsfeld at the Pentagon. "But I'm the decider, and I decide what's best." To be sure, Bush was hardly one to shy away from making hard decisions, which he had done throughout his life with reflexive ease, utterly assured in his ability to collate information, however contradictory, get to what was important, and arrive at a resolution. He took pride in the tough decisions he made in his presidency and the principled reasoning behind them, believing they reflected the best of American ideals. As he put it six years after he left office, "The fact that there was any doubt in anyone's mind about who the president was blows my mind," adding that Cheney and Rumsfeld "didn't make one fucking decision." And just as it miffed 41 when those in his son's administration wondered of him, "Why didn't you take care of Saddam Hussein when you had the chance?" 43 came to resent the armchair quarterbacking over Iraq from those of his father's administration. "People around George H. W. Bush and other [former] presidents say, 'We were for freedom,'" he said. "It's one thing to say you're for freedom and it's another to implement reforms that yield freedom."

Still, why *hadn't* 43 further sought his father's advice on Iraq? "I was content with the informed advice I was getting," he said, "and it's not like I wasn't getting advice on both sides. The decision-making process was fairly lengthy and at times hectic. I was getting ample advice, and maybe it didn't occur to me to ask him because circumstances had changed. He had never been confronted with an issue like 9/11." Forty-three surmised that his father didn't openly question his Iraq policy because his "disclose, disarm, or face serious consequences" ultimatum made clear his intention.

A lesson he taught me was, if you say something, you'd better mean it. And I meant it. So, when Saddam Hussein defied not just us but a coalition, like the one we formed in the Gulf War, it was easy for [my father] to see. Maybe if I had gone headlong in without trying to put together a coalition, he might have called and said, "Don't you need allies on this deal?" [But] to get the congressional authorization, I didn't call him on the phone and ask, "Do you think I should get congressional authorization?" I had plenty of capable people advising me. Some said no, some said yes; I made up my mind to go. Should I go to the United Nations? Some said yes, some said no.

As the 2004 presidential election neared and 43's approval rating fell below the 50 percent mark, 41 did offer his son some political advice. Without specifically counseling him to dump his vice president, he suggested that he might consider "shaking up the ticket" by tapping a new running mate. Forty-three considered it—just as his father had considered his suggestion that he replace Quayle in 1992—but chose Cheney again when he couldn't think of a better replacement. But while Cheney would remain Bush's vice presidential pick, his influence would wane. Throughout the balance of 43's presidency, as he settled further in the office, the "decider" would move in a decidedly different direction.

48

THE LAST CAMPAIGN

NEARLY THIRTY-EIGHT YEARS TO THE day before her future husband was inaugurated as the forty-third president, Laura Welch was among the twenty thousand who queued up at the Lyndon Baines Johnson Presidential Library in Austin, Texas, to pay their respects to LBJ, who lay in state the day after his death from a heart attack. As the twenty-five-year-old University of Texas graduate student approached the thirty-sixth president's flag-draped coffin, she was greeted by his widow, Lady Bird Johnson, who graciously shook her hand and thanked her for being there. At the time, Laura thought she had nothing in common with the former first lady except for the fact that they were two women from Texas. Serene and stolid, Lady Bird had been her husband's closest confidante throughout their tumultuous White House years marked by national tragedy, political division, and a sustained and increasingly unpopular war. Laura would eventually find she would have more in common with Mrs. Johnson than she could have imagined.

Laura Welch Bush had eased gracefully into the role of first lady, having been the first lady of Texas for six years and having watched her mother-in-law fulfill the roles of second lady and first lady of the United States for a collective twelve years. Still, she didn't feel she had hit her stride until well into her first year, late in the summer of 2001. With a conference on early childhood development at Georgetown University and her first state dinner under her belt, and the successful September 8 launch of the National Book Festival, a take-off on the Texas Book Festival she had brought to her home state as the governor's wife, she was finding her "place in the world of Washington." When 9/11 befell the nation, her life changed every bit as much as her husband's. Though she couldn't see it at the time, the opportunity to make her mark would be much greater than before.

An old friend, her college roommate at Southern Methodist University, put things in perspective for her. When Laura became first lady, incurring all the scrutiny that went with it, she told Laura she didn't envy her at all. But after 9/11 she said she felt "a little jealous"—at a time when Americans were wondering what they could do for their scarred country, Laura had a platform. Like Lady Bird Johnson, she used it. She reminded parents to hug their children after the terrorist attacks, and in November became the first first lady to step in for her husband in his weekly radio address, generating awareness of the oppressive plight of Afghan women and children under the Taliban regime, a cause she would continue to pursue. Shortly after giving the address, during a shopping excursion in Austin while visiting Jenna at the University of Texas, a store clerk expressed her gratitude. "Thank you so much for speaking for the women of Afghanistan," she said. As Laura recalled it, "It was the first time I thought, 'Hey, they heard me.'"

Her husband heard her, too. After his "Wanted Dead or Alive" remark about Osama bin Laden, she gently admonished him by ribbing, "Bushie, you gonna get 'im?" Afterward, he toned down his rhetoric. George W. Bush appreciated that he had a full partner in Laura, someone who was going to slog through the worst with him when the possibility of another homeland attack lingered well after the smoke cleared over Manhattan. "I know he knows it," she said later, "he could have been married to someone who said, '[Before 9/11] it was fine, but not now . . .'" Like his son, George H. W. Bush marveled at the first lady. "What surprised me was that years ago she didn't want to give a speech, didn't want to be involved," he said admiringly, "and now she's captured the imagination and the hearts of the country."

Laura was one of the few around the Bush White House who eluded controversy as the country fell into a red state–blue state political chasm, fostered by the media, during the latter part of Bush's first term. When the Bushes traveled to New York City in late August 2004, where the president would accept his party's charge for a second term at the Republican National Convention, Laura saw the divide. "Being in the city reminded me of how completely our country came together after 9/11," she wrote, "when for a time personal passions were put aside and we had a common care and a common purpose. Now I began to feel the country separating along new seams."

The extended Bush family all came together with the first couple in New York. The twins, Barbara and Jenna, now twenty-two-year-old college graduates, had shunned participation in their father's previous campaigns. But this one was different. Earlier in the year, Bush received a letter from Jenna, in her last semester at college, conveying her desire to be involved, a sentiment shared by Barbara.

> You and Mom have taught us the meaning of unconditional love . . . At age twenty-two, I finally have learned what that selfless pain must have felt like.
>
> I hate hearing lies about you. I hate when people criticize you. I hate that everybody can't see the person I love and respect, the person that I hope to someday be like.
>
> It is because of all of these reasons that I have decided that if you want me to I would love to work full-time for you in the fall . . .
>
> I want to try to give you something for the twenty-two years you have given me . . .
>
> I love you and am so proud of you.

It may well have been a letter Bush would have written to his own father, and a welcome acknowledgment that the unconditional paternal love he received—"the greatest gift a father can give a child," Bush would often call it—he, in turn, had given to his daughters. Jenna and Barbara spoke in support of their father at the convention in a joint prime-time appearance, and took to the stump for him in the fall. As Bush later wrote, "My last campaign would be their first."

In his acceptance speech, Bush served up red-state twang while admitting to some of his foibles. "Some folks look at me and see a certain swagger, which in Texas is called 'walking,'" he said. He also allowed that "now and then I come across as a little too blunt," adding while pointing to his seventy-nine-year-old mother in the audience, "For that, we can all thank the white-haired lady sitting right up there." But he offered no apologies for his decision to go to war in Iraq. Alluding to the false assurances of Saddam that he had no WMD, Bush asked the party faithful, "Do I forget the lessons of September 11, and take the word of a madman, or do I take action to defend our country? Faced with that choice, I will defend America every time." He talked of second-term domestic goals—reform in Medicare, Social Security, and immigration while continuing No Child Left Behind, passed by a bipartisan vote in 2001, and faith-based initiatives—but it was his role as the unyielding commander in chief in the war on terror that he most touted as he took to the campaign trail in the fall. "This election is going to come down to who knows how to lead, who will take on the big issues, and who can keep American safe," he said privately during a quail hunting respite in South Texas in early January.

Standing in the way of Bush's second term was Massachusetts senator John Kerry. A month before the GOP convention, at their own gathering in Boston, the Democrats gave their presidential nod to Kerry, who had in turn chosen his former challenger, North Carolina senator John Edwards, as his running mate.

The sixty-year-old Kerry also aimed at projecting the image of a strong leader. Two years earlier, Bill Clinton counseled fellow Democrats that "when people are insecure, they'd rather have somebody who's strong and wrong than somebody who's weak and right." A recipient of the Silver Star, Bronze Star, and three Purple Hearts, Kerry served with distinction in Vietnam as a U.S. Navy Swift boat captain in the late sixties, before returning stateside to become a leading voice in the antiwar movement as a civilian. "How do you ask a man to be the last man to die for a mistake?" he asked poignantly before a Senate subcommittee in 1971. But it was the former experience as a decorated soldier that he emphasized in his acceptance speech, opening with, "I'm John Kerry, and I'm reporting for duty," and noting later that he had "defended this country as a young man and I will defend it as president." He went on to promise to fight "a smarter, more effective war on terror," positioning himself as a "commander in chief who wouldn't mislead us into war." "In these dangerous days," he dug at his opponent, "there is a right way and a wrong way to be strong. Strength is more than tough words."

Once again, Bush ran a disciplined campaign, characterizing Kerry as a liberal "flip-flopper," who had said earlier in the year of Bush's $87 billion package to fund the war on terror, "I actually did vote for the $87 billion before I voted against it." Kerry sustained a more devastating blow from a series of attack ads from a nonprofit group called Swift Boat Veterans for Truth. The campaign featured Vietnam veterans who accused Kerry of lying to earn his Bronze Star and one of his Purple Hearts and making exaggerated claims about the conduct of American soldiers in Vietnam in his testimony to Congress. "He dishonored his country," a fellow Vietnam veteran said in one of the ads, "and more importantly the people he served with." The ads were roundly criticized as grossly unfair and misleading. Bush drew his own controversy for ignoring calls to denounce them until a week after they stopped running, when he commented from the ranch that so-called 527 ads, those financed by nonprofit groups attempting to influence political opinion, should be abolished as being "bad for the system." Mostly, they were bad for Kerry.

Regardless of Kerry's setbacks, it was the senator who looked as though he had the edge going into election night, which, like the election of 2000, would be a nail-biter. John Adams and John Quincy Adams, the only other father-and-son presidents, offered a dim historical precedent. After the elder Adams became the first president to fail to win reelection upon his defeat to Thomas Jefferson in 1800, John Quincy Adams became the second, losing the presidency in 1828 to Andrew Jackson, his principal rival four years earlier, by a margin of 178 electoral votes to Adams's 83. (Martin Van Buren, the last sitting vice president to win the presidency before George H. W. Bush in 1988, became the third.)

The elder Adams, who died over two years before the election of 1828, was spared his son's drubbing. The elder Bush, however, was very much alive as his son's fate hung in the balance on Tuesday, November 2, 2004, and every bit as nervous as he had been four years earlier. He and Barbara were on hand at the White House with the first couple to monitor the evening's results as exit polls, even in the reddest of red states, tilted toward Kerry. Earlier in the day, after Karl Rove gave 43 a bleak heads-up on Air Force One as he flew back to Washington from Crawford, where he and Laura cast their votes, the president felt as if he had been "punched in the stomach." Memories of 41's painful reelection defeat hung tacitly in the air. "I knew life would go on, as it had for Dad," Bush recalled thinking as he "moped around" the Treaty Room. "But the rejection was going to sting."

It was going to sting every bit as much for his father. Condoleezza Rice recalled the elder Bush ducking his head into her West Wing office, looking "the most devastated" she had ever seen him. "It looks kind of bad, doesn't it?" he asked dolefully early in the evening. Rice invited him into her office as the two huddled at her computer to study election data on the website *RealClearPolitics*.

"There is a glimmer of hope about the character of the exit polls," she said, citing the fact that they were "unbalanced" in a way that was advantageous to 43. It was enough to bolster her former boss, whose demeanor changed palpably. He followed the returns with hope, as the election came down to the outcomes in swing states Iowa, New Mexico, Nevada, and Ohio. The networks, mindful of their premature declarations four years earlier, cautiously held off on announcing a winner, as vote totals were pointed increasingly and unexpectedly toward a Bush victory. Ohio, with its mother lode of twenty electoral votes, was looking more promising and decisive as the night turned to morning. Still, there was no call from the networks, nor a concession from Kerry.

Regardless, Bush's advisers in the early morning hours urged him to declare victory. Laura thought otherwise. "George, you can't go out there," she advised her husband, for whom patience was not always a strong suit. "Wait until you've been declared the winner." He agreed. At 5:00 a.m., Andy Card told the press, "President Bush decided to give Senator Kerry the respect of more time to reflect on the results of this election. We are convinced that President Bush has won reelection with 286 electoral votes."

It was 286 electoral votes exactly, to Kerry's 251, with Bush eking out just over 50 percent of the popular vote. Before noon on Wednesday, November 3, Kerry called to congratulate the president on his victory. Bush became the first presidential candidate to win a majority of the popular vote since his father yielded 53 percent in 1988, racking up more ballots than any other president in history to that point, and the Bush father and son had now surpassed the Adams father and son by achieving a collective third term. Only the Roosevelt family, with Theodore's White House tenure of just under eight years and his fifth cousin Franklin's at a month over twelve years, had a greater hold on the nation's highest office than the Bushes.

George and Barbara Bush left the White House for Texas the

morning of the third, before George W.'s win was official. Prior to their departure, 41 and 43 had a chance to visit in the Oval Office. Afterward, the president, who had gotten just two hours of sleep, asked his father to accompany him on the putting green on South Lawn. As he watched his son drain putts, fished out of the holes by his Scottish terrier, Barney, he thought about what lay ahead for him in his next term.

Will God continue to give him the strength and perseverance he needs? What comes next in the fight against terror, in working with Congress, in bringing our divided country together? What can he do to lower the decibels of hatred and anger?

I wish I could help this son of ours. I wish I could do something to help ease the burden . . .

But I cannot. I am an old guy. My experiences are out of date. Perhaps my instincts aren't as good as I once thought they were.

A few minutes later, he and Barbara boarded a limousine en route for Reagan National Airport, where they went by air to Houston, leaving the forty-third president at the White House with the sobering challenge of leading a fractured nation forward for the next four years.

"BROTHER FROM ANOTHER MOTHER"

IF 9/11 WAS THE DEFINING moment of his presidency, it was the righteous cause of freedom with which the forty-third president most wanted to be defined as he prepared to take the helm for a second term. Upon leaving his first cabinet meeting after his reelection, Bush spotted his chief speechwriter, Michael Gerson. His upcoming inauguration address on his mind, Bush gave Gerson a simple direction. "I want this to be a freedom speech," he said.

After working closely with Gerson on twenty-two drafts, a freedom speech is what Bush got. On January 20, 2005, before a crowd of 400,000—at least 146 of whom were members of the sprawling Bush clan—Bush spoke for twenty-one minutes using the words freedom, free, or liberty just shy of fifty times. "At this second gathering, our duties are defined not by the words I use, but by the history we have seen together," he said. "For a half a century, America defended our own freedom by standing watch on distant borders. After the shipwreck of Communism came years of relative quiet, years of repose, years of sabbatical—and then there came a day of fire." The global dissemination of liberty, he asserted with echoes of Woodrow Wilson more than eight decades earlier, was "the calling of our time" and "the best hope for peace." "Some, I know, have questioned the global appeal of liberty," he continued, "though this time in history—four decades defined by the swiftest advance of freedom ever seen—is an odd time for doubt." Regardless, it was a tough sell for the many Americans wary of the festering aftermath of the wars in Iraq and Afghanistan and who saw some national security measures as the draconian trampling of civil liberties.

George H. W. Bush watched his son from the inaugural platform, once again doing double duty as a father and as part of a contingent of former presidents who gather quadrennially to observe the inauguration of a presidential successor, a further manifestation of the peaceful transition of power as the hallmark of American democracy. With him were fellow "formers," Jimmy Carter and Bill Clinton. A dozen years earlier, Bush had cordially and obligingly turned the office over to Clinton, but there had been a chill between them then. How could there not have been? The elbows had been sharp in the '92 campaign during which Bush had called Clinton and Al Gore "two bozos," as Clinton successfully depicted his opponent as out of touch and indifferent.

Presently, though, as the elder Bush and Clinton sat together against the cold of the day, just a notch above freezing, a thaw had set in. Against all odds perhaps, the two men had become friends. The relationship had sprouted with a recent entreaty by the president who asked that they team up in the name of American public relations, international diplomacy, and money-raising in the wake of one of the most devastating natural disasters on record. On December 26, 2004, a 9.1 magnitude earthquake shook beneath the Indian Ocean triggering a massive tsunami that overwhelmed coastal areas throughout Southeast Asia. The death toll would measure over 230,000, in fourteen different countries in the region. Initially, the U.S. responded by offering up $15 million in disaster relief, later upped to $35 million. At a time when the American image was tarnishing due to the war in Iraq and when the U.S. ranked last among developed nations in foreign aid as a percentage of gross domestic income, the American contribution was looked upon as paltry.

As the White House scrambled to show American compassion, Andy Card suggested tapping 41 and Clinton to encourage donations in the private sector and to tour the most ravaged areas of

Southeast Asia. It was a big statement—two ex-presidents and former rivals coming together in the name of philanthropy and international goodwill. Accepting the president's request, the twosome, like a mismatched duo in a buddy movie, set off in mid-February from Los Angeles for Thailand in a blue-and-white Boeing 757 with the words "United States of America" emblazoned on its fuselage. As they got to know each other better on the trip—which included stops in Thailand, Indonesia, Sri Lanka, and Maldives—a genuine friendship blossomed.

Bush conceded that before the trip he "did not really know" his former rival. He gathered that Clinton would have scored poorly in his grade school report card in the area of "Claims no more than his fair share of time and attention in the classroom," and lamented that "Clinton time" nullified his instinctive punctuality. But as he spent time with Clinton, he found himself liking him. The fifty-eight-year-old Clinton treated him with deference and consideration, including his insistence that his eighty-year-old elder take the plane's only bed in the stateroom while he slumbered in a seat in the plane's main cabin, despite recently being laid up with quadruple-bypass surgery. Bush also appreciated the fact that Clinton "went out of his way not to criticize" his son, the president.

Putting the two together "wasn't 'Let's have an experiment in character,'" 43 said later. "It was really focused on how to get money from the private sector to show the world that not only does the government of the U.S. care but so do the citizens—to help deal with a major catastrophe. And so we put them together, [and] it was the beginning of a really unique relationship."

Soon, 41 and 42, the "odd couple" as they became known, were constant companions, side by side on their Asian tour, appearing together at the Super Bowl and on the links at a charity golf tournament in Florida, and on the airwaves in joint commercials soliciting donations and in news interviews. Together they raised more than $12 million, which their offices directed to outside organizations that channeled the funds into relief efforts. When 43 asked his father and Clinton to represent him as part of a U.S. delegation at the funeral of Pope John Paul II in April, 41 encouraged his new friend to accept. "Come on," he said, "it will be better with you along." Throughout the year, the pair became symbols of bipartisanship lacking in national discourse. "We found that when we traveled abroad, people said this couldn't have happened in their country," Bush said in June that same year. "The equivalent of a Republican and a Democrat—this never would happen. Well, it doesn't have to be that way." *Time* magazine named the pair "Partners of the Year," as an honorable mention for their annual "Person of the Year" designation. "I think people look at George and me and they say, 'This is the way it ought to work,'" Clinton said in an interview with the magazine.

Nearly twenty-five years earlier, Gerald Ford had formed an unlikely friendship with his former rival, Jimmy Carter, under similar circumstances. When Ronald Reagan asked Ford and Carter, along with Richard Nixon, to represent the United States at the 1981 funeral of assassinated Egyptian president Anwar Sadat, the two bonded during the long plane rides to and from Cairo, precipitating a relationship that Carter described as "almost like brothers." While 41 and Clinton were also part of the fraternity of former presidents, the pair were less like brothers and more, in the eyes of those who observed them, like father and son. Calling the friendship a "dividend" in his life, Bush, too, mused, "Maybe I'm the father he never had," and Barbara saw it the same way.

Forty-one's affection for Clinton was evident in the invitation he extended for Clinton to join the Bush family for a late-June sojourn in Kennebunkport. During the visit, the Bush progeny took to calling Clinton "brother from another mother," or simply, "bro." "We all like him," Barbara said of the family's newest friend. The feeling was mutual. "[Clinton] would say, 'Your dad means a lot [to me],' or 'I love your dad,' and then I realized how sincere he was," said 43, who recognized "the enormous capacity of George Bush to become a father figure in people's lives." The relationship surprised him as much as anyone and enhanced his admiration for his father. "It's a unique quality of a person to be able to put aside [his] defeat and see [something] larger," he said. At the same time, he developed his own kinship with Clinton, picking up the phone "just to chat" at times, especially during his second term. "I talked to Clinton quite a bit," he said. "I like being around him. I don't think there's any obligation for any president to talk to former presidents, but if you like [the] person and they're open-minded, you can learn . . . And if there was an issue [he knew] he would say, 'Here's my experience with that,' and I'd be very interested."

Eight months after the tsunami swept over Southeast Asia, America saw nature's fury on its own shores as Hurricane Katrina, propelled by 127-mile-an-hour winds, met landfall as a category four storm between Grand Island, Louisiana, and the mouth of the Mississippi River, at 6:00 a.m. on Monday, August 29. Katrina's wrath, mainly in Louisiana, Mississippi, and Alabama, would make it the most destructive natural disaster in American history—resulting, according to the Federal Emergency Management Agency (FEMA), in a total of $108 billion in damages and just over eighteen hundred deaths. A million people were left homeless in the storm's wake. Eighty percent of New Orleans, where levees holding back Lake Pontchartrain had failed, was underwater, as ten thousand of the city's displaced residents, mostly poor and predominantly African American, sought shelter in the Superdome, which lacked adequate accommodations. The city's Lower Ninth Ward, a poor, low-lying section, was decimated.

Bush was at his Crawford ranch when Katrina struck. The day after, he traveled to San Diego for a scheduled visit, then returned to his ranch for the night before heading back to Washington on the thirty-first. As Air Force One glided over New Orleans, Bush pressed his face to the window and looked out at the submerged, shattered city below, giving the impression of his detachment and indifference.

In general, the administration appeared flat-footed. Bush worried about federal overreach in alleviating the situation, convinced by aides that his powers were limited by law and unwilling to override them. Donald Rumsfeld, incurring Bush's ire, opposed sending in the Eighty-Second Airborne, which he feared would convey the image of martial law. Neither was Bush pleased with the sluggish pace of the FEMA relief effort, though as cameras rolled, Bush praised Mike Brown, the organization's director, by saying, "Brownie, you're doing a heck of a job." Bush conceded later that he should have "recognized deficiencies sooner and intervened faster."

To be sure, there was a dearth of local leadership in Louisiana. The city's mayor, Ray Nagin, and the state's governor, Kathleen Blanco, haplessly flailed during the crisis as chaos reigned throughout New Orleans, where reports of lawlessness, some later proven false, abounded. A committee comprised of eleven Republican members of the House of Representatives looking into the response to Hurricane Katrina issued a universally damning report in February of the following year. "Our investigation revealed that Katrina was a national failure," it read, "an abdication of the most solemn obligation to provide for the common welfare. At every level—individual, corporate, philanthropic and governmental—we failed to meet the need that was Katrina." But Bush was the media's main target as criticism of him and his administration hit a fevered pitch. During a star-studded NBC telethon to raise money for disaster relief, rapper Kanye West blurted, "George Bush doesn't care about black people." Katrina marked, Bush wrote later, "the worst moment" of his presidency.

As always, George H. W. Bush felt the sting as acutely as his son did, perhaps even more. It was a natural instinct for a family member of a politician to lament a public image that contradicted the person they knew, but even more so for Bush, who was one of the handful of living Americans who knew the presidency first hand. In a September 2 letter to his friend Hugh Sidey, Bush wrote:

> I am really down about the way the President has been attacked. Over and over again the networks attack him. First for being late in moving. Then for flying over Louisiana on the way back to Washington. Then on the snaillike pace of relief.
>
> One story in yesterday's NY Times suggested that the President didn't care because Katrina's main victims were African American . . .
>
> These attacks by politicians and news media remind me of those I weathered back in 1992 when Hurricane Andrew hit South Florida. We could do nothing right. Every person who lost a roof or a house aimed their fire at me personally.
>
> Now my son is under this kind of blistering, mean-spirited attack. People assign to him the worst possible motives. They do not recognize how complex the recovery is. They do not want to say that it was impossible to foresee the extent or even the types of the damage. And they also seem to feel that these gun toting, knife wielding thugs should get a pass.
>
> The critics do not know what is in 43's heart, how deeply he feels about the hurt, the anguish, the losses affecting so many people, most of them poor . . .

In fact, the elder Bush helped mitigate the situation when 43 once again asked him and Clinton, the "A-Team" by estimation of the *New York Post*, to head up disaster-relief efforts. "I see you've reunited your father and stepbrother," his mother quipped afterward. David Letterman, referring to a struggling NBC *Friends* sitcom spin-off, cracked in a *Late Night* monologue, "That's what they do now—whenever there's trouble, they send in Presidents Clinton and

Bush. Earlier today they arrived on the set of *Joey*." The Clinton-Bush team dutifully set out to the worst-hit areas, just as they had in Southeast Asia, and encouraged philanthropy from the private sector. By October the Katrina fund in their names had generated over $120 million, more than any other nonprofit raised except the Red Cross and Salvation Army.

The joint effort further deepened the bond between the forty-first and forty-second presidents. Earlier in the year, Clinton had joked that his unofficial adoption into the Bush clan showed the lengths the family would go "to have another president in the family," expressing hope that he could "get them to adopt Hillary." That may have been one Clinton too many. Anticipating the future ambitions of members of the Bush and Clinton families, 41 predicted in early 2006 that his amity with Clinton "might become a little strained when Hillary runs [for president]," adding, as if it were necessary, "I won't be her campaign manager." Still, if his friendship with Mr. Clinton came as a surprise to many in 2005, his vote for Mrs. Clinton eleven years later as she squared off against her Republican challenger, Donald Trump, in the election of 2016—Bush's first ballot for a Democratic presidential nominee—would be far bigger.

50

"THE FIRST JEWISH PRESIDENT"

KATRINA WAS JUST ONE OF George W. Bush's domestic woes as his second term played out. Cheney's chief of staff and closest adviser, Lewis "Scooter" Libby, resigned in the fall of 2005 after being indicted for perjury and obstruction of justice and sentenced to thirty months in prison resulting from a twenty-two-month CIA investigation around the leak of the identity of covert CIA operative Valerie Plame. In early 2006, Cheney accidently shot a hunting companion while hunting quail in Texas, angering Bush by not disclosing the accident for a day and a half. Gas prices soared and the economy, hampered by debt, sputtered.

Even Bush's domestic successes came under fire. No Child Left Behind effectively held public schools accountable for achievement, putting in place standardized tests in reading and math, as fourth-grade literacy and math skills and eighth-grade math skills climbed to their highest levels in history. Critics, however, held that schools were "teaching to the test," leaving students behind in other areas of study. The Medicare Modernization Act, signed by Bush at the end of 2003, gave the thirty-eight-year-old federal health insurance program a much-needed overhaul that offered sweeping new benefits including prescription drug coverage, but costs were far greater than anticipated. "By the end of 2005," Bush wrote later of the ebb of his presidency, "much of my political capital was gone."

Abroad, the rebuilding of Afghanistan was tough going. Insurgency and the failure of multilateralism in stabilizing the country necessitated an expansion in U.S. troops from twenty thousand to thirty thousand, along with a twofold increase in funding the effort. Likewise, the rebuilding of Iraq continued to be a drain on U.S. military and financial resources. In December 2005, Americans were warmed by the news that nearly two-thirds of Iraq's registered voters went to the polls in a free election to determine members of the nation's 275-seat parliament, many voters proudly lifting purple ink-stained index fingers as a symbol of the exercise of their democratic right. It was a positive sign that democracy had taken hold. But as violence continued to plague the country with no exit strategy in place for U.S. involvement, Americans were increasingly losing patience.

So was Bush. For two years, he thought about asking for Donald Rumsfeld's resignation. Twice he had been talked out of it by Cheney, Rumsfeld's closest ally in the administration. In the fall of 2006, though, Bush had firmly made up his mind—it was time for Rumsfeld to go. Cheney and Rumsfeld had been close since the Ford administration, when Cheney had served as deputy chief of staff under Rumsfeld, whom he supplanted as chief of staff when Rumsfeld moved to the Pentagon in his first turn as secretary of defense. "I disagree with your decision, Mr. President," Cheney told Bush. "But it's your call. You're the president." When Bush held firm, Cheney offered to be the one to break the news to Rumsfeld. "I owe Don an awful lot and he should hear [it] from me," he said.

Bush's first choice to replace Rumsfeld was Jim Baker—and it was 41 who paved the way. "[Forty-one] called and asked me, 'Would you consider this?'" recalled Baker. While the seventy-five-year-old erstwhile secretary of state and secretary of the treasury agreed to talk to 43 about the position, he ultimately demurred due to "the toll" he knew the job would take. Eventually, 43 settled on a like-minded alternative, Robert Gates, who had served as director of the CIA in 41's administration and went on to become dean of the Bush School of Government and Public Service at Texas A&M University, then president of the university at large. Gates was "strongly advocated" by 41 and Baker, whose pragmatic, multilateral approach to foreign policy aligned with that of Gates, an implicit departure from the unilateralism favored by Cheney and Rumsfeld. During a conversation with Bush about the job, Bush preemptively addressed the elephant in the room—the issue of Cheney's influence on his foreign policy going forward. "Cheney?" he asked on Gates's behalf, answering, "He is one voice, an important voice, but only one voice."

In fact, while Bush heard Cheney's voice throughout the balance of his second term, he often passed it over for the more moderate voice of Condoleezza Rice. Rice would become Bush's most prominent adviser, leading a chorus of other like minds including Gates and Hadley that one National Security Council official called the "revolt of the radical pragmatists." It showed in decisions Bush made contrasting those he had made in the first term. He was receptive to European involvement in Iraq and was willing to make diplomatic overtures—ultimately unsuccessful—to Iran and Syria relating to Iraq's future. Both were signs that 43 was taking a different course as he became more settled in the presidency and more confident in his own ability to read foreign affairs.

While the course 43 took in his second term, by extrapolation, looked a lot more like the pragmatic, diplomatic line his father would have walked in his place, they didn't see eye to eye on everything. In an interview with *Time* magazine just prior to the end of calendar year 2004, 41 was asked to express his hope for his son's second term. "Peace," he replied. "Clear, positive solutions to Iraq and the Middle East. What I'd like to see is the President's view of a Palestinian state." But the views between father and son on how to achieve peace differed discernably, representing generational and ideological differences. Conservative thinker William Kristol, who worked in the Bush 41 administration as chief of staff for Dan Quayle, said, "Bush the father was from a certain generation of political leaders and foreign policy establishment types. He had many years of dealings with leading Arab governments; he was close to the Saudi royal family." As president, 41 opposed Israeli settlements on occupied land—even if a less pro-Israeli stance further hurt his cause with Christian conservatives at home—and maintained close ties with Arab leaders, using his diplomatic facility and stature in the region after the Gulf War to help broker negotiations in nuanced regional disputes.

Bush the son had far greater affinity for Israel, sparked on a formative expedition as governor of Texas in 1998. During a helicopter tour of the country with Israeli foreign minister Ariel Sharon, Bush was taken by how small and vulnerable it was, a narrow slice of rugged land a shade larger than the state of New Jersey. *We have driveways in Texas bigger than that*, he thought. Just after becoming president, in a private March 2001 Oval Office meeting with Sharon, who had advanced to become Israeli prime minister, Bush gave him his unsolicited assurance that if necessary he was prepared to use U.S. force to protect Israel. Bush's determination intensified after 9/11. Bolstering his support among evangelicals, Bush stuck by Israel during a military conflict with the Hezbollah fighters in Lebanon lasting just over a month in the summer of 2006, despite international calls for a cease-fire and wide condemnation of Israel after a bomb attack on a Lebanese apartment complex killed twenty-eight civilians.

"How's the first Jewish president doing?" Barbara Bush greeted her son in a phone call at the time, reflecting skeptics in the Bush 41 camp who held 43's Israeli policy in question. "[My parents] were concerned—and the chattering class was concerned—about how I was handling the Israeli situation," 43 said, "because they just weren't that informed about my strategy, which was first and foremost making it clear to the Middle East that the United States would not abandon Israel—and that's necessary to get Israel to consider peace and for the Arab world not to exploit differences." He added, "Frankly, what I never got credit for among the chatterers was [convincing] Ariel Sharon that the two-state solution was the only solution for peace." In 2014, Bush looked at the state of Israel as vindication of his policy. "To the extent that the debate has changed considerably in Israel, the strategy worked," he said.

Arab leaders would occasionally use their relationship with Bush's father to try to change 43's policy—including using 41 as an intermediary to lobby on their behalf. "I suspect that some of these leaders would send messages trying to get me to change my foreign policy," 43 said. "I guess they didn't understand my nature. I had made up my mind." Among them was Egyptian president Hosni Mubarak, who had held the office since 1981 and had resisted 43's pressure for Egypt to hold open elections. Forty-three related a conversation in which the Egyptian president invoked his relationship with the elder Bush.

> With President Mubarak, for example, it was a fairly typical line, "Well, your father and I are very close," or, "Your father this," or, "Your father that." And, of course, he was uncomfortable with my change in policy. And basically, it was this:
> [Bush]: "Why don't you have open elections?"
> [Mubarak]: "Because the Muslim Brotherhood will win."
> [Bush]: "Oh, really, well why don't you find out why the Muslim Brotherhood will win and do a better job of appealing to the people as a change in foreign policy?"

Shortly after the exchange with Mubarak, 43 heard from his father that the Egyptian president was "unsettled." If 41 was relaying Mubarak's feedback as an indirect means of challenging 43's Middle East policy, it fell on deaf ears. "Look," he told his father, "this is what we're doing in order for there to be peace." Forty-three wasn't looking to him for advice or approbation. "When I would discuss this with Dad," he explained, "it wasn't like I was trying to feel my way along after I made policy decisions, it was that I believed it was the best course for the country. It unsettled people, but when you really think about what unsettled them, it betrays our values. 'How dare you think people can self-govern?'" He saw in 2011 Mubarak's ouster by popular uprising, spurred by the Arab Spring movement, as vindication of his stance toward the oppressive Egyptian president. "Turns out I might have been right because he then gets overthrown because he didn't listen to the people on the street," 43 said. For that matter, Bush saw the Arab Spring, the independent antigovernment movement that swept through the Arab states starting in 2010, as being "influenced" by the kindling of democracy in Iraq.

Still, dissention between Bush and his parents was a rarity. The bulk of their interaction was not about politics but family; it's what had always bound them together. Early in Bush's second term, Jean Becker, 41's chief of staff, placed a call to Marvin Bush in which she told him about a health matter affecting his mother that had eluded her doctors. "I think your mother is really struggling right now," she said. Less than fifteen minutes later, she received word that the president was on the line from a flustered receptionist who connected them. "He was calling to tell me that he was the oldest and if there was ever anything about his parents of concern, that I need to call him [first]," she recalled. Knowing the enormity of the demands he shouldered, she gently pushed back.

"Mr. President," she said, "as long as you're in the Oval Office, I'm not going to call you because you're so busy. I would call Andy Card, but I'm not going to call you."

"Yes, you will," the president said firmly. "You *will* call me."

Even during his eight years in the White House, George W. Bush never stopped being the First Son.

51
THE TOUGHEST DECISION

DURING HIS MOST TRYING DAYS in the White House, it wasn't the forty-first president that George W. Bush conjured but the sixteenth. Throughout his tenure, Bush read no less than fourteen biographies of Abraham Lincoln, who like him presided over the nation during a time of division and war. It wasn't unusual. All presidents draw from Lincoln. Bush's father, alluding to the fraternity of presidents, called him "everybody's favorite." The martyred Lincoln is the presidential standard, the one who suffered, risked, and ultimately achieved more than any, the one whom every president looks to for inspiration and guidance—but none more so than George W. Bush. "Always bear in mind," Lincoln wrote a friend in 1855, offering leadership counsel that would live well beyond the moment, "that your own resolution to succeed is more important than any [other] thing." As the 2007 calendar year opened, Bush's resolution to succeed in the Iraq War drove "the toughest and most unpopular decision" of his presidency.

The clamor around folding in Iraq, or at least substantially reducing American troops, grew daily in Bush's second term. In September 2006, Senate majority leader Mitch McConnell urged Bush to "bring some troops home from Iraq," warning him that his "unpopularity is going to cost us control of the Congress." His call was prescient. The 2006 midterm election, by and large a referendum on the war and Hurricane Katrina, proved disastrous for the GOP; Democrats picked up thirty seats in the House and six in the Senate, giving the party control of both houses of Congress for the first time since the election of Bill Clinton in 1992. In December, the Iraq Study Group, a ten-member nonpartisan panel appointed by Congress and led by Jim Baker and former Democratic representative Lee Hamilton, released its report recommending a heightened effort to train Iraqi troops and the withdrawal of all American combat units by the spring of 2008. "We're not winning the war, we're losing," Bush confessed to the *Washington Post* later the same month, contradicting his insistence before the election that "absolutely, we're winning."

Wary of the lessons of another war president, Lyndon Johnson, Bush had resisted micromanagement of the war, generally deferring decisions on war strategy to Rumsfeld and his commanders. But early in 2007, rejecting the advice of Condoleezza Rice and his top military advisers, Bush took the war in another direction. He doubled down.

In a televised speech on January 10, he announced "a new way forward," ordering the deployment of five additional brigades—more than twenty thousand troops—to Iraq, principally in Baghdad, as a means of curtailing insurgent violence and stabilizing the country. The reaction was brutal. A *Los Angeles Times*/Bloomberg poll conducted several days later showed that only 34 percent of Americans agreed that the surge in troops would "now allow the U.S. and Iraqi forces to defeat the insurgency and win the war in Iraq."

Shortly afterward, he called to the Oval Office a group of aides who had worked on the strategy and pointed to a bronze bust of Lincoln. "Thinking about what Lincoln went through lends some perspective to things," he told the group of twelve. "You know, I am no Lincoln, but I am in the same boat." He then gave his reasoning behind the decision. "People say, 'Bush needs to see the world as it is,'" he said. "Well, I've been here for six years now and I see the world as it is, maybe better than most. The world as it is, that world needs America to lead. Because nothing happens if we don't lead."

Congress resisted the surge, putting forward a bill that continued funding for the war contingent on a substantial troop withdrawal later in the year. Bush quickly vetoed it, signing a bill in May that allowed for funding of the war and no deadline for troop withdrawal. The surge went forward. The same month, 146 American soldiers died in Iraq, the highest rate in the war's four-year history. But in short order there were positive signs that the strategy was working, as the increased presence of U.S. troops meant greater interaction with Iraq's civilian population and the creation of coalitions resulting in intelligence on brewing insurgent activity. It paid off. "We are no longer headed toward a catastrophic defeat," wrote the *Wall Street Journal* in March, "and may be on the path to a remarkable victory." In May of 2008, the number of U.S. troops killed in action was nineteen. A Pentagon report a month later reflected a 40 to 80 percent decline in violence after the surge including Iraqi civilian deaths, which fell by over two-thirds during the same period. In the summer of 2008, when Bush ordered the withdrawal of the five brigades that composed the surge in troops, fatalities in Iraq were at their lowest since the war had begun five long years earlier.

It is inevitable that 43's Iraq policy will be held up in history to his father's, along with the supposition, based on extrapolation, that the elder Bush would have played a different hand in Iraq if he had stood in his son's place. Regardless, 43 believed that history would eventually reflect well not only on his father's policy in Iraq but also, ultimately, his own. Historians, he contended in his postpresidency, would judge his father's decision not to invade Iraq after the liberation of Kuwait as "just right," an assessment with which he agreed. But he also believed that 9/11 "changed the equation" for Iraq, necessitating a change in the policy paradigm from that of his father. "I'm very comfortable that when people fully analyze my decisions in the proper context," he asserted, "they will understand why my foreign policy—not in the principles of U.S. leadership but in the application—was different."

Forty-one had alluded to the mistakes his son had made in the war, telling historian Jon Meacham, "The big mistake that was made was letting Cheney bring in his own State Department. I think they overdid that. But it's not Cheney's fault, it's the president's fault." The buck, in other words, stopped with his son. But asked in May of 2014, in his modest office at Walker's Point, if the Iraq War was worth its sum in blood and treasure, 41 responded, "Well, in the final analysis, it will be seen as the right thing to do. Saddam Hussein was a bad guy." Pausing, he added, "I think history will be okay with it." Would he have made the same choices his son made in Iraq? "Hard to tell," he replied laconically, "but I think so." Just as hard to tell was whether the elder Bush was speaking as a dispassionate former president or a loyal, loving father supporting the toughest decision of his son's onerous presidency.

Upon the death of President Zachary Taylor in 1850, Lincoln wrote, "The Presidency, even to the most experienced politicians, is no bed of roses; and Gen. Taylor like others, found thorns within it." While the surge had shown what Lincoln may have called Bush's "own resolution to succeed," Bush would find little absolution in it as 2008 and his presidency wound down. There were more thorns yet to come.

52

LAST DAYS

"IS THIS THE WORST CRISIS since the Great Depression?" President Bush somberly asked Ben Bernanke, the chairman of the Federal Reserve bank in late September 2008.

It was a reasonable question. A spiral in the American housing market had precipitated a series of calamitous events that had sent the economy, heavily steeped in mortgage-backed securities, reeling. Corporate behemoths began falling like dominoes. Mortgage giants Freddie Mac and Fannie Mae collapsed and, deemed "too big to fail," were taken over by the government on September 7; a week later, Lehman Brothers filed for Chapter 11, becoming the largest bankruptcy in American history; the following day, AIG, the world's largest insurance company, took an $85 million bailout from the government in exchange for an 80 percent stake. Other corporate aftershocks would follow, including the closing of Washington Mutual Bank, the largest bank failure in history, and the cratering of the domestic automobile industry, as the economy wobbled on the brink of free fall. Bush's presidency was, in Laura's words, bookended by two Septembers, "September 11, when the nation was devastated by forces without, and the September of 2008, when it was threatened to collapse from within."

"Yes," replied Bernanke as he considered Bush's question, "In terms of the financial system we have not seen anything like this since the 1930s, and it could get a lot worse."

In fact, it might have been *worse* than the Great Depression. Bernanke, a foremost scholar on the Depression, later stated that "September and October of 2008 was the worst financial crisis in global history, including the Great Depression," adding that, of the thirteen "most important financial institutions in the United States, twelve were at risk of failure within a period of a week or two."

On September 19, as the crisis escalated, Bush addressed the nation from the Rose Garden. "There will be ample opportunity to debate the origins of this problem," he said. "Now is the time to solve it." But blame was heaped inevitably on Bush. As he knew well, it was part of the territory for a president, especially for a two-termer at the end of a reign rife with crises. Even well before the market nosedived, Bush had felt the rabid bite of capricious public opinion. The previous April, seven years after Bush garnered the highest approval rating in presidential history at 90 percent, a Gallup poll reflected that 69 percent of Americans disapproved of his performance, a low since Gallup began posing the question in 1938. "To say that Bush is unpopular only begins to capture the historic depths of his estrangement from the American public," wrote the *New York Times Magazine* a little over a week before the crisis erupted. "He is arguably the most disliked president in seven decades." If it got to Bush, he didn't let it show. He made the decisions he thought were right, with little regard for public approbation. "I would like to be remembered as president as somebody who did not compromise principles in order to be loved or liked," he said later.

Was Bush to blame for the economic crisis? The market had erupted due to the cumulative effects of a trend toward financial deregulation that had started in the 1970s, cresting during the Clinton administration, which helped clear the way for the trading of derivatives, complex financial instruments that would become the chief culprit of the crisis. Additionally, Clinton had pushed to repeal the 1933 Glass-Steagall Act, relaxing restrictions on commercial banks to act as investment banks. Though Bush had advocated unsuccessfully for greater regulation of Fannie Mae and Freddie Mac, the administration also, like Clinton's, promoted greater home ownership while generally upholding the climate of deregulation that saw overheated lending institutions run amok, ultimately sinking under the weight of loan defaults as the housing market crashed.

While Bush had channeled Lincoln during the surge, it was the most iconic Democratic president he looked to as he approached the financial meltdown. "If we're looking at another Great Depression," he told aides after his meeting with Bernanke, "you can be damn sure I'm going to be [Franklin] Roosevelt, not Hoover." The federal government, in other words, would step forward to manage and mitigate the crisis as much as possible. To that end, Bush pushed for the creation of the Troubled Asset Relief Program—TARP—which authorized $700 billion in federal money, the largest federal expenditure in American history, to buy up toxic assets from banks to ensure their solvency and stabilize the economy. Congress's rejection of the proposal sent the Dow Jones Industrial Average spiraling 778 points, the largest one-day drop in its history. Four days later, Congress hurriedly passed a revised version. The Department of the Treasury, Federal Deposit Insurance Company, and Federal Reserve scrambled to reduce the strain on banks and other major corporations with guarantees, funding access, and capital. Heading up Bush's efforts was Henry Paulson, whom Bush had lured from the top spot at Goldman Sachs to become his secretary of the treasury in 2006. Bush wrote later that Paulson's efforts during the crisis rivaled "those of Henry Morgenthau under FDR or Alexander Hamilton at the founding of the country."

The efforts of the administration were enough to stave off economic free fall but not a deep recession. By the end of 2008, the Dow Jones Industrial Average was down by over a third, and unemployment rose to 7.2 percent—a 2.3 percent lift over the previous December. That the ramifications could have been far worse came as little consolation to Americans who struggled with the effects of a crippled economy.

As the situation worsened, Bush kept the two presidential candidates, Republican John McCain and Democrat Barack Obama, abreast of developments. Due to the "historic" nature of the crisis, McCain decided in late September to suspend his campaign to involve himself in the negotiation of federal bailouts and further monitor the situation from Washington, suggesting that Obama was ignoring the matter's severity by remaining on the stump. The move backfired. "It's my belief that this is exactly the time that the American people need to hear from the person who in approximately 40 days will be responsible for dealing with this mess," Obama said. "It's going to be part of the president's job to deal with more than one thing at once." It proved a turning point in the campaign. As a GOP strategist said afterward, "We don't need to quit the game. We need to *change* the game." Even beforehand, a *Washington Post*–ABC News poll showed Americans believed Obama would do a better job on the economy by a margin of two to one, a gap McCain wasn't able to overcome. Campaigning on a platform of hope and change, an implicit and often direct indictment of the Bush administration, Obama managed a decisive win over McCain on November 4, with just under 53 percent of the popular ballot, taking 365 electoral votes to McCain's 173. The first African American had been elected to the nation's highest office.

Taking his father's lead with Clinton, Bush handled the presidential transition with grace. He invited his successor, fifteen years his junior, to the White House for a series of cordial meetings. On January 7, he hosted an unprecedented White House reception for the president-elect and all the former presidents—Carter, Clinton, and his father—to welcome Obama into their small club, soon to be integrated. Standing side by side in the Oval Office, which Obama would inherit in less than a fortnight, the quintet looked like an animated would-be Mount Rushmore. "One message that

I have, and I think we all share," Bush told Obama before the five men adjourned for a private lunch, "is that we all want you to succeed. Whether we're Democrat or Republican, we care deeply about this country."

Bush's last days were filled with tasks big and small. The economic crisis was the main priority, but there were also last-minute details to attend to, including presidential pardons. Cheney argued persistently that his former chief of staff, Scooter Libby, should be pardoned for any wrongdoing in the investigation into compromising the identity of CIA covert operative Valerie Plame during Bush's first term. "I felt we were leaving a good man wounded on the battlefield," Cheney said later. "It was a major strain on our relationship, obviously a major source of friction. The president had the power to fix it and make the right decision and chose not to." Bush had commuted Libby's thirty-month prison sentence several months after it was handed down, but told Cheney he was unwilling to overturn the jury's verdict.

In the wake of a heated exchange with Cheney over the matter, Bush uncharacteristically wrestled anxiously over the decision during what was to be a celebratory last weekend at Camp David with family, friends, and a few staffers in mid-January. "Don't let this be a pall over your last days as president," Condoleezza Rice counseled Bush. "You deserve better . . . and this shouldn't be the way that you spend your last hours as president." It was emblematic of the change that had taken place in Bush's second term, as Rice supplanted Cheney as Bush's chief adviser and distance grew between the president and vice president. Regardless, Bush would leave the White House believing that Cheney had served him well. "I understand the way the system works, somebody has to be the bad guy," he said of his vice president, adding, "When I made decisions he didn't agree with, he didn't immediately trod out into the public and leak. He was always straightforward; there were no hidden agendas."

On the morning of January 20, 2009, Bush went to the Oval Office as president for the last time. Observing the tradition that Reagan had begun by leaving a handwritten note for 41, 43 wrote out best wishes to his successor.

Dear Barack,

Congratulations on becoming our President. You have just begun a fantastic new chapter in your life.

Very few have had the honor of having the responsibility you now feel. Very few will know the excitement of the moment and the challenges you will face.

There will be trying moments. The critics will rage. Your "friends" will disappoint you. But, you will have an Almighty God to comfort you, a family who loves you, and a country that is pulling for you, including me.

No matter what comes, you will be inspired by the character and compassion of the people you now lead.

Sincerely,

George

After placing the letter in a manila White House envelope and sticking a Post-it Note on it, on which he wrote "44," Bush left the Oval Office, much like he would the presidency, with little reservation and without looking back.

A little before 10:00 a.m., he arrived at the residence where he and Laura stood outside on the North Portico in 28-degree air to greet Barack and Michelle Obama as their limousine glided through the west gates. "Welcoming Barack Obama was a historic moment and I was looking forward to it," Bush recalled. "I was looking forward to going home—eight years is a lot." As he did almost invariably during the momentous times of his public life, Bush thought about his dad and wondered "what he must have been going through" when he went through the same ritual with Clinton under different circumstances sixteen years earlier—and it occurred to him that no one at the time thought to ask.

His father was on hand at the Capitol with Clinton and Carter to watch, along with 1.8 million others who made up the record crowd that stretched from the south side of the building well beyond the Washington Monument. Fending off the winter chill in a trapper's hat and purple scarf that looked as though it had been snatched from a Minnesota Vikings fan, 41 walked gingerly with Barbara beside him, his legs suffering from a vascular disorder that would soon render him unable to walk. It would be the last inauguration the elder Bushes would witness as guests.

After bidding the Obamas farewell from the Capitol's front steps, George W. and Laura boarded Marine One, the presidential helicopter, where George H. W. and Barbara awaited them. "The love of the Bush family had come full circle," Laura wrote later, "the pride George had felt for his parents, they felt in return for their son. They too had made this journey we were about to begin, and had found unexpected joys in the years beyond." Boos could be heard from the dispersing crowd as the helicopter thwacked over the Mall toward Andrews Air Force Base. Emotions around the Bush presidency still seethed—the wars in Iraq and Afghanistan, Hurricane Katrina, the financial crisis—even as the jubilant masses invested their hope in Barack Obama. But as he left Washington, looking down at his nation beneath him, Bush could feel due satisfaction around what *hadn't* happened—no major act of terrorism had befallen America after 9/11. He had kept the nation safe just as he had vowed to do.

PART VIII

The Last Republicans

53

"IT WASN'T ME"

IN 2007, TWO YEARS BEFORE his son left the presidency, George H. W. Bush wrote an essay for the Kennebunkport Conservation Trust, a nonprofit environmental group working to preserve Maine's coastline, in which he waxed reflectively. "I have had an exciting, full life," he wrote.

> With the help of friends . . . across our land, I climbed the highest political mountain in the world. I knew good times and times of great disappointment, but when my body and soul cried out for a calm and relief my thoughts came right back here to Kennebunkport.

George W. Bush may have written the same words save for the last, which for him would have been *Texas*. A day after leaving the presidency, the forty-third president awoke with Laura before the sun at their Crawford ranch. Nearly two decades of his life had been spent on the East Coast—in New England for prep school, college, and business school; in Georgia for flight training; in Washington to help his father achieve the presidency, then for his own turn in the White House. Now he was back home in his beloved state. For good.

After reading the Bible, he scanned the headlines of the *Dallas Morning News* and *Waco Tribune-Herald* and asked himself instinctively of the issues of the day, "What are we going to do about this?"

then realized, "It wasn't me. It was my successor." They were all Barack Obama's problems now. The sense of responsibility that had become "ingrained" in Bush for eight years began to dissipate.

After breakfast, he gathered up his Scottish terriers, Barney and Miss Beazley, hopped in his Ford pickup truck, drove to his small office on the ranch grounds, and began writing about the presidency he had left the day before, generating material for the book that would become his 2010 bestselling memoir *Decision Points*. The process of writing the book was cathartic for Bush, helping him to channel his inherent restlessness while providing "finality" to his presidency, allowing him to put it neatly behind him. "By focusing on your presidency," he said, "it helps you to realize that you're no longer president."

As with nearly every other epoch of his life, his father had been there before him, offering a model. "He moved on with his life," Bush said of his dad. "He didn't linger. I learned from him that when it's over, it's over." George W. was content to be out of the presidential spotlight, and, given his respect for the dignity of the office, to let his successor find his way in the role without carping from the sidelines. It was, of course, easier for him to move on than it had been for his father. While he had left office with a low approval rating, he had been reelected and compelled to leave the White House in adherence to the Constitution's Twenty-Second Amendment, limiting a president to two terms in office. It was easier in another sense, too. More than 41, his time in politics was "a chapter," not his life. He had no desire to be back in the arena—or "the swamp," as he called it—and after giving his best in two White House terms, he felt he had nothing left to prove.

The Bushes tapped seamlessly back into a community of loyal friends in Dallas, where they moved into a $3 million 8,501-square-foot home in Preston Hollow, not far from the home where they lived with the twins prior to moving into the Texas governor's mansion in 1995. Friends had always been a staple in the Bushes' lives, including their White House years, helping to ground them. Laura recalled a visit from several of their friends early in Bush's presidency, in which, during a tour of the West Wing, one of them said, "Man, Bush, I can't believe I'm here in the Oval Office," whereupon he looked quizzically at Bush as the group chortled at the whims of fate.

As always, Bush stayed fit, mountain biking and golfing, a game that helped him to focus and practice discipline. Like his father, he took his turn on the lucrative speaking circuit, something he anticipated before leaving the presidency when he announced his intention to "replenish the ol' coffers." For fourteen years in the political arena, he had spoken at gatherings for free; now he was commanding $100,000 to $150,000 in honoraria for light duty behind a microphone. "It didn't take long to adjust," he said jokingly. In his first two years after leaving the presidency, Bush would rack up an estimated $15 million in speaking fees. As his mother quipped of her eldest son's riches within the Bush family, "W. could buy and sell us all."

He and Laura immersed themselves in planning the George W. Bush Presidential Center, a twenty-five-acre complex on the grounds of Southern Methodist University, Laura's alma mater, complete with a fifteen-acre urban park planned by Laura featuring native Texas vegetation. Among other things, the library would house a museum, an archive for the eighty terabytes of digital records of the Bush administration, and the George W. Bush Institute, focusing on the areas of interest to Bush in his presidency including education reform, support of veterans, the international empowerment of women, and the fostering of human freedom. (Additionally, Bush would later team up with Bill Clinton to inaugurate the Presidential Leadership Scholars program, aimed at nurturing leaders from the nonprofit, military, public, and private sectors by drawing on the lessons of presidential leadership through their own presidential libraries and those of 41 and Lyndon Johnson.) The Bush Institute was a means of keeping Bush's hand in policy. "I don't view the post-presidency as burnishing any legacy," he said in 2013, just prior to the library's opening. "The challenge when you're a former president is how do you use your God-given talents? People will tell you that you deserve to rest. Well, you never deserve to rest."

Yet Bush did gravitate toward more reflective pleasures. Throughout his life, he had a way of surprising those around him. His passionate pursuit of oil painting several years into his post-presidency took many aback, including Laura. "If someone said, 'One day you will be writing a foreword for a book that includes George's paintings,'" Laura wrote in the foreword to her husband's 2016 bestselling book, *Portraits of Courage: A Commander in Chief's Tribute to America's Warriors*, "I would have said, 'No way.'" Bush often exuded, even cultivated at times, the image of a philistine more apt to clear brush than clean brushes. Nonetheless, Bush found himself drawn to the canvas after Yale historian John Lewis Gaddis recommended Winston Churchill's book *Painting as a Pastime* during a 2012 visit. His diligent attempt at honing his artistic ability, including frequent lessons from local Dallas talent, paid off. Writing for the *New York Times Book Review*, Jonathan Alter called Bush's work "evocative and surprisingly adept." Even in his midsixties, Bush was expanding his horizons.

In late spring of 2009, less than five months after his son left the White House, George H. W. Bush had a simple wish for his eighty-fifth birthday—the same one he had for his seventy-fifth and eightieth, and would for his ninetieth: He wanted to make a parachute jump. On June 12, his eighty-fifth birthday, Bush leaped from a plane over Kennebunkport at just under eleven thousand feet in a tandem jump with a member of the Army's Golden Knights, plunging toward earth at 100 miles per hour, before landing safely on the grounds of St. Ann's Church, which Barbara joked would be the perfect destination if the jump had gone wrong, allowing for a quick burial. After being greeted by his wife of sixty-four years, Bush declared, "It's a great, exhilarating feeling. I don't feel a day over eighty-four."

But age was catching up to the Bush patriarch. By the next year, vascular Parkinsonism had rendered Bush's legs unable to move, relegating him to a wheelchair or motorized scooter. It was a dramatic change for Bush, who had led life in constant motion. "I kept hoping that things would change for the better and that I'd get back in the game," he said in 2012, "but you just face reality and make the best of it."

Later the same year, on November 23, Bush was admitted to Houston Methodist Hospital after a chronic cough was diagnosed as a case of bronchitis. The family rallied around him. George W. and Laura visited and found him in good spirits. When George W. asked him how he was doing, he replied, "It's not the cough that carries you off, it's the coffin you go off in." In December, as his condition declined to pneumonia, George W. and Laura, anticipating the worst, returned to the hospital with Barbara and Jenna, the latter who was five months pregnant with her first child. Her grandfather put his hand on her protruding stomach. "There's death and there's new life," he said, as the family quietly wept.

Death came close just days later. On December 23, Bush was admitted to the intensive care unit where he would remain through the holidays. After one particularly harrowing night, Bush awoke the following morning and asked Jean Becker, "Did I almost buy the farm last night?" In fact, his condition was more dire than the public knew; when Becker later asked Bush's doctor to rate the seriousness of Bush's condition between one and ten, she replied, "Eleven." But against all odds, Bush left the hospital on January 14, weak but in stable condition. George W. believed it was family that pulled his father through, just as it had sustained him in his darkest hours—in World War II, as he paddled desperately in the Pacific to evade

Japanese capture after being shot down; upon losing Robin to cancer; after his presidential defeat in '92.

As he began his slow convalescence at home in Houston, his eldest son reminded him that 43's presidential library was opening in April. "I'll be there, son," he said. The old man made good on his promise. On April 25, 2013, he was in attendance in Dallas, along with 43, Jimmy Carter, Bill Clinton, and Barack Obama, to cut the ribbon on the nation's thirteenth presidential library under the auspices of the National Archives. It was the first time the quintet had been together since 43 hosted them at the White House, welcoming Obama two weeks before his inauguration. Now Obama, over four years into the job, stood gray-haired among four peers whom he called "a support group," having carried the burdens of the most powerful office in the world just as they had.

Support is what the group gave George W. Bush as they each spoke during the ceremony. Forty-one spoke only briefly, thanking all those who brought the occasion about but stopping short of expressing his teeming pride in his son due, likely, to his worry that it would result in a flood of tears. Though he offered the fewest words, he got the greatest response from the audience, which offered a groundswell of applause. "Too much?" he asked his son after delivering his cursory remarks. Carter, who had been openly critical of 43 during his presidency—much to the ire of 41—found common ground with him around his commitment to PEPFAR. Clinton also lauded 43's efforts on AIDS relief while talking affectionately about his friendship with the Bush family, and Obama praised, among other things, Bush's "strength and resolve" after 9/11, efforts to pass comprehensive immigration reform, and "commitment to reach across the aisle." There were no mentions of the war in Iraq, Hurricane Katrina, the financial collapse, or other issues that complicated Bush's legacy. Those were topics for another day. Instead, the occasion was a time to express gratitude to Bush, who, like the other men who graced the stage, loved his country and had done his best.

Afterward, the presidents adjourned with their wives and other VIPs to a private reception in the library's courtyard. At its center were nine-foot bronze statues depicting the Bush father and son: 43 standing confidently, his hands at his sides, looking as though he is about to say something; 41 standing mutely a half a step behind him, his left hand in his pocket, offering steady support—as always.

54

TRUMPED

AS IF HE NEEDED IT, Donald J. Trump was encouraged early on to consider politics by a man he had never met, Richard Nixon. After Nixon's wife, Pat, saw Trump on *The Phil Donahue Show* in December 1987, the former president, now largely rehabilitated as a foreign policy and political sage, wrote Trump a short missive from his office in New York City's Federal Plaza.

Dear Donald,

I did not see the program, but Mrs. Nixon told me that you were *great* on the Donahue Show.

As you can imagine, she is an expert on politics and she predicts that whenever you decide to run for office you will be a winner!

With warm regards,

Sincerely,

RN

Not all former Republican presidents, however, saw Trump's charm. Sometime in the 1990s, well before Donald Trump reached a new level of fame through reality TV, George H. W. Bush found himself waiting in the noncommercial terminal at Boston's Logan Airport while the private plane on which he would travel was being serviced for a mechanical issue. As he passed the time by reading a newspaper, Jean Becker, his chief of staff, was approached by an airport attendant who told her that Trump Air was about to land. "Would President Bush like to say hello to Mr. Trump?" she asked Becker, who dutifully presented the opportunity to Bush. "God, no," he replied as he peered over his reading glasses. "Is he coming in here?" Becker indicated that she didn't know, whereupon the former president raised the newspaper to shield his face in the event Trump and his entourage passed. Aside from a grip-and-grin photo op with Trump at a New York City fundraiser in 1988, the same year Bush roundly rejected Trump's unsolicited overture to be considered as Bush's running mate when it was brought to him by Lee Atwater, it was the only time Bush found himself in the orbit of Donald Trump. No such luck for Jeb Bush, for whom Donald Trump would become a political gadfly twenty-seven years later.

After Barack Obama's decisive 2012 reelection victory against GOP challenger Mitt Romney, pundits began looking ahead at the political landscape in the next presidential election with the prospect of a matchup between the Bush and Clinton political dynasties—Jeb Bush going head-to-head with Hillary Clinton—a familial reprise of the '92 presidential campaign. Both Jeb and Clinton were logical establishment candidates, the former boasting two solid terms as Florida governor, a high policy profile on issues like immigration and school choice, and a reliably conservative record. "[Jeb Bush] is going to be a very effective candidate if he runs," surmised Charlie Black, a Republican lobbyist, reflecting the views of untold members of the GOP establishment, "because he is going to talk about the future without backing down or pandering to the Tea Party side." But Jeb, who had said he would run only if he could do so with "joy" in his heart, was on the fence about a campaign for the White House. Part of his concern was the zeitgeist. "I think our politics is a mirror of our culture," he said astutely in 2014, portending the challenges he would face. "It's us. And what we are today is less humble, more arrogant, more coarse, more vulgar, more accepting of crap. We've moved down this slow slide. Politics is more like 'The Kardashians.'"

His parents were split on the matter of him mounting a challenge for the presidency, with his father strongly in favor and his mother declaring on the *Today* show in 2013, "We've had enough Bushes." George W. aligned with his father. Hillary Clinton would set a precedent for a dynastic choice. "What's the difference if it's Bush, Clinton, Bush, Obama, Clinton," George W. asked rhetorically of presidential succession, "or Bush, Clinton, Bush, Obama, Bush?" Still, while George W. hoped Jeb would run, he didn't pretend to know his brother's intentions on a presidential bid, nor did he expect Jeb would consult him. Jeb had been loyal to his brother even when members of their own party turned their backs on 43. "Till death do us part," he told CNN in 2010. But it didn't mean the brothers were political confidants.

Surveying the field of GOP "would-be's," in early 2015, 43 said he was impressed by New Jersey governor Chris Christie, whom he felt to be underrated, and was complimentary of Kentucky senator Rand Paul, whom he characterized as "interesting" and far more thoughtful than the Tea Party extremists. He believed that Paul Ryan was "too stiff" to be a successful national candidate, adding, "That's one guy you won't find farting on Air Force One." But there was no doubt in his mind that his brother would be the best candidate for the party, and the best president for the nation.

After much hemming and hawing, Jeb quietly decided over the Thanksgiving holiday in 2014 in Mexico with Columba and their three children, George P., Noelle, and Jeb Jr., that he would make a go of the presidency. In the months afterward, the Bush machine, dormant since the aughts, went into overdrive. Barbara Bush, retracting her initial pronouncement said, "What do you mean there are too many Bushes, I changed my mind." Later she elaborated. "I thought that in a country this size, if we couldn't find more than three families . . ." she said. "But now I've decided that, truthfully, we need him. We need someone who doesn't think compromise is a dirty word; we need someone who isn't flip-floppable; we need someone who immediately will talk to Democrats *and* Republicans." The smart money went with Jeb, who seemed to have every advantage. Eventually, his campaign would amass a war chest of $150 million. In any other year all that may have been enough.

Jeb made his candidacy official on June 15, 2015, stating at a Miami rally that the country was "on a bad course," and pledging to "campaign as I would serve, going everywhere, speaking to everyone, keeping my word, facing the issues without flinching, and staying true to what I believe." He continued, "I will take nothing and no one for granted. I will run with my heart. I will run to win."

The following day, on June 16, Donald Trump glided down the shiny escalator of Manhattan's gilded sixty-eight-floor Trump Tower with his wife, Melania, to declare his own intentions to seek the presidency. He said nothing about his heart but, like Bush, was taking nothing for granted. Less a rally than a rant, Trump set the tone for his populist campaign. "Our country is in serious trouble," he said, positioning himself—a successful, rich outsider who owed nothing to a broken, impotent Washington—as the only one who could fix it. His decidedly dystopian view heaped blame on foreign influences, necessitating mass deportation, bans on Muslim immigration, and shutting down mosques. In the speech's most indelible line, Trump called many Mexican immigrants "rapists," adding as an aside, "And some, I assume, are good people." He promised to remedy illegal Mexican immigration by building "a great, great wall on our southern border." The kicker was that Mexico would pay for it.

The speech raised hackles. Citing his "derogatory statements" regarding immigrants, NBC—the network behind Trump's high-rated reality game show, *The Apprentice*, which debuted in 2004—ended its "business relationship" with Trump later the same month. Nonetheless, Trump, due largely to his fame from the hit show, entered the race with unmatched name recognition and by July 1, yielded 12 percent of likely Republican voters to Jeb's 19 percent in a CNN/OCR poll, with other candidates, including Florida senator and one-time Jeb protégé Marco Rubio, Texas senator Ted Cruz, brain surgeon Ben Carson, and Chris Christie pulling single digits.

But as the election cycle played out, Trump continued to gain traction, tapping into a raw nerve of anger that permeated much of forgotten America. By turns channeling circus showman P. T. Barnum, professional wrestler Gorgeous George, and ruthless political-fixer and one-time Trump mentor Roy Cohn, Trump inexplicably dominated the news coverage. George H. W. Bush had surged ahead in the 1988 GOP primaries when his chief rival, Bob Dole, uncivilly snarled that Bush should "stop lying" about his record. But none of the traditional rules of politics seemed to apply to the New York billionaire. Trump's offhand and off-the-wall hyperbole and tweets frequently had Republicans slapping their heads and Democrats slapping their knees, though it didn't seem to matter; Trump hurled invective with relative impunity, saying of John McCain, a revered Vietnam POW, "He's not a war hero because he was captured," and of Fox News' Megyn Kelly, after she aggressively moderated a GOP debate, she "had blood coming out of her eyes, out of her wherever." At a campaign rally in January, he proclaimed, "I could stand in the middle of Fifth Avenue and shoot somebody and I wouldn't lose voters," words that themselves might have sunk candidates in other times but that figuratively captured Trump's ostensible invulnerability. Yet his antics were a boon for the struggling news media outlets. By March, he had yielded an estimated $2 billion in free media.

Trump's attacks against Jeb were especially damaging. During a series of GOP debates in the fall, his characterization of Jeb as "low energy" hit home. Suddenly, Jeb's own campaign slogan—*Jeb!*—mocked him. His father's son, Jeb had obeyed the rules and behaved civilly—decorum, along with his wonkishness, that seemed woefully out of fashion. When he did hit back, as he did in a mid-December debate, claiming that Trump's ban on all Muslims would preclude a coalition to destroy ISIS and presciently warning that Trump "is a chaos candidate and he'd be a chaos president," his blows wreaked little havoc; Trump's eye-rolling reactions got just as much attention, his reality-star wattage blinding not only Bush but the other candidates as well.

By the time primary season heated up in January, polls showed Jeb relegated to the field's also-rans as Trump surged to the top of the heap trailed by Cruz and Rubio. The January Iowa caucuses later the same month gave Cruz 37 percent of the vote over Trump's 35 percent, as Jeb limped in with under 3 percent. Jeb did a little better in the New Hampshire primary in early February, reaping 11 percent of the vote—with Trump topping the field with 33 percent—though his fourth place showing wasn't enough to reflect a momentum shift in his campaign. The South Carolina primary, on February 20, became do or die for Jeb. Pulling out all the stops, he tapped his older brother in Texas to stump with him.

The call came as a last resort. Jeb struggled to be his own man in the campaign, but just as 41's legacy had hovered over George W. Bush's campaign in 2000, George W. shadowed his brother in 2016. Torn by fraternal loyalty and political practicality, Jeb had stumbled awkwardly over the issue of the war in Iraq early in the contest before categorically distancing himself from his brother's decision by stating, "Knowing what we know now . . . I would not have gone into Iraq." Regardless of the war's toxicity, 43 remained popular in South Carolina, which had saved his own campaign in 2000 and was where Jeb hoped his brother might have a rub-off effect on him.

To that point, 43 had respectfully kept his distance in his brother's campaign, limiting his involvement to appearing at fundraisers for his brother and calling him "a couple of times" with the names of potential supporters who expressed interest in helping him. But now, a few days before the primary, the former president, Laura by his side, was happy to do his part, speaking ebulliently on Jeb's behalf for twenty minutes at a North Charleston rally. "These are tough times," he said.

> And I understand that Americans are angry and frustrated. But we do not need someone in the Oval Office who mirrors and inflames our anger and frustration. We need someone who can fix the problems that cause their anger and frustration, and that's Jeb Bush.

The event proved a double-edged sword for Jeb, pulling in a crowd of three thousand, far larger than he had seen in earlier stops, but highlighting his own contrast with his brother. As one attendee, buying into Trump's characterization of Jeb, put it, "I'm a big W. fan . . . Jeb would be fine, if he could get some energy."

Trump seemed to relish taking on not only Jeb in the campaign but also George W., who Trump had implied was responsible for 9/11 and suggested should be impeached for sending troops to Iraq. During a South Carolina debate, just prior to George W.'s campaign appearance, Trump had railed against 43's administration, claiming, "They lied. They said that there were weapons of mass destruction and there were none, and they knew there were none." Jeb rose passionately to his brother's defense. "While Donald Trump was busy building a reality-TV show," he responded, "my brother was busy building a security apparatus to keep us safe. And I'm proud of what he did." Still, Jeb's more forceful debate performance, along with his brother's campaign reinforcement, was too little, too late. Trump took the primary with ease, earning 33 percent of the vote, with Bush distantly tying for fourth place with just 8 percent.

The writing on the wall, Jeb pulled out of the race after the results came in. "In this campaign, I have stood my ground, refusing to bend to the political winds," he said in an emotional address to supporters. The same winds blustered Trump on to secure the Republican nomination. He accepted the party's bid at the GOP National Convention, held in Cleveland, where the Bush family was conspicuously absent.

Forty-one was stung by his son's defeat, just as he had been after Jeb's failed gubernatorial effort in 1994. He rejected the notion that it was due to "too many Bushes," attributing it instead to "timing." "I think he would've been the best president if elected," he said. "So, it was a personal disappointment. He went all out. He did everything he could do. It just wasn't meant to be."

Forty-three believed Trump's success at his brother's expense was fueled by the anger resulting from a moribund economy. The reality star's antiestablishment fury better fit the country's mood, trumping his brother's conservative ideas, he believed. "You can either exploit the anger [and] incite it," he said, "or you can come up with ideas to deal with it. [Jeb] came up with solutions . . . but it didn't fit with the mood . . . If you're angry with the powers that be, you're angry with the so-called establishment, and there's nothing more established than having a father and brother that have been president." As for whether Jeb could have done anything in his campaign differently, 43 replied, "I don't know . . . It's hard to tell."

The antiestablishment rage Trump fomented, an indictment of the very party that had made him its standard-bearer, raised the question, Was Trump a Republican? In 2016, conservative analyst and journalist William Kristol, invoking conservative icon William F. Buckley, wrote in the *National Review*, "Hasn't Donald Trump been a votary merely of wealth rather than freedom? Hasn't he been animated in the art of the deal rather than the art of self-government? . . . Isn't Trump a two-bit Caesarism of a kind that American conservatives have always disdained? Isn't the task of conservatives today to stand athwart Trumpism, yelling Stop?" If so, Trump yelled louder. The GOP, long in disarray and absent a definitive, unifying doctrine beyond opposition to Barack Obama, was ripe for insurgency.

At a reunion of aides from his administration in Dallas, George W. Bush said privately, "I'm worried that I will be the last Republican president." It was a legitimate concern. Modern Republicanism, at its best, was rooted in America's engagement in the world, enhancing its international power and prestige through its leadership role by fostering greater openness and human freedom, and representing the essence of opportunity and democratic ideals. It was epitomized in Nixon's historic opening of China in 1972, his most important foreign-policy achievement, and in Reagan's overriding quest to wipe out Soviet oppression. Reagan's most resonant oration was delivered at the Berlin Wall's Brandenburg Gate in 1987, when he demanded of his Soviet counterpart, "Mr. Gorbachev, tear *down* this wall!" America stood for knocking walls down, not building them.

The Bushes continued the tradition, as 41, driven by humanitarian concern and economic interests, led the charge to drive Iraqi captors out of Kuwait, and to catalyze the reunification of Germany as a democratic stronghold despite European opposition. Like his father and Reagan, 43 saw the U.S. as a "force for good." Even after 9/11, 43—while enacting security measures many considered extraconstitutional—declined to take the political path of least resistance by yielding to xenophobia, instead visiting a mosque, proclaiming, "Islam is peace," and later pushing for progressive changes in immigration policy. Though the war in Iraq was ill-founded and lacked a clear nation-building strategy, its intent, in addition to better ensuring national security, was a Wilsonian push to promulgate democracy.

Well before the 2016 election cycle began, 43 expressed disquiet about the country's growing isolationism and protectionism as "twins of negativity." In 2010, he said, "I worry about a nation withdrawing and saying, 'It's too hard. Let's forget what's going on over there.' I also worry about protectionists, which is another way of saying less competition, less trade around the world." By 2016, Bush was reminded of the "America First" movement waged by Charles Lindbergh to keep America out of World War II; the 1930 Smoot-Hawley Tariff Act, which put stiff tariffs on imported goods and exacerbated the Depression; and the National Origins Act of 1924, which placed rigid quotas on non–Western European countries. "Those sentiments are alive again," he lamented.

He had similar concerns about his party's presidential nominee, who fanned the flames of nativism. When Trump entered the 2016 race, Bush thought, *Interesting; won't last.* After Trump emerged as the GOP's lead candidate, 43 was as surprised as anyone else. "When you're not out there and you're not with the people, you don't get a good sense of [the mood]," he said. But his reservations about the candidate took hold. When asked what people should look for in a president, Bush often replied, "humility," a virtue that he believed should include "recognizing your limitations and surrounding yourself with people who know what you don't know." When Trump said, "I'm my own adviser," Bush thought, *Wow, this guy really doesn't understand the job of president.* Bush also saw as a poor sign Trump's inability to "poke fun" at himself. "As you know from looking at my family," he continued, "[humility] is a certain heritage, that's what they expect, and we're not seeing that [in Trump]."

Indeed, his father *did* expect humility in American leadership— and he didn't see it in Trump, either. "I don't like him," George H. W. Bush said bluntly of Trump in May 2016, after Trump had become the party's presumptive nominee. "I don't know much about him, but I know he's a blowhard. And I'm not too excited about him being a leader." When asked what he thought Trump was seeking in running for the presidency, it wasn't Trump's desire to serve that Bush cited, as it had been for him and members of his family, but that Trump had "a certain ego." Could Trump unite the country if elected? "Yes," 41 replied, but it would require "humility," making it a greater challenge for Trump. Asked in the same sitting whom he would vote for in a matchup between Trump and Hillary Clinton, he responded, "I think I'd probably be for Hillary over Trump, if that was the choice," a vote he confirmed after Election Day. "I'm just down on Trump."

The Bushes' resistance to Trump was no great surprise. During the height of the Watergate scandal, in the letter to his sons that George W. said set a political "standard" for him and Jeb, 41 warned presciently, "Power accompanied by arrogance is very dangerous. It's particularly dangerous when men with no real experience have it— for they can abuse our great institutions."

Forty-three, however, wasn't much more enthusiastic about Clinton. Less than a month before the election, he stated, "The question for the country to decide—on both candidates, by the way—is to what extent should we be insisting upon integrity and solid character." While he "knew nothing" about Trump before he threw his hat in the ring in 2015, 43 had spent some time with Clinton. "In my presence, she was polite . . . thoughtful," he said, but alluding to her using a private email server as secretary of state, he added, "obviously tangled up in bad judgment. This email thing, putting confidential information out there in a world where all kinds of people can figure out how to get your emails was not good judgment." It was, he said understatedly, "a strange election year," in which both candidates were "among the few in the country" who made each other "viable." Neither Donald Trump nor Hillary Clinton would earn the vote of the forty-third president. As he said later, "I voted 'None of the Above' for president, and Republican down ballot in 2016."

When the votes were tallied on Election Day, Americans, many holding their noses, gave the popular vote to Clinton by 2.9 million ballots, but handed the presidency to Trump, who took 304 electoral votes to Clinton's 227. With one guerilla campaign and a trail of scorched earth, Trump had felled two political dynasties—the Bushes and the Clintons. It was a different world. And, for the moment at least, it was Trump's world.

55

"WHAT ABOUT GEORGE?"

JOHN QUINCY ADAMS WAS GIVEN to apoplexy after his defeat at the hands of Andrew Jackson in 1828. Throughout his one-term presidency, Adams had been excoriated and then rejected by the American people. "Posterity will scarcely believe . . ." he vented bitterly in his last days in the White House, "the combination of parties and of public men against my character and reputation such as I believe never before was exhibited against any man since this Union existed." Two years after retiring to his native Quincy, he was unexpectedly drafted back into public service by local constituents who elected him as their representative in Congress. Adams's post-presidency was consumed by eighteen fruitful years in the House of Representatives, where he earned the reputation of venerable elder statesman along with the nickname "Old Man Eloquent." After his passing on the House floor at age eighty, an outpouring of grief followed not seen since the death of his father and other prominent founding fathers. Newspapers across the country effused their praise for his service to the nation as thousands lined up to pay their respects while he lay in state in the Capitol and as his body was carried by rail to Boston and then taken back to Quincy, where he was buried next to the elder Adams at the United First Parish Church.

History is like that. Only with the passage of years do passions begin to fade, allowing for more detached reflection. It takes at least a generation to assess a president's legacy with any degree of objectivity. Simple mob judgment gives way to more nuanced views as complexities are acknowledged and weighed dispassionately with the benefit of hindsight. Events that play out after a president's reign often show the effects and consequences of the actions a president took and the policies he put into place. Did he do the greatest good for the greatest number? Did he contribute toward the lasting betterment of the nation? Did he elevate America's position in the world, making it a stronger and more influential and prosperous force? Those questions and others take a while to sort out.

Harry Truman left office in 1953, with an approval rating of 34 percent, blamed for the lingering war in Korea. Worn down by Vietnam, Lyndon Johnson opted not to run for reelection in 1968, leaving office with the dark cloud of the war balefully obscuring his transformational domestic legacy, including civil rights. A year after leaving office in 1989, Ronald Reagan was seen in a decennial poll among prominent historians to be a "below average" president, placing him in the fourth quintile in ranking the thirty-nine presidents from Washington forward. For the legacy of each, time offered perspective and clear-eyed reflection after the myopia of contemporaneous appraisal. Historians would come to rank Truman, Johnson, and Reagan either as "high average" or "near great" presidents—just below the pantheon reserved for Washington, Lincoln, and FDR. Even Gerald Ford, while widely placed in the "average" category, would be vindicated for his damning pardon of Richard Nixon, which is now widely seen as a courageous, healing act that allowed America to move forward.

Time would be a friend to George H. W. Bush as well. Just two decades after leaving office, no longer in the shadow of the iconic Reagan, Bush would begin to be recognized for his sheer competence as president during a seminal time, credited for his incisive foreign-policy mind, diplomatic facility, and steady, prudent hand as commander in chief. Even his broken "No new taxes" pledge, his political undoing, was exonerated. While Americans tossed him out of office due to the image of him toiling indifferently in the Oval

Office as the economy limped along, economists would generally come to believe that the tax hikes Bush was compelled to sign into law helped pave the way toward the prosperity of the Clinton years, which saw the biggest economic expansion since the post–World War II boom.

Moreover, in a barbed, self-aggrandizing age when passion all too often overcame reason, Americans came to value 41's character. In February 2011, Barack Obama draped the Medal of Freedom around Bush's neck in a White House East Room ceremony. "Like the remarkable Barbara Bush," Obama said, paying tribute to his predecessor, "his humility and his decency reflect the very best in the American spirit. *This* is a gentleman." The following month, *Newsweek*, which may have landed the unkindest blow to Bush with its 1987 cover story "Fighting the 'Wimp Factor,'" provided its own repudiation in a feature article titled "A Wimp He Wasn't." "Qualities once branded as vices," it read, "[Bush's] civil tone, willingness to reach across the aisle, even his sway with Mideast strongmen—suddenly seem more like virtues in a world weary of attack politics and confronting a cascading series of global crises." Two years later, after the octogenarian former president shaved his head to show solidarity with the two-year-old leukemia-stricken son of a member of his Secret Service detail, a photo of Bush and the boy, both bald as cucumbers, spread like a balm throughout social media. The forty-first president had become a beacon of decency. In the unexpected warmth of his winter years, the public servant who called for a "kinder, gentler" nation got a little of it back.

Kindness and gentleness found its way to George W. Bush, too. In June 2014, Bush's approval ratings had risen above Barack Obama's, exceeding the 50 percent mark for the first time in almost a decade. By the time the Trump era swept vociferously across the American landscape, 43's policies didn't look quite as harsh as they did when he was in office. The month after Trump's inauguration, where he reportedly commented of Trump's fiery speech, "That was some weird shit," Bush, in an interview on the *Today* show, implicitly repudiated Trump. Responding to Trump's view of the media as "the enemy of the people," Bush called it "indispensable to democracy," adding: "We need an independent media to hold people like me to account." Asked about immigration, Bush replied that he was for "policy that is welcoming and upholds the law," and responding to allegations that Trump had ties to Russia, claimed, "We all need answers." Suddenly, Bush was a font of reason. Shortly after the interview, the *National Review* published an article under the headline "George W. Bush, Liberals' New Hero," which included a tweet from Glenn Greenwald, an author and left-leaning journalist specializing in national security issues, that encapsulated its essence.

2005: George W Bush is a pillaging, torturing war criminal who let a city drown.
2017: I may have disagreed with Bush but he was A Good Man™.

But not enough time had passed for history to cast anything but diffuse light on 43's legacy. Given the complications of the age in which he ruled and the messy aftermath of the decisions he made, a dispassionate assessment would likely take far longer than it had for his father—especially with untold documents relating to his administration's security measures unprocessed or yet to be declassified, and with the morass of the war in Iraq, the dominant part of his legacy, still too thick for a clear view. Only time would tell.

On February 21, 2017, George H. W. Bush and Jean Becker had lunch in the Grille, a cozy, elegant dining room at Houston's Forest Club, next door to Bush's office on Memorial Drive. Forty-one was now back in good health and good spirits after a bout with pneumonia that had landed him in the hospital for over two weeks in January. Less than a week after his release, on February 5, he was well enough to toss the coin at the Super Bowl in the Houston Astrodome, where the former president earned a standing ovation from a crowd of more than seventy thousand, which included Mike Pence, the vice president of just over two weeks. As 41 dug into a prodigious slice of apple cobbler with vanilla ice cream, Becker talked of how beloved he was. "You've become an icon," she would often tell him, and the old man would roll his eyes. When asked how he would like to be remembered, he would say repeatedly, "Let history be the judge." Now history's indebted nod was clear. "I'm glad that the judgment of history has come in your lifetime," she told him as he enjoyed his dessert.

At that moment, George H. W. Bush's thoughts were less about his own presidency than that of his eldest son. George W. Bush hadn't concerned himself with his legacy while he was in the White House, nor did he have illusions that he would see a binding verdict in his lifetime. One of the lessons from his father that helped to guide his decisions in the White House was "History will ultimately sort things out, so one shouldn't worry about legacy." But George H. W. Bush, whose now-lauded presidency was stunted by the verdict of American people who he believed didn't know his heartbeat, *was* worried about his son's legacy.

"What about George?" the forty-first president asked Becker plaintively, his heartbeat as palpable as at any point in his ninety-two years. "I want this for George."

EPILOGUE

THE TRAINING SESSION for members of the Texas State Republican Executive Committee and county chairs in a generic, windowless meeting room of Austin's Wyndham Garden Hotel was a low-key, business-casual affair. On Saturday, August 6, 2016, thirteen weeks before Election Day, Republicans from around the state sat at round tables prepared for an afternoon of PowerPoint presentations and roll-up-your sleeves strategy sessions around fundraising and voter turnout. Presiding over the gathering, dressed in an untucked white polo shirt was Texas GOP victory chairman George P. Bush.

The dark-haired, olive-skinned forty-year-old was a rising star in Texas politics. Two years earlier, flashing a lustrous smile that looked as much Kennedy as Bush, he put a political stake in the ground, handily winning election as Texas land commissioner as the only Latino in the election cycle to win a statewide office. Long before his election, George P., the oldest of Jeb and Columba Bush's three children, oozed political promise, emerging as the likeliest to carry the Bush family torch into the next generation. After George W.'s election as president, when the family began calling him and his father "43" and "41" respectively, they jokingly referred to George P. as "44." During the 2004 convention, his father remarked that the family gave him a hard time "just to make sure he keeps all this in perspective."

Eleven years later, in 2015, who could have foreseen that his father, riding an early wave of establishment support and the windfall of cash that came with it, would be the odds-on favorite to win the party's presidential nomination the following year—and hold the possibility of a third Bush landing in the White House in eighteen years? Bush 45? And who would have predicted that he would soon be routed by political neophyte Donald Trump? Jeb's doomed prospects came as much of a surprise to George P. as anyone. So had Trump's raging populist rise. Even in May 2016, after accepting the position to plot the GOP's state election strategy as victory chairman but still stung by his father's loss, George P. held deep reservations about Trump. "I, along with others," the Spanish-speaking scion of a Mexican immigrant told the *Austin American-Statesman*, "are not in a position to endorse [Trump] at this time because of concerns about his rhetoric and inability to create a campaign that brings people together." Instead, George P. explained, he would focus on the state's down-ballot races.

But now, in early August 2016, several weeks after the party's national convention in Cleveland, it was time to get real. Like it or not, Trump was the Republican standard-bearer. Pragmatism and prudence, Bush family political instincts, dictated George P.'s loyalty to the party and support of the man at the top of the ticket, however objectionable. "It's time to put it aside," he said of his opposition to the New York billionaire and reality-TV star, as the room clapped its approval. "From Team Bush, it's a bitter pill to swallow, but you know what? You get back up and you help the man who won, and you make sure we stop Hillary Clinton." With that, George P. Bush became the first and only member of the Bush family to endorse the presidential candidacy of Donald J. Trump.

As he stood behind the podium, microphone in hand, one could almost hear his grandfather, father, and Uncle George urging him on, encouraging him to go forward without being constrained by those who went before him. He honored the Bush family not by being bound to their legacies but by serving. "Chart your own course," they whispered in his ear. "No one will ever question your loyalty. It's your turn now in the family."

ACKNOWLEDGMENTS

This book simply wouldn't have happened without the forty-first and forty-third presidents, George H. W. Bush and George W. Bush, giving generously of their time to offer firsthand glimpses into their remarkable lives. I am deeply indebted to them and to members of their family, including Barbara Bush and Laura Bush, and to members of their staffs and foundations: Catherine Branch, Brian Cossiboom, Logan Dryden, Ken Hersh, Hutton Hinson, Catherine Jaynes, David Jones, Kristin King, Holly Kuzmich, Nancy Lazenby, Mike Meece, Laura Pears, Evan Sisley, and Tobi Young. Special thanks to Jean Becker and Freddy Ford.

I am grateful to others who agreed to be interviewed for the project—James Baker, Ben Barnes, Jean Becker, Barbara Bush, Jeb Bush, Andy Card, Dick Cheney, Clay Johnson, Ron Kaufman, David Hume Kennerly, Mary Matalin, Condoleezza Rice, Karl Rove, and Chase Untermeyer—and to those I interviewed independent of the book but whose feedback factored into these pages: Laura Bush, Jimmy Carter, Eric Draper, the late Gerald Ford, Robert Gates, Steve Hadley, Karen Hughes, Henry Kissinger, Dan Rather, Bob Schieffer, Brent Scowcroft, and Hugh Sidey, a friend and mentor who offered a standard for writing on the presidency.

Many have helped to guide and support this endeavor, including my agent and friend, Jim Hornfischer, and, at HarperCollins, Adam Bellow, Jonathan Burnham, Chris Goff, Kate D'Esmond, John Jusino, Eric Meyers, and my editor, Eric Nelson. Isabel Saralegui, Jake Solomon, and Ben Stein diligently transcribed interviews between their school studies.

Our presidential libraries, under the auspices of the National Archives and Records Administration and the direction of archivist of the United States, David Ferriero, and director of presidential libraries, Susan Donius, are national treasures. The George Bush Presidential Library and the George W. Bush Presidential Center are no exceptions. At the George Bush Presidential Library, my thanks to director Warren Finch, Mary Finch, Robert Holzweiss, and Cody McMillian; and at the George W. Bush Presidential Center, the directors Alan Lowe and Pat Mordente, Brook Clement, Eric Mc-Crory, and Emily Robison.

I am thankful to a number of accomplished friends and colleagues who provided insight, inspiration, editorial advice, or a willing ear, including Ken Adelman, John Avlon, Peter Baker, Michael Beschloss, Philip Bobbitt, Bill Brands, Doug Brinkley, Christopher Buckley, Ken Burns, Marc Burstein, Don Carleton, Matthew Dowd, Michael Duffy, Jeffery Engel, Frank Gannon, Nancy Gibbs, Mike Gillette, Doris Kearns Goodwin, Steve Harrigan, Will Imboden, Bob Inman, Tom Johnson, Mark Lawrence, Jim Magnuson, Jon Meacham, Todd Purdum, Cokie Roberts, Richard Norton Smith, Robert Schenkkan, Bat Sparrow, Larry Wright, and Bob Woodward.

Valuable perspective also came from those who served in either the Bush 41 or 43 administrations: John Bridgeland, Randy Erben, Margaret Hoover, Anita McBride, Tim McBride, Margaret Spellings, and Kevin Sullivan. At the LBJ Presidential Library and the LBJ Foundation, my thanks to Claudia Anderson, Amy Barbee, Ben Barnes, Adam Brodkin, Elizabeth Christian, Jennifer Cuddeback, Rodney Ellis, Wayne Gibbens, Jay Godwin, Mary Herman, Bill Hobby, Luci Johnson, Tom Johnson, Sarah McCracken, Cappy McGarr, the late Harry Middleton, Suzanne Mirabal, Marge Morton, Kassie Navarro, Lyndon Olson, Lynda Robb, Roy Spence, Larry Temple, and Anne Wheeler; and at the National Medal of Honor Museum Foundation, Carlyle Blakeney, Marilyn Buist, Bill Phillips, Darwin Simpson, Peter Stent, and Jim Taylor.

Thank you to old and loyal friends: Craig Barron, Marty Dobrow, Amy Erben, Bill Gurney, Steve Huestis, David Hume Kennerly, Jim Popkin, Lee Rosenbaum, Nick Segal, Hal Stein, and Ray Walter, who died far too soon in 2014. And I couldn't ask for a better sister and brother-in-law than Susie and Glenn Crafford, or in-laws in Kim and Richard Storm, and Sandy and Skip Wood, or family members in Jim and Wendy Krombech, Herbert Krombech, Cindy Kaskey, Jeff and Loretta Kaskey, and Mike and Andria Kaskey.

Frequently, as I worked on these pages about a father and son, I was reminded of my own father, John Jacob Updegrove, a good and decent man, and a great dad, who died in 2014. I owe so much to him and to my beloved mother, Naomi, both of whom labored, scrimped, and sacrificed to give their children every advantage.

My wife, Amy, my dearest friend and greatest love, has been with me at every step, lending support and encouragement as I wrote this book, which is dedicated to her. She and our blended family—Isabel, Charlie, Mateo, and Tallie—make me lucky beyond measure.

MKU
Austin, Texas

NOTES

INTRODUCTION

4   "buy no more": Chuck Wills, *Jack Kennedy: The Illustrated Life of a President* (San Francisco: Chronicle Books, 2009), 99.

5   "What people can't": Mark K. Updegrove, "Bush 2.0," *Texas Monthly*, December 2010.

5   "love letter": George W. Bush (hereafter "GW Bush"), author interview.

6   "people have certain": Bill Minutaglio, *First Son: George W. Bush and the Bush Family Dynasty* (New York: Three Rivers Press, 1999), 230.

7   "Are you going": Skip Hollandsworth, "Born to Run," *Texas Monthly*, May 1994.

7   "[I]t's part of": Peter Schweizer and Rochelle Schweizer, *The Bushes: Portrait of a Dynasty* (New York: Doubleday, 2004), xiv.

7   "a higher father": William Hamilton, "Bush Began to Plan War Three Months Before 9/11," *Washington Post*, April 17, 2004.

7   "Any nuance of": George H. W. Bush (hereafter "GHW Bush"), author interview.

7   "W. avenged his dad": Tatiana Morales, "Bushworld: Enter at Your Own Risk," CBC News, August 10, 2004.

9   "He's probably the": Nick Glass, "Jeb Bush Recounts Conversation with Mitt Romney," *Politico*, January 5, 2016.

CHAPTER 1: "A FLASH OF LIGHT"

13   "bullshit": GW Bush, author interview.

13   "chased a lot of pussy": Ibid.

14   "Why should I talk": Jean Becker, author interview.

14   *New York Times*: GW Bush, author interview.

14   "All these questions": Jeb Bush (hereafter "J. Bush"), author interview.

14   "a man of few words": Jean Becker, author interview.

14   "I've run out of things to say": Ibid.

15   "more than a Yale linebacker": Barbara Bush (hereafter "B. Bush"), author interview.

15   "She put me in the back seat": Ibid.

15   "a flash of light": Ibid.

15   3,288,672 American children: Halbert L. Dunn, *Vital Statistics of the United States: 1946* (Washington, DC: Federal Security Agency, 1948).

16   "Of course, George couldn't carry me": B. Bush, author interview.

17 *We had better do something*: James Bradley, *Flyboys: A True Story of Courage* (New York: Little Brown, 2003), 80.

17 "an imposing presence": Ibid., 86.

17 "No, sir, I'm going in": Ibid.

17 His father nodded: GHW Bush, *All the Best, George Bush: My Life in Letters and Other Writings* (New York: Scribner, 1999), 25.

17 It was the first time: Schweizer and Schweizer, *The Bushes*, 69.

18 "the youngest guy": Bradley, *Flyboys*, 86.

18 Prescott Bush cried: Ibid.

18 "glory of being": GHW Bush, *All the Best*, 45.

18 "I hope John and Buck": Ibid.

19 "family and survival": Paula Zahn, "Story of George H. W. Bush World War II Experience," *CNN Presents*, aired December 20, 2003.

19 "As you get older": Thomas Ferraro, "Bush Heads to Pearl Harbor Commemoration," UPI, December 6, 1991.

CHAPTER 2: "BAR"

21 Dressed in white tie: B. Bush, *Barbara Bush: A Memoir* (New York: Scribner, 1994), 16.

21 "the most beautiful girl": Jim McGrath, *Heartbeat: George Bush in His Own Words* (New York: Scribner, 2001), 267.

22 "humiliated": Schweizer and Schweizer, *The Bushes*, 67.

22 "secretly engaged": B. Bush, *Barbara Bush*, 20.

22 "Bar" after Barsil: Ibid., 19.

22 "Although my childhood": GHW Bush, *All the Best*, 59.

22 "I think about those guys": "WWII 'Flyboys' Remembered," *Washington Times*, September 28, 2003.

23 "George loved that baby": B. Bush, author interview.

24 "the enforcer": GW Bush, author interview.

24 "My mother hated to be": B. Bush, *Barbara Bush*, 27.

CHAPTER 3: GO WEST

25 its Capitol dome in Austin: "It's True: Texas Capitol Stands Taller Than Nation's," *Orlando Sentinel*, January 14, 1999.

25 "To make it on my own": GHW Bush, author interview.

26 "didn't want to go to Wall Street": Ibid.

27 Even after S.P. died: Schweizer and Schweizer, *The Bushes*, 36.

27 Pres asked his grandchildren: Walter Isaacson, "George W. Bush: My Heritage Is Part of Who I Am," *Time*, August 7, 2000.

27   "a very dignified person": Ibid.

27   "He'll say, 'My father was a great man'": Todd Purdum, "41 + 43 = 84," *Vanity Fair*, September 2006.

27   "a real fighter": GHW Bush, *All the Best*, 91.

27   "the most competitive woman": Isaacson, "My Heritage," *Time*.

28   "Don't be a braggadocio": GHW Bush, author interview.

28   "He talked all the time about": Brent Scowcroft, author interview for Mark K. Updegrove, *Second Acts: Presidential Lives and Legacies After the White House* (Guilford, CT: Lyons Press, 2006).

28   "charm the fangs off a cobra": Schweizer and Schweizer, *The Bushes*, 95.

28   "Right now I am bewildered": GHW Bush, *All the Best*, 61–62.

29   "capitalize completely on": Ibid., 62.

29   "too confining": Ibid.

29   "touch and feel": B. Bush, author interview.

29   "would have followed George Bush anywhere": Ibid.

29   "I've always wanted to live in Odessa": Minutaglio, *First Son: George W. Bush and the Bush Family Dynasty*, 25.

29   "It was the first time in our lives": Schweizer and Schweizer, *The Bushes*, 97.

29   62,249,000 barrels of oil: Texas State Historical Association, https://tshaonline .org/handbook/online/articles/hce02.

30   "Trust fund babies don't share duplexes": GW Bush, author interview.

30   "Never heard of it": B. Bush, *Barbara Bush*, 33.

30   "as different from Rye, New York": Ibid., 32.

30   "It was the people": GHW Bush, author interview.

30   "You should see Georgie now": GHW Bush, *All the Best*, 65.

CHAPTER 4: "THE SKY'S THE LIMIT"

32   "You're a product of who raised you": GW Bush, author interview.

32   "at the very bottom of the corporate ladder": GHW Bush, *All the Best*, 63.

33   the town was home to 21,713 souls: Robert L. Martin, *The City Moves West: Economic and Industrial Growth in Central West Texas* (Austin, TX: University of Texas Printing Division, 1969), 156.

33   "I didn't feel any sense of being different": GW Bush, author interview.

33   "a big name, no money": Ibid.

34   "my dad went to Greenwich Country Day School": Hollandsworth, "Born to Run," *Texas Monthly*.

34   "Georgie has grown to be a near-man": GHW Bush, *All the Best*, 70.

34   "In Midland, the sky sits overhead": Laura Bush (hereafter "L. Bush"), *Spoken from the Heart* (New York: Simon & Schuster, 2010), 48.

35   "There are no trees": GW Bush, author interview.

35   embodied the American Dream: Ibid.

35   The Bush family first settled: "Bush Family History," The George W. Bush Child-
     hood Home, http://www.bushchildhoodhome.org/family_history.

35   Even with the housing upgraded: Ibid.

36   "There is a plainness": L. Bush, *Spoken from the Heart*, 121.

36   His father saw it in his free-spiritedness: GHW Bush, author interview.

36   "a square-jawed, curly-headed man": R. W. Apple Jr., "The 1994 Campaign:
     Texas; In Texas Race, a Bush Avoids Father's Errors," *New York Times*, Novem-
     ber 7, 1994.

36   though he had left Texas: GW Bush, speech in Midland, Texas, January 20,
     2009.

CHAPTER 5: "MORE THAN TONGUE CAN TELL"

37   "a heroic figure": GW Bush, author interview.

37   "Mother adored him": Ibid.

37   "George was a great father": B. Bush, author interview.

37   "I was busy, though": GHW Bush, author interview.

37   "Keep in mind": GW Bush, author interview.

37   "so-called camping trip": Ibid.

38   "Son, you've arrived": Ibid.

38   His starkest memory: GW Bush, *A Charge to Keep: My Journey to the White House*
     (New York: Harper Perennial, 1999), 14–15.

38   "I don't know what to do this morning": B. Bush, *Barbara Bush*, 59.

39   $350,000 in investment capital: Schweizer and Schweizer, *The Bushes*, 101–2.

40   "He didn't forgive us for quite a while": B. Bush, author interview.

40   Their posture reflected their own hope: Schweizer and Schweizer, *The Bushes*, 103.

40   she "fell apart" upon her return to Midland: B. Bush, *Barbara Bush*, 103.

40   "Mother's reaction was to envelop herself": Schweizer and Schweizer, *The
     Bushes*, 108.

40   "I was thinking I was there for him": Ibid.

40   *She was the epitome of innocence*: McGrath, *Heartbeat*, 270–71.

40   "For forty years, I wasn't able to talk about it": Schweizer and Schweizer, *The
     Bushes*, 107.

41   "I'll bet she can see the game better": B. Bush, *Barbara Bush*, 46.

41   "family clown": Schweizer and Schweizer, *The Bushes*, xi.

41   "After Robin died": B. Bush, author interview.

42   "There is about our house a need": GHW Bush, *All the Best*, 81.

42   "every child more valuable": GHW Bush, author interview.

42   "I love you more than tongue can tell": B. Bush, author interview.

43   "I stopped being a sissy about it": McGrath, *Heartbeat*, 170.

43   In regular "check-in" telephone conversations: GW Bush, author interview.

CHAPTER 6: AGGRAVATION AND PRIDE

44 In his late eighties: GHW Bush, author interview.

44 "He was hands-off": GW Bush, author interview.

44 "I think all of us would tell you": Ibid.

44 "I can remember screwing up": J. Bush, author interview.

45 "George [W.] aggravates the hell out of me": GHW Bush, *All the Best*, 79.

45 "It's not that W. rebelled": Isaacson, "My Heritage," *Time*.

45 "a rambunctious lad": GHW Bush, author interview.

45 "a wonderful, incorrigible child": Hollandsworth, "Born to Run," *Texas Monthly*.

45 "Hiya, little lady": Isaacson, "My Heritage," *Time*.

45 "swaggered in as though he had done": Kathryn Moore, *The American President: A Complete History* (New York: Barnes & Noble, 2007), 622.

45 "I don't remember [my dad] ever": J. Bush, author interview.

46 "I have my father's eyes": Susan Page, "Bush's 'Love Letter' to Dad, and Message to Jeb: Run," *USA Today*, November 14, 2014.

46 "easier to vent with": GW Bush, author interview.

46 "George W. Bush is the most disciplined": Andy Card, author interview.

46 "He was all excited": Ibid.

46 "He does these amazingly loving things": B. Bush, author interview.

CHAPTER 7: "A CERTAIN SORT OF EXPECTATION"

48 "an obligation": GW Bush, author interview.

48 "cold and distant": Moore, *American President*, 613.

49 "For George [W.], everywhere he went": Schweizer and Schweizer, *The Bushes*, 150.

49 "Everybody knew that this was the guy": Minutaglio, *First Son*, 61.

49 "There was very little reference to [his dad]": Clay Johnson, author interview.

49 "If you're humble and successful": GW Bush, author interview.

50 "The greatest inheritance I got": Ibid.

50 "I didn't see any desire [on his part]": Clay Johnson, author interview.

50 "He was already who he was": Minutaglio, *First Son*, 61.

50 "He was a flip, in-your-face kind of kid": Unnamed source, author interview.

50 "one of the cool guys": Minutaglio, *First Son*, 62.

51 something, he conceded later, he didn't exactly brag about: GW Bush, author interview.

51 would later become George W.'s favorite movie: Nancy Gibbs and Michael Duffy, *The Preacher and the Presidents: Billy Graham in the White House* (New York: Center Street, 2007), 327.

51 "He became a master at it": Schweizer and Schweizer, *The Bushes*, 150.

51 After his grades weren't up to snuff: GW Bush, *Charge to Keep*, 20.

52 "had this fear that generation after generation": Nicolas D. Kristof, "The 2000

Campaign: The Cheerleader; Earning A's in People Skills at Andover," *New York Times*, June 10, 2000.

52 His grades went up: Minutaglio, *First Son*, 261.

52 "Why don't you try walking in his shoes": Peter Baker, *Days of Fire: Bush and Cheney in the White House* (New York: Doubleday, 2013), 24.

52 "I was fully prepared to go to Texas": GW Bush, author interview.

52 "just two guys wandering around": Ibid.

53 Yale accepted him even though his board scores: Nicolas D. Kristof, "A Boy and His Benefits," *New York Times*, January 24, 2003.

53 "wasn't dying to go": GW Bush, author interview.

53 "there had to be something there": Ibid.

53 "What are you going to Houston for": Clay Johnson, author interview.

54 "What the hell is this": Helen Thorpe, "Go East, Young Man," *Texas Monthly*, June 1999.

CHAPTER 8: "TO DO SOMETHING OF SERVICE"

57 "I have in the back of my mind": GHW Bush, *All the Best*, 67.

57 "I knew what motivated him": "George H. W. Bush," *American Experience: The Presidents*, PBS, aired May 5, 2008.

58 "dangerous divisions among": Ibid.

58 "I'm not going to vote for 'nother": Ibid.

58 "George Bush's instinct politically": Ibid.

59 "He admired his dad": GW Bush, author interview.

59 "my reasons for not supporting the bill": GHW Bush, *All the Best*, 88.

59 "I want to win": Ibid.

59 "politically inspired and is bad legislation": Steven A. Holmes, "The Nation; When the Subject Is Civil Rights, There Are Two George Bushes," *New York Times*, June 9, 1991.

59 "I took some far right positions": Moore, *American President*, 568.

60 Little consolation came from the fact: Ibid.

61 "Watching him on the campaign trail": GW Bush, author interview.

61 "If life is to be led to the fullest": Ibid.

61 "in the arena": *The Almanac of Theodore Roosevelt*, "The Man in the Arena," www .theodore-roosevelt.com/trsorbonnespeech.html.

CHAPTER 9: YOUNG AND FOOLISH

62 "I was a pretty independent guy": GW Bush, author interview.

62 didn't feel the "weight": James Carney, "George W's Love-Hate Affair with Yale," *Time*, May 23, 2001.

62 "I knew your father": Evan Thomas, "War Stories," *Newsweek*, February 22, 2004.

63   "George probably knew 1,000": Schweizer and Schweizer, *The Bushes*, 167.

63   "He was third generation Yale": Ibid.

64   "only a cigarette burn": Maureen Dowd, "Liberties; President Frat Boy?" *New York Times*, April 7, 1999.

64   "borrowing" a Christmas wreath: Larry Margasak, "Alcohol an Issue in the Bush Campaign," *Washington Post*, November 3, 2000.

64   "I liked to drink": GW Bush, author interview.

64   "He was a little wild": J. Bush, author interview.

64   "I don't ever remember my father": GW Bush, author interview.

65   "sometimes was sort of sensitive": Minutaglio, *First Son*, 97–98.

65   "Take your clubs": B. Bush, author interview.

65   "turned around to pick up": Ibid.

65   "the pick of the litter": Schweizer and Schweizer, *The Bushes*, 100.

66   "The whole process": Minutaglio, *First Son*, 123.

66   as close to "undoing" George W.: Ibid.

66   "just sort of panicked": Lois Romano and George Lardner Jr., "Bush: So-So Student but a Campus Mover," *Washington Post*, July 27, 1999.

66   "wasn't ready": GW Bush, author interview.

66   "He believed that his father's position": Romano and Lardner, "So-So Student," *Washington Post*.

67   Last weekend our son came home: GHW Bush, *All the Best*, 119–120.

67   "intellectual arrogance": Carney, "Love-Hate," *Time*.

67   "I still believe": Ibid.

67   "wrote him off": Isaacson, "My Heritage," *Time*.

67   The year George W. was enrolled: Jerome Karabel, "Legacy of Legacies," *New York Times*, September 13, 2004.

68   "felt guilty about their lot in life": Minutaglio, *First Son*, 119.

68   "This is what's going to work": Ibid., 104.

68   "To me, that was sort of symbolic": Ibid.

69   "It's not one of those things": GW Bush, author interview.

69   "He hung out with my family": Romano and Lardner, "So-So Student," *Washington Post*.

69   "It wasn't said in passing": Ibid.

CHAPTER 10: "A WHOLE DIFFERENT LIFE"

70   "do justice" to Zapata's shareholders: GHW Bush, *All the Best*, 93.

71   "young black soldiers fighting": Ibid., 107.

71   I'll vote for the bill: Ibid.

72   "With one of the most conservative": Ibid., 108.

72   "by far the most meaningful": "Interview with George W. Bush," moderated by Brian Lamb, C-SPAN, aired September 4, 1988.

72   "a whole different life": B. Bush, *Barbara Bush*, 65.

72   "kept the family together": GW Bush, author interview.

72   "She handled all the moves": Ibid.

73   "I think you should work": Ibid.

73   When Barbara told her seventeen-year-old son: Ibid.

74   "He called me down to his office": Ibid.

CHAPTER 11: UP IN THE AIR

75   "activity in study": Minutaglio, *First Son*, 115.

75   "We didn't have the luxury": Herón Márquez, *George W. Bush* (Minneapolis, MN: Twenty-First Century, 2007), 35.

75   "felt that in order not to derail": Minutaglio, *First Son*, 116.

76   "we could not explain the mission": GW Bush, *Charge to Keep*, 50.

76   George W. discovered through friends that there were openings: George Lardner Jr. and Lois Romano, "At Height of Vietnam Bush Picks Guard," *Washington Post*, July 28, 1999.

76   "I want to be a fighter pilot": Minutaglio, *First Son*, 120.

76   Spots in the "Champagne Unit": Lardner and Romano, "At Height of Vietnam, Bush Chooses Guard," *Washington Post*.

76   Despite scoring a 25 percent: Ibid.

77   "There is absolutely no way": Ben Barnes and Lisa Dickey, *Barn Burning, Barn Building: Tales of a Political Life from LBJ through George W. Bush and Beyond* (Houston, TX: Bright Sky Press, 2006), 120.

77   as many as two thousand names ahead of Bush's: Ben Barnes, author interview.

77   "Sid Adger was an oilman": Ibid.

77   "pay back a dividend or two": Barnes and Dickey, *Barn Burning*, 109.

77   just as he had for "dozens": Ben Barnes, author interview.

77   "a total lie": Mary Jacoby, "George W. Bush's Missing Year," *Salon*, September 2, 2004.

77   "Nobody's come up": Ibid.

78   "Don Evans reported your conversation": Joe Hagan, "Truth or Consequences," *Texas Monthly*, May 2012.

78   with nicknames like "Fly": Schweizer and Schweizer, *The Bushes*, 193.

78   "a lively individual": Moody Air Force Base website, http://www.moody.af.mil/.

78   "How would you like to fly": George W. Bush, *41: A Portrait of My Father* (New York: Crown, 2014), 93.

79   "Being a swashbuckling pilot": Ibid., 94.

79   "Get any": Ibid., 95.

79   George Walker Bush is one of the members: Suzanne Goldenberg and Oliver Burkeman, "George's War," *Guardian*, February 11, 2004.

80   "The difference between": GHW Bush, author interview for *Second Acts*.

81   "Being in politics": B. Bush, author interview.

81   "The funniest part": Minutaglio, *First Son*, 132.

81   "The loss hurt a lot more": GW Bush, *41*, 92.

81   "What Will George Bush Do Next": Jon Meacham, *Destiny and Power: The American Odyssey of George Herbert Walker Bush* (New York: Random House, 2015), 150.

81   "The future—I don't know": GHW Bush, *All the Best*, 128–129.

82   "He was lost": Minutaglio, *First Son*, 134.

CHAPTER 12: "LISTEN TO YOUR CONSCIENCE"

84   "broad, interesting, fantastic": GHW Bush, *All the Best*, 147.

84   They took a Chinese delegation: B. Bush, *Barbara Bush*, 92.

84   When journalist Dick Schaap: McGrath, *Heartbeat*, 332.

85   He also invited some: Ibid.

85   "I am continually amazed": GHW Bush, *All the Best*, 142.

85   "The Senator dreaded": Meacham, *Destiny and Power*, 161.

85   "My Dad was": GHW Bush, *All the Best*, 161.

86   "good Nixon man—first": Moore, *American President*, 571.

86   it fell to Bush to gently tell Dole: Ibid.

87   "Dad helped inculcate into us": GHW Bush, *All the Best*, 181–86.

88   "He couldn't have realized it": GW Bush, *41*, 105.

88   "battered and disillusioned": GHW Bush, *All the Best*, 185.

88   "Both of our fathers": B. Bush, author interview.

88   "the whole goddamned thing": GHW Bush, *All the Best*, 183.

88   "one of unreality": Meacham, *Destiny and Power*, 171.

89   "I'm sure there will be impeachment": Meacham, *Destiny and Power*, 174.

89   "We ought to wait": Ibid.

89   "expeditiously": Ibid.

89   "It is my considered judgment": GHW Bush, *All the Best*, 193.

89   "I'm glad dad's not alive": Schweizer and Schweizer, *The Bushes*, 231.

90   dolefully concluded was amoral: Meacham, *Destiny and Power*, 171.

CHAPTER 13: FAMILY VALUES

91   "Can't you keep a steadier hand": James Traub, *John Quincy Adams: Militant Spirit* (New York: Basic Books, 2016), 26.

92   "all gone to smash": Robert Dallek, *An Unfinished Life: John F. Kennedy 1917–1963* (New York: Back Bay Books, 2003), 107.

92   "You know how much": Ibid.

92   "You know my father wanted": Robert Dallek, *Camelot's Court: Inside the Kennedy White House* (New York: HarperCollins, 2013), 6.

92 "For the Kennedys": Chris Matthews, *Jack Kennedy: Elusive Hero* (New York: Simon & Schuster, 2012), 122.

92 "My father was totally honest": *Today*, NBC, aired September 28, 1999.

92 "I never thought any of them": B. Bush, author interview.

92 "There was no game plan": GW Bush, author interview.

93 "All five of them": B. Bush, author interview.

93 a "beacon": GW Bush, author interview; and J. Bush, author interview.

93 "very close . . . some would say": Dorothy Bush Koch, *My Father, My President: A Personal Account of the Life of George H. W. Bush* (New York: Warner, 2006), 295.

93 "love, loyalty, faith, friends": GHW Bush, author interview.

94 "If they had had IQ tests": B. Bush, author interview.

94 "My memories of my childhood": J. Bush, author interview.

94 "always be ready to give you": James Baker, author interview.

95 "grinded myself to a place": J. Bush, author interview.

95 "I actually lowered": Ibid.

95 struck by lightning: Alex Altman and Zeke J. Miller, "The Bush Identity," *Time*, March 16, 2015.

95 "She thought he was the son": B. Bush, author interview.

95 "How I worry about Jeb": Altman and Miller, "Bush Identity," *Time*.

95 "It took me a while": B. Bush, author interview.

96 "blowin' and goin'": J. Bush, author interview.

96 "It's not that I wasn't aspirational": Ibid.

96 "When they were growing up": Gibbs and Duffy, *Preacher and Presidents*, 328.

96 "You look at second-generation kids": J. Bush, author interview.

97 "I'm not good at psychoanalysis": GW Bush, author interview.

CHAPTER 14: HORIZONS

98 "The events of our life": GW Bush, author interview.

98 hold three jobs: Schweizer and Schweizer, *The Bushes*, 193.

98 "nomadic years": Isaacson, "My Heritage," *Time*.

99 "I was a badass back then": Hagan, "Truth or Consequences," *Texas Monthly*.

99 "He had a couple of girls": Lardner and Romano, "At Height of Vietnam, Bush Picks Guard," *Washington Post*.

99 "active lifestyle": J. Bush, author interview.

99 "share a view which few others": GHW Bush, *All the Best*, 130–31.

99 "Dad was shy": Minutaglio, *First Son*, 9.

99 more "sedate" side: Ibid., 182.

100 He dropped the idea: Isaacson, "My Heritage," *Time*.

100 "We hope he'll feel settled": Meacham, *Destiny and Power*, 167.

100 "a political thing": B. Bush, author interview.

100 "dull coat and tie job": Hollandsworth, "Born to Run," *Texas Monthly*.

100 to do "some good": GW Bush, author interview.

100 was "an eye-opening experience": Ibid.

100 "very frustrating": B. Bush, author interview.

101 replied simply, "No": GHW Bush, author interview.

101 I worry about the family situation: GHW Bush, *All the Best*, 143.

101 "What's sex like after fifty": Schweizer and Schweizer, *The Bushes*, 305.

101 "I got drunk": GW Bush, author interview.

101 "disgraceful behavior": B. Bush, author interview.

102 "You want to go mano a mano": Purdum, "41 + 43 = 84," *Vanity Fair*.

102 "He never said that": B. Bush, author interview.

102 Neither George nor George W.: GHW Bush, author interview; and GW Bush, author interview.

102 a "holy shit" moment: J. Bush, author interview.

102 "Instead of saying the obvious": Ibid.

102 "He didn't have to tell me": GW Bush, author interview.

102 "slinked out of the room": Ibid.

102 "a juvenile delinquent": Maureen Dowd, "Poppy Bush Finally Gives Junior a Spanking," *New York Times*, November 7, 2015.

102 "I don't know if he was trying to": James Baker, author interview.

102 created "a myth": GW Bush, author interview.

103 "That was better than caning him": B. Bush, author interview.

103 "I don't think George [W.] could drink": Undisclosed source, author interview.

103 "he was becoming a real boozer": Schweizer and Schweizer, *The Bushes*, 235.

104 his "biggest accomplishment": GW Bush, author interview.

104 His family knew nothing: GW Bush, author interview.

104 "I wasn't sure": Ibid.

104 "No, I'm not really sure": Ibid.

104 "I had no possessions": Ibid.

104 "planted a seed": Ibid.

104 "You know something, George": Ibid.

105 "At a place like Harvard Business School": Mica Schneider, "George W.'s B-School Days," *Bloomberg*, February 15, 2001.

105 Under 8 percent of Cambridge's fifty thousand: Minutaglio, *First Son*, 154.

106 "He was not a star": John Solomon, "Bush, Harvard Business School and the Making of a President," *New York Times*, June 18, 2000.

106 "George's leadership was not based on": Ibid.

106 "turning point": GW Bush, *Charge to Keep*, 60.

106 "energy and entrepreneurship": Ibid., 56.

106 Bush "could smell something happening": Hollandsworth, "Born to Run," *Texas Monthly*.

107 the $20,000 in seed money: GW Bush, *Charge to Keep*, 56.

107 "Cambridge, Massachusetts": Minutaglio, *First Son*, 163.

107 "your horizons suddenly expand": Ibid.

CHAPTER 15: GO EAST

109 "plainly [the choice] should be Bush": Gerald R. Ford, *A Time to Heal: The Autobiography of Gerald R. Ford* (New York: Harper & Row, 1979), 143.

109 "Well, there are some": GW Bush, *41*, 110.

109 "not yet ready to handle": Ford, *A Time to Heal*, 142–43.

109 "nationally and internationally known": Dick Cheney, author interview.

110 "It hurts so much": Schweizer and Schweizer, *The Bushes*, 237.

110 "I indicated that way down the line": Meacham, *Destiny and Power*, 177.

110 "An important, coveted position": Ibid., 178.

110 "cut off from the day to day news": GHW Bush, *All the Best*, 202–3.

111 "pace in my life": Ibid.

111 "It doesn't matter": Meacham, *Destiny and Power*, 182.

111 "We found them very like us": B. Bush, author interview; "An Exclusive Conversation with President and Mrs. Bush," *Parade Magazine*, July 15, 2012.

112 Great talk with Bar: GHW Bush, *All the Best*, 21.

112 "My visit underscored": GW Bush, *Charge to Keep*, 61.

113 "He is off to Midland": GHW Bush, *All the Best*, 239.

113 Countries of little strategic: Meacham, *Destiny and Power*, 186.

114 "with a frankness": GHW Bush, *All the Best*, 431.

114 "Right away": Minutaglio, *First Son*, 165.

CHAPTER 16: COMING HOME

115 "toxic waste dump": George W. Bush, *Decision Points* (New York: Crown, 2010), 24.

115 "Harvard, at first": Schweizer and Schweizer, *The Bushes*, 255.

116 "began to fit in": Ibid.

116 "was replaced by": Minutaglio, *First Son*, 163–64.

116 "I was collecting": GW Bush, *Decision Points*, 27.

117 "He didn't dress down": Nicholas D. Kristof, "Learning How to Run: A West Texas Stumble," *New York Times*, July 27, 2000.

118 "He wasn't obsessed with politics": Eric Pooley and S. C. Gwynne, "How George Got His Groove," *Time*, June 14, 1999.

118 "a total shock": GHW Bush, *All the Best*, 233–34.

118 "The CIA!": B. Bush, *Barbara Bush*, 131.

118 "He called back": Ibid.

118 "I look forward": GHW Bush, *All the Best*, 236.

119 Henry, you did not know: Ibid., 233–34.

119   "I do not have politics": Ibid.

119   "graveyard for politics": B. Bush, *Barbara Bush*, 132.

120   "Man, Rummy got your ass": Meacham, *Destiny and Power*, 192.

120   "There was some feeling": GHW Bush, author interview.

120   "like two cocks": David Hume Kennerly, author interview.

120   "I had a husband": B. Bush, *Barbara Bush*, 135.

121   nearly two-thirds of Americans disapproved: Joseph Carroll, "Americans Grew to Accept Nixon's Pardon," Gallup News Service, May 21, 2001.

CHAPTER 17: "DEFEAT ISN'T DEFEAT"

123   "It was great to be back": GHW Bush, *All the Best*, 271.

123   "Well, this is your big mistake": Meacham, *Destiny and Power*, 210.

123   "There is a missing": GHW Bush, *All the Best*, 271.

124   "I think I want to run": Ibid.

124   "I had just gotten out": GW Bush, author interview.

125   "Was it [Carter's] policy alone": Ibid.

125   "You can't win": Herbert S. Parmet, *George Bush: The Life of a Lone Star Yankee* (New York: Scribner, 1997), 207.

125   "I wonder if George Bush Jr.": Meacham, *Destiny and Power*, 212–13.

125   "You're not going to win": GW Bush, author interview.

125   "I mean if you're an expert": Ibid.

126   "And I ran anyway": Ibid.

126   "[O]ut of nowhere": Lois Romano and George Lardner Jr., "A Run for the House," *Washington Post*, July 29, 1999.

126   "passing on of tradition": Parmet, *Lone Star Yankee*, 207.

126   "George is off": GHW Bush, *All the Best*, 272–73.

127   "He was at the age": Romano and Lardner, "A Run for the House," *Washington Post*.

127   "At that particular moment": L. Bush, *Spoken from the Heart*, 96.

127   "two complementary souls": L. Bush, author interview, The Texas Book Festival, October 16, 2010.

127   "late miscarriages": Ibid.

128   "He has a steadfastness": Ibid.

128   "They are perfect": Meacham, *Destiny and Power*, 213.

128   "Animal House": Koch, *My Father, My President*, 146.

129   "In years to come": L. Bush, *Spoken from the Heart*, 97.

129   "We spent nearly a year": Ibid., 99.

129   "We don't need Dad": Nicholas Kristof, "Learning to Run," *New York Times*.

130   "I'm not interested": Ibid.

130   "are pretty darn good": GHW Bush, *All the Best*, 274.

130   a war chest of $400,000: Kristof, "Learning to Run," *New York Times*.

131   "a lot of class": GHW Bush, *All the Best*, 277.

131   "You know how close": Ibid.

131   "The guy rolled": Kristof, "Learning to Run," *New York Times*.

131   "Daddy and granddad": Romano and Lardner, "A Run for the House," *Washington Post*.

131   "Maybe it's a cool thing": Ibid.

131   "Kent lives here": Ibid.

132   "When someone attacks": GW Bush, *Charge to Keep*, 175.

132   Bush's friends later saw: Romano and Lardner, "A Run for the House," *Washington Post*.

132   "One of the things": GW Bush, author interview.

CHAPTER 18: "THE BIG MO"

135   "up and down": Ronald Reagan, Address on Behalf of Senator Barry Goldwater: "A Time for Choosing," American Presidency Project, University of California at Santa Barbara, http://www.presidency.ucsb.edu/ws/?pid=76121.

136   "so low that in many": GW Bush, *41*, 128.

136   "a lifelong Republican": "Presidential Announcement Statement of Ambassador George Bush," accessed at http://www.4president.org/speeches/1980/georgebush 1980announcement.htm.

136   "progressive Republican policies": Marc Fisher, "GOP Platform Through the Years Shows Party's Shift from Moderate to Conservative," *Washington Post*, August 28, 2012.

137   "The age thing": Schweizer and Schweizer, *The Bushes*, 279.

137   "George Bush had the knowledge": GW Bush, interview with Michael Gillette, January 2, 1990.

138   "We campaigned in backyards": B. Bush, *Barbara Bush*, 157.

138   "I look back": Koch, *My Father, My President*, 151.

138   "Jeb bawl[ed] him out": Schweizer and Schweizer, *The Bushes*, 275.

139   "I've got the 'Big Mo'": Andrew Glass, "The 'Big Mo': It Can Change with Time," *Politico*, April 17, 2007.

140   "Brilliant strategy": Craig Shirley, "Fast Times at Nashua High," *National Review*, October 19, 2009.

140   "Goddamn it, you guys": Ibid.

140   "Is this on": Ibid.

141   "may have won": Ronald Reagan, *An American Life* (New York: Simon & Schuster, 1990), 213.

141   "It cost him": GW Bush, interview with Gillette.

141   "like second-class citizens": Adam Clymer, "Reagan Easily Defeats Bush and Baker in New Hampshire," *New York Times*, February 25, 1980.

141   "Bush stiffed us": Ibid.

141 "but that's something": Ibid.

141 "Looking back at it": GW Bush, *41*, 134–35.

142 "What went wrong": "Money Is So Hard to Get," *Time*, June 2, 1980.

CHAPTER 19: THE CALL FROM REAGAN

143 "What's it going to be like": Alan Peppard, "Command and Control: Tested Under Fire," *Dallas Morning News*, May 18, 2015.

143 "[T]his guy had": Minutaglio, *First Son*, 187.

144 "I have strong reservations": Parmet, *Lone Star Yankee*, 245.

144 "They needed a putty knife": Thomas M. DeFrank, *Write it When I'm Gone: Remarkable Off-the-Record Conversations with Gerald R. Ford* (New York: Penguin Group, 2007), 111.

144 even though Ford rejected: Gerald Ford, author interview for *Second Acts*.

145 *Wait a minute*: Reagan, *An American Life*, 215.

145 "As far as Ronnie was concerned": Meacham, *Destiny and Power*, 249.

146 "It's not fair": GHW Bush, author interview.

146 "It was a dream ticket": Mark K. Updegrove, *Second Acts: Presidential Lives and Legacies After the White House* (Guilford, CT: Lyons Press, 2006), 130.

146 "Hello, George": Meacham, *Destiny and Power*, 253.

147 the "Western" man: Minutaglio, *First Son*, 198.

147 "a man of decisive action": Ibid.

147 "little chance": GW Bush, *A Charge to Keep*, 176.

147 "Surprised and thrilled": GW Bush, *41*, 138.

147 "If anyone wants to know": Hedrick K. Smith, "Reagan Wins Nomination and Chooses Bush as Running Mate After Talks with Ford Fail," *New York Times*, July 17, 1980.

147 "out of the clear blue sky": Ibid.

CHAPTER 20: A HEARTBEAT AWAY

148 Preparing for their own: "Bush Poor at Investing but Doubles Wealth Political Perks, Advisers Help President Prosper," *Money*, July 18, 1989.

149 "work, work, work": HK Smith, "Reagan Wins," *New York Times*, July 17, 1980.

149 "I know I satisfied the Reagans": Koch, *My Father, My President*, 175.

150 "a warm bucket": Patrick Cox, "John Nance Gardner on the Vice Presidency—In Search of the Proverbial Bucket," Briscoe Center for American History, https://www.cah.utexas.edu/news/press_release.php?press=press_bucket.

150 "a stuck pig at a screwing match": Harry Middleton, author interview for *Indomitable Will: LBJ in the Presidency*.

150 "Presidents don't get to do that very much": Kenneth T. Walsh, "Reagan and Bush's Admirable Partnership." *U.S. News & World Report*, March 17, 2011.

150   "He would be forceful": GW Bush, author interview.

150   "Reagan expected Bush": Chase Untermeyer, author interview.

150   "dead to the world": Lou Cannon, *Ronald Reagan: The Role of a Lifetime* (New York: Public Affairs, 1991), 172.

151   "There was no doubt": Peppard, "Command and Control," *Dallas Morning News*.

151   "I guess every Vice President": GHW Bush, *All the Best*, 316.

151   "a small fluttering": Reagan, *An American Life*, 258.

152   "Recommend that you return": Peppard, "Command and Control," *Dallas Morning News*.

152   "Constitutionally, gentlemen": Richard V. Allen, "When Reagan Was Shot, Who Was in 'Control' at the White House?" *Washington Post*, March 25, 2011.

153   "Something about it": Ibid.

153   "with perfect equanimity": Ibid.

153   "tiny and afraid": Ibid.

153   "Honey, I forgot to duck": Susannah Cahalan, "I Forgot to Duck," *New York Post*, February 11, 2011.

153   "I hope you're all Republicans": Paul Bedard and Mallie Jane Kim, "Reagan's Hospital One-Liners Inspired by Hollywood," *U.S. News & World Report*, August 1, 2011.

154   "All things considered": Godfrey Sperling Jr., "The Good-Humored Reagan," *Christian Science Monitor*, April 6, 1981.

154   "the will of the people": Moore, *American President*, 554.

154   $110.7 billion: Edward Cowan, "Deficit in Reagan Budget at Record $110.7 Billion," *New York Times*, October 27, 1982.

154   plunged to 35 percent: Frank Newport, Jeffrey M. Jones, and Lydia Saad, "Ronald Reagan from the People's Perspective: A Gallup Poll Review," Gallup News Service, June 6, 2004.

CHAPTER 21: ARBUSTO OR BUST

155   "[W]e were outliers": L. Bush, *Spoken from the Heart*, 112–13.

155   "Thank you for teaching": Ibid., 113.

155   "My heart was filled": GHW Bush, *All the Best*, 325.

155   "George would call young George": Schweizer and Schweizer, *The Bushes*, 328–29.

156   When Reagan was shot: Minutaglio, *First Son*, 200.

156   "You die, I fly": GHW Bush, *All the Best*, 197.

156   "Laura and I didn't know": GW Bush, *Charge to Keep*, 185.

156   "He changed diapers": L. Bush, *Spoken from the Heart*, 111.

156   "most important responsibility": GW Bush, author interview.

157   "Running came up a lot": Minutaglio, *First Son*, 200.

157   "The politician was in him": Pooley and Gwynne, "Got His Groove," *Time*.

157 "George was an easy sale": Lois Romano and George Lardner Jr., "Bush's Move Up to the Majors," *Washington Post*, July 31, 1999.

157 "company maker": Pooley and Gwynne, "Got His Groove," *Time*.

157 "I had some success": GW Bush, author interview.

157 Of the ninety-five wells: Pooley and Gwynne, "Got His Groove," *Time*.

158 "totally inebriated": Hollandsworth, "Born to Run," *Texas Monthly*.

158 mocked by some Midland oil insiders: Minutaglio, *First Son*, 202.

158 "doodah days": Pooley and Gwynne, "Got His Groove," *Time*.

158 "The competitive nature": Andy Card, author interview.

158 "It's the first-son syndrome": Pooley and Gwynne, "Got His Groove," *Time*.

CHAPTER 22: "WHY DON'T YOU COME TO WASHINGTON?"

160 "Under President Reagan's leadership": Ronald Reagan TV ad, video retrieved on October 25, 2016, from: https://www.youtube.com/watch?v=EU-IBF8nwSY.

160 Afterward, Reagan's lead: David Weigel, "When Reagan Blew a Presidential Debate and Dropped Seven Points in the Polls," *Slate*, October 12, 2012.

160 "I won't let age": "Reagan-Mondale Debate: The Age Issue," video retrieved on October, 28, 2016, from: http://www.youtube.com/watch?v=LoPu1UIBkBc.

161 "We did not, repeat": Robert Busby, "The Scandal That Almost Destroyed Ronald Reagan," *Salon*, February 4, 2011.

161 "A few months ago": "Address to the Nation on the Iran Arms and Contra Aid Controversy," Ronald Reagan Presidential Library & Museum website, retrieved on November 5, 2012, https://reaganlibrary.archives.gov/archives/speeches/1987/030487h.htm.

162 "formidable" competitor: GW Bush, interview with Gillette.

162 "Why should we": Ibid.

163 "loyalty yardstick": Ibid.

163 "Let me rephrase": Ibid.

163 "That was some tough": Ibid.

163 "I want y'all": Ibid.

163 "palace intrigue": J. Bush, author interview.

164 "did not want his life's": GW Bush, interview with Gillette.

164 "At the time": Ibid.

164 "Midland was suffering": GW Bush, *Charge to Keep*, 64.

164 "I don't know where": Pooley and Gwynne, "Got His Groove," *Time*.

164 Bush himself was given: Hollandsworth, "Born to Run," *Texas Monthly*.

164 Not bad considering: Ibid.

165 "His name was George Bush": Márquez, *George W. Bush*, 53.

165 Harken stock amounting to $312,000: Lois Romano and George Lardner Jr., "Bush's Life-Changing Year," *Washington Post*, July 25, 1999.

165 "enormous worry that": GW Bush, interview with Gillette.

165 In politics, access is the key: Ibid.

CHAPTER 23: AWAKENING

167 "was on the road": Pooley and Gwynne, "Got His Groove," *Time*.

168 "mistakes": Laurie Kellman, "Bush Once Pleaded Guilty to DUI," *Washington Post*, November 3, 2000.

168 "four B's": Craig Unger, "How George Bush Really Found Jesus," *Salon*, November 8, 2007.

168 Laura recalled later: L. Bush, *Spoken from the Heart*, 118.

168 "I wasn't going to": Ibid.

169 "Kemp and Indiana Senator": Mark Crispin Miller, *Bush Dyslexicon: Observations on a National Disorder* (New York, W. W. Norton, 2002), 50.

169 "You no good": Minutaglio, *First Son*, 208–9.

169 "a priority": GW Bush, author interview.

169 "I realized that alcohol": Romano and Lardner, "Bush's Life-Changing Year," *Washington Post*, July 25, 1999.

169 "He looked in the mirror": Minutaglio, *First Son*, 210.

169 "There was a strong feeling": Schweizer and Schweizer, *The Bushes*, 331.

170 father was "frustrated": GW Bush, author interview.

170 "We didn't know": B. Bush, author interview.

170 "It was a big deal": GW Bush, author interview.

170 "He couldn't trust": "The Jesus Factor," *Frontline*, PBS, aired April 29, 2004.

171 "I encouraged him": Arthur Blessitt, *The Official Website of Aurthur Blessitt*, retrieved November 29, 2016, www.blessitt.com/praying-with-george-w-bush.

171 "It was just two people": Gibbs and Duffy, *Preacher and Presidents*, 328.

172 "They say you have to": GW Bush, author interview.

172 *Wow, that's a good answer*: Ibid.

172 "saddled up": Ibid.

172 "'Do you have the right'": Gibbs and Duffy, *Preacher and Presidents*, 330.

172 "The man is powerful": Ibid., 331.

172 "planted a seed": P. Baker, *Days of Fire*, 37.

172 "George was always": Pooley and Gwynne, "Got His Groove," *Time*.

173 "Fuck 'em. You can't": Jeff Shesol, "What George H. W. Bush Got Wrong," *New Yorker*, November 13, 2015.

173 "the religious guy": GW Bush, author interview.

173 "unbelievable gift of grace": Ibid.

173 "semi-surrogate, special voter": GW Bush, interview with Gillette.

174 "Don't worry, son": GW Bush, author interview.

174 "a little too large": Ibid.

CHAPTER 24: JUNIOR

175 "The only person": Mary Matalin, author interview.

176 "George W. didn't learn": Pooley and Gwynne, "Got His Groove," *Time*.

176 I think George [W.] Bush: GHW Bush, *All the Best*, 353.

176 the "project": GW Bush, interview with Gillette.

176 "If you're the candidate's son": Ibid.

176 "chew his ass out": GW Bush, author interview.

176 "You think this is bad?": Ibid.

176 "fiercest of warriors": Ibid.

177 the "guard dog at the gate": Jean Becker, author interview.

177 "Give me one good reason": Ibid.

177 "Just doing my job": Hollandsworth, "Born to Run," *Texas Monthly*.

177 "My role evolved": GW Bush, interview with Gillette.

177 "the clerks": "How George Got His Groove," *Time*, Asia edition, July 12, 1999.

177 "Even people that": Mary Matalin, author interview.

177 "[E]veryone in the campaign": GW Bush, interview with Gillette.

178 "They're just not true": Schweizer and Schweizer, *The Bushes*, 341.

178 "impugned the character": GW Bush, *41*, 161–62.

178 "Lance the boil": Minutaglio, *First Son*, 221.

178 "The answer to the big": GW Bush *41*, 161.

178 "one of the great campaign": GW Bush, author interview.

178 "I would intentionally be": GW Bush, interview with Gillette.

178 "a little fear": Ibid.

179 "just be yourself": GW Bush, author interview.

179 importance of "decompression": Ibid.

179 "He was his father's confidant": James Baker, author interview.

179 "When he knows": GW Bush, interview with Gillette.

179 "A strong [leader] makes": Ibid.

CHAPTER 25: "BE TOUGH"

182 "If I had sat there": Meacham, *Destiny and Power*, 302.

182 "Most importantly, president": Ibid., 299–300.

182 "I'm one of the few": Ibid., 299.

182 "Dad, what is this": GW Bush, interview with Gillette, April 30, 1990.

183 "blind trust": "Garry Trudeau on George W. as College Boy," *Slate*, September 9, 1999.

183 "as he traipses": Curt Smith, *George H. W. Bush: Character at the Core* (Norman, OK: Potomac Books, 2014), 85.

183 Bush, who formally declares: Margaret Garrard Warner, "Bush Battles the 'Wimp Factor,'" *Newsweek*, October 19, 1987.

183 "[T]hat was about": GW Bush, interview with Gillette.

183 That *Newsweek* story: GHW Bush, *All the Best*, 369.

184 "This is the beginning": E. J. Dionne Jr., "Dole Wins in Iowa, with Robertson Next," *New York Times*, February 9, 1988.

184 "Even before tonight's": Ibid.

185 "You do, Dan": Meacham, *Destiny and Power*, 319–20.

185 "knocked it out of the park": Ibid.

185 "Senator Straddler": E. J. Dionne Jr., "Bush vs. Dole: Behind the Turnaround." *New York Times*, March 17, 1988.

185 "Stop lying about": Ibid.

186 "George Bush's success": Ibid.

CHAPTER 26: NOT SO KIND, NOT SO GENTLE

187 Dukakis boasted of 800,000 new jobs: Nicholas M. Horrock, "Dukakis' 'Miracle' Losing Luster," *Chicago Tribune*, September 4, 1988.

187 "They are losing": Ibid.

187 "What's fourteen inches long": Purdum, "41 + 43 = 84," *Vanity Fair*.

188 "Let 'em stay where they belong": Timothy Naftali, *George H. W. Bush* (New York: Times Books, 2007), 58.

188 "After listening to George Bush": "Past Democratic Convention Speeches," C-SPAN, September 2, 2012.

188 The only hiccup came: "Governor Bill Clinton Endorsing Mike Dukakis at the 1988 DNC," video retrieved on November 10, 2016, from: https://www.youtube .com/watch?v=vvTRvTII40o.

188 "with the wind": Edward Walsh, "Dukakis Accomplished Convention Goals," *Washington Post*, July 23, 1988.

189 "Dukakis, a native": Ibid.

189 Donald Trump, the self-promoting: Kathy Burke, "New Book: Trump Wanted to Run with Bush 41," *Newsmax*, November 5, 2015.

190 "In naming the staunchly": Jack Nelson and Richard E. Meyer, "Bush Selects Quayle as His Running Mate: Calls Indiana Senator, 41, Future Leader," *Los Angeles Times*, August 17, 1988.

190 "The thing that's important": Minutaglio, *First Son*, 231.

191 "Quayle was a generational statement": GW Bush, interview with Gillette.

191 "George, go out and win": "Republican National Convention: President Reagan's Address, 8/15/88," video retrieved on December 28, 2016, from: https://www .youtube.com/watch?v=oi1JJUe7h5Y.

191 "about decency and honor": GHW Bush, author interview.

191 "I don't hate government": George Bush: Address Accepting the Presidential Nomination at the Republican National Convention in New Orleans, American

Presidency Project, retrieved on December 1, 2016, from: http://www.presidency
.ucsb.edu/ws/?pid=25955.

193   "I served with Jack Kennedy": Joshua Green, "The Story Behind the Best Debate
Zinger Ever," *Bloomberg Businessweek*, October 3, 2012.

193   A *Saturday Night Live* skit: "George Bush Debate," *Saturday Night Live*, accessed
on January 2, 2017, https://www.hulu.com/watch/4117.

194   "backwater towns": GW Bush, interview with Gillette.

195   "in the trenches": Roberto Suro, "The 1992 Campaign: President's Family;
One of Bush's Campaign Advisers Is Also His Son," *New York Times*, April 26,
1992.

195   "It was a wonderful experience": Pooley and Gwynne, "Got His Groove,"
Time.

195   "[Junior] was the first": Mary Matalin, author interview.

195   "If there was competition": Paul Burka, "The W. Nobody Knows," *Texas Monthly*,
June 1999.

196   "George went up": Pooley and Gwynne, "Got His Groove," *Time*.

196   "go-to prayer guy": GW Bush, author interview.

196   "Many of us will begin": B. Bush, *Barbara Bush*, 248–49.

197   "Scrub Team": Veda Boyd Jones, *Modern World Leaders: George W. Bush* (New
York: Chelsea Books, 2007), 62.

CHAPTER 27: "A NEW BREEZE"

201   "the man who made": Jones, *Modern World Leaders*, 62.

202   AIDS had taken fifty-five thousand lives: National Geographic Society, *National
Geographic Eyewitness to the 20th Century: An Illustrated History* (Washington,
DC: National Geographic, 2004), 362.

202   "America is never wholly herself": George Bush: Inaugural Address, American
Presidency Project, retrieved on June 17, 2016, from: http://www.presidency.ucsb
.edu/ws/?pid=16610.

204   "One of the worst things": Doug Wead, *All the Presidents' Children: Triumph
and Tragedy in the Lives of America's First Families* (New York: Atria Books,
2003), 1.

204   "Two things the media": Pooley and Gwynne, "Got His Groove," *Time*.

205   "Daddy's coattails": Ibid.

205   "What's the boy": Ibid.

CHAPTER 28: "COMFORTABLE IN THE JOB"

206   "I feel comfortable": GHW Bush, *All the Best*, 411.

206   "knew where the keys": GHW Bush, author interview.

206   "in no small measure": Ellen Goodman, *Washington Post* Writers Group, "The Re-
       freshing New Age of the 'Silver Fox,'" *Chicago Tribune*, January 29, 1989.
206   "My mail tells me": Ibid.
206   "We approach everything": GHW Bush, author interview for *Second Acts*.
206   "the loneliest job": Ibid.
207   "So far": Meacham, *Destiny and Power*, 359.
207   fifteen of forty-one delegations: Minutaglio, *First Son*, 245.
207   "Tonight's pair will": B. Bush, *Barbara Bush*, 414.
208   "work hard, study": James Baker, author interview.
209   "captured the imagination": GHW Bush, *All the Best*, 429.
209   "anguished letter": Ibid., 428.
209   "convey to the Chinese": Ibid., 431.
210   "possible for every citizen": Mary Elise Sarotte, "How the Fall of the Berlin Wall
       Really Happened," *New York Times*, November 6, 2014.
210   "I'm not an emotional": Lesley Stahl, *Reporting Live* (New York: Touchstone,
       1999), 355.
210   "could cry at": James Baker, author interview.
210   Bush left it to Marvin: GHW Bush, *All the Best*, 443.
211   "The worst thing": Frederick Kempe, "Brent Scowcroft on the Fall of the Berlin
       Wall," Atlantic Council, November 2, 2009.
211   "I've got Bush": Michael Beschloss, ed., *The American Heritage: Illustrated History
       of the Presidents* (New York: Crown, 2000), 495–96.
211   "a grab of Noriega": GHW Bush, *All the Best*, 441.
211   "hardest days of": B. Bush, *Barbara Bush*, 323.
212   "I couldn't have": Ibid., 322.
212   "It's been some year": GHW Bush, *All the Best*, 451.
212   80 percent approval rating: Gallup, "Presidential Approval Ratings—George W.
       H. Bush," accessed on June 15, 2015, from: http://www.gallup.com/poll/116500
       /presidential-approval-ratings-george-bush.aspx.
212   "one of the greatest": GHW Bush, *All the Best*, 452.

CHAPTER 29: A WHOLE NEW BALL GAME

213   as "the increasingly likely": Minutaglio, *First Son*, 235.
213   "He went to Greenwich": Patricia Kilday Hart, "Don't Call Him Junior," *Texas
       Monthly*, April 1989.
213   "I want to affect": Ibid.
214   "This could be": GW Bush, *Charge to Keep*, 198.
214   But though he had $606,000: Ibid.
214   "like a pit bull": Ibid.
215   "There is no question": Lois Romano and George Lardner Jr., "Bush Earned
       Profit, Rangers Deal Insiders Say," *Washington Post*, July 31, 1999.

215 An investment group: Ibid.

216 "I've seen this before": Schweizer and Schweizer, *The Bushes*, 388.

216 "peeing in the same": Minutaglio, *First Son*, 240.

216 "being the president's son": Schweizer and Schweizer, *The Bushes*, 390.

216 "I don't want": Ibid., 387.

216 "When you make": George Lardner Jr. and Lois Romano, "Tragedy Created Bush Mother-Son Bond," *Washington Post*, July 26, 1999.

CHAPTER 30: COMMANDER IN CHIEF

217 "We're proud of you": B. Bush, *Barbara Bush*, 322.

217 "I thought—I sent her son": GHW Bush, *All the Best*, 456.

217 "Bar is gone . . . I sit here": Ibid., 467.

218 "given up what": David Lauter and William J. Eaton, "Bush Abandons Vow, Backs a Tax Increase," *Los Angeles Times*, June 27, 1990.

218 "reform populist conservatism": Naftali, *George H. W. Bush*, 116.

219 "I would be naïve": Sharon LaFraniere, "The Savings and Loan Scandal—Neil Bush Stands in the Shattered Ruins of Denver Thrift," *Washington Post*, August 1, 1990.

219 "Neil Bush Stands in": Ibid.

219 "His whole problem": B. Bush, *Barbara Bush*, 325.

219 "[I'm] . . . worried about": Meacham, *Destiny and Power*, 419.

219 "spirits needed a lift": GW Bush, *41*, 223.

220 "it looks very bad": George [H. W.] Bush and Brent Scowcroft, *A World Transformed* (New York: Vintage Books, 1998), 303.

220 "worst fears": Ibid.

220 he had murdered thousands: BBC News, "On this Day, March 16, 1988."

221 "The Most Dangerous Man": "The World's Most Dangerous Man," *U.S. News & World Report*, June 4, 1990.

221 "I would not discuss any": Schweizer and Schweizer, *The Bushes*, 393.

221 "The truth is": Meacham, *Destiny and Power*, 424.

221 "There is little the U.S.": GHW Bush, *All the Best*, 476.

221 "naked aggression": Opinion, "Iraq's Naked Aggression," *New York Times*, August 3, 1990.

221 "If Iraq wins": Meacham, *Destiny and Power*, 426.

222 "American people might": GHW Bush and Scowcroft, *A World Transformed*, 328.

222 "My first objective": Ibid., 328.

222 "smoldering intensity to him": "George H. W. Bush," *American Experience: The Presidents*, PBS, aired May 5, 2008.

222 "This will not stand": Jean Edward Smith, *George Bush's War* (New York: Henry Holt, 1992), 89.

222 "[T]he enormity of Iraq": GHW Bush, *All the Best*, 476.

CHAPTER 31: "YOU CAN'T GIVE IN"

223 "vision thing": Howard Gleckman, "It's Not Too Late for Bush to Get 'The Vision Thing,'" *Bloomberg*, August 24, 1992.

223 "new world order": George Bush: Address Before a Joint Session of the Congress on the Persian Gulf Crisis and the Federal Budget Deficit, American Presidency Project, retrieved on December 12, 2015, from: http://www.presidency.ucsb.edu /ws/?pid=18820.

224 "all means necessary": "The Long Road to War, Chronology," *Frontline*, PBS.

224 "Bush, unlike Ronald Reagan": George J. Church, "The Two George Bushes," *Time*, January 7, 1991.

225 "It is only the United States": Meacham, *Destiny and Power*, 452.

225 "[W]hat do we do": Ibid., 451.

225 "If I don't get": Ibid., 453.

225 "Principle must be": GHW Bush, *All the Best*, 498.

226 "pro-41, anti-43": GW Bush, author interview.

226 "It's conventional wisdom": Jon Meacham, "The Hidden Hard-Line Side of George H. W. Bush," *Politico Magazine*, November 12, 2015.

227 "I was in those meetings": Dick Cheney, author interview.

227 "I had spent much": Hugh Sidey, *Portraits of the Presidents: Power and Personality in the Oval Office* (New York: Time Books, 2000), 160.

228 "the final chapter": GHW Bush and Scowcroft, *World Transformed*, 184.

228 "Kuwait is liberated": Ann Devroy and Dan Baltz, "Kuwait Is Liberated," *Washington Post*, February 28, 1991.

228 which stood at 89 percent: "Presidential Approval Ratings—Gallup Historical Statistics and Trends," retrieved on January 3, 2016, from: http://www.gallup .com/poll/116677/presidential-approval-ratings-gallup-historical-statistics-trends .aspx.

CHAPTER 32: COMING OF AGE

229 "living life to the fullest": GW Bush, author interview.

229 "It was an accomplishment": Ibid.

229 "seemed to be almost": Church, "Two George Bushes," *Time*.

229 "You start out with Mr. Rogers": Jacob Rubin, "Why Dana Carvey's George H. W. Bush is the Best Impression of All," *Slate*, March 17, 2015.

230 "home run for conservatism": Jeremy Rabkin, "The Sorry Tale of David Souter, Stealth Justice," *Weekly Standard*, November 6, 1995.

231 He had never seen: GW Bush, *41*, 234.

231 "He was beginning": GW Bush, author interview.

232 "It was a big honor": Ibid.

232 "I have asked [my] son George": "Unhappy Warrior," *Newsweek*, October 31, 1992.

232 "I felt that whenever": Andy Card, author interview.

232 "a symbol of the": Andrew Rosenthal, "Sununu Resigns Under Fire as Chief Aide to President; Cites Fear of Hurting Bush," *New York Times*, December 4, 1991.

232 "There was a lot of angst": GW Bush, author interview.

233 "You're isolated": Meacham, *Destiny and Power*, 493.

233 "good listener," GW Bush, author interview.

233 "Who's going to tell": GW Bush, *41*, 227.

233 "George can't do": B. Bush, author interview.

233 "that's not true": GW Bush, author interview.

233 "George W. was given": Andy Card, author interview.

233 "a drag": Rosenthal, "Sununu Resigns," *New York Times*.

234 "close confidant": GW Bush, *41*, 226.

234 "maybe created a new": GW Bush, author interview.

234 Dear Dad, This past week: Letter from GW Bush to GHW Bush, December 4, 1991, George Bush Presidential Library, College Station, Texas.

235 "caught up in real": GHW Bush, *All the Best*, 543.

235 "At this special time": Ibid.

235 "There was something": Ibid.

CHAPTER 33: "DEFEAT WITH DIGNITY"

237 "reaped the seeds": Robin Toner, "Bush Jarred in First Primary; Tsongas Wins Democratic Vote," *New York Times*, February 19, 1992.

237 "In November, we will win": Ibid.

238 "The economy, stupid!": Michael Kelly, "The Democrats—Clinton and Bush Compete to Be Champion of Change; Democrat Fights Perceptions of Bush Gain," *New York Times*, October, 31, 1992.

238 "Do I bring a different perspective": Suro, "President's Family," *New York Times*.

239 "The American people": Memo from GW Bush to GHW Bush, May 12, 1992, George Bush Presidential Library.

239 "I want to tell you": Minutaglio, *First Son*, 254.

239 "like watching the disintegration": GW Bush, *41*, 232.

240 "it's time to take out": Anthony Bennett, *The Race for the White House from Reagan to Clinton: Reforming Old Systems, Building New Coalitions* (New York: Palgrave, 2013), 136.

240 "turned sour": GHW Bush, *All the Best*, 361.

240 "I think he was driven": Tierney Sneed, "Exclusive Clip: Perot Driven by 'Personal Dislike,' Bush 41 Says," *U.S. News & World Report*, September 10, 2013.

240 "seen as a weirdo": Naftali, *George H. W. Bush*, 144.

240 "He'll be back": GW Bush, *41*, 232.

241 Benefiting from the Perot: R. W. Apple Jr., "The 1992 Campaign: Overview; Poll Gives Clinton a Post-Perot, Post-Convention Boost," *New York Times*, July 18, 1992.

241 Bush lugged *Truman*: Meacham, *Destiny and Power*, 519.

241 "You know, I think I'm going": James Baker, author interview.

241 "didn't want to leave": Ibid.

242 George W. was struck: GW Bush, *41*, 238.

242 "The agenda Clinton and Gore": Timothy Stanley, *The Crusader: The Life and Tumultuous Times of Pat Buchanan* (New York: Thomas Dunne, 2012), 3.

242 "convey a kinder": GW Bush, *41*, 238.

242 "Quayle was right": Memo from GW Bush to GHW Bush, May 12, 1992, George Bush Presidential Library.

242 But he also thought: P. Baker, *Days of Fire*, 42.

243 sixty-five deaths and over $26 billion: "1992—Hurricane Andrew," Hurricanes: Science and Society website, retrieved on February 12, 2016, from: http://www.hurricanescience.org/history/storms/1990s/andrew/.

243 "We're going to win": Meacham, *Destiny and Power*, 519.

243 "We're gonna lose": Matalin, author interview.

243 "It was George W. Bush": Ibid.

244 "would my father have loved": GHW Bush, *All the Best*, 571.

244 "I've given it my best": Ibid., 571–72.

244 "It looks like": GW Bush, *41*, 243.

244 "My country didn't": GHW Bush, *All the Best*, 583.

244 Yet he was convinced: Ibid.

CHAPTER 34: "FINISH STRONG"

245 "Be strong, be kind": GHW Bush, *All the Best*, 572.

245 "tired old lady": Ibid., 469–70.

246 "Hurry up, George": B. Bush, author interview.

247 tried to "psychobabble": GW Bush, author interview.

247 Dear Bill, When I walked: George Stephanopoulos, "Exclusive—Bush's Oval Office Letter to Clinton," ABCNews.com, January 11, 2011.

247 "Don't worry": GHW Bush, *All the Best*, 578.

247 29 percent the previous summer: "Presidential Approval Ratings—Gallup Historical Statistics and Trends," retrieved on January 23, 2017, from: http://www.gallup.com/poll/116677/presidential-approval-ratings-gallup-historical-statistics-trends.aspx.

247 "I've done what I could": Updegrove, *Second Acts*, 209.

248 "Remember," he said: R. W. Apple Jr., "The New Presidency: Last Day in the White House; 'Over and Out,' Bush Says, Then Goes to Walk the Dog," *New York Times*, January 20, 1993.

248 "I've asked him": Peter Applebome, "Bush Set Out to Prove He Can Go Home Again," *New York Times*, January 24, 1993.

248 "I've been in public life": Ibid.

CHAPTER 35: "THE L-WORD"

251 "white collar crime": GHW Bush, author interview for *Second Acts*.

251 "We have much": GHW Bush, *All the Best*, 586.

251 "All presidents go through": GW Bush, author interview.

252 "Barbara is way out": GHW Bush, *All the Best*, 586.

252 "On January 20": B. Bush, *Reflections: Life After the White House* (New York: Scribner, 2003), 6.

252 "Aren't you Barbara Bush": B. Bush, *Barbara Bush*, 525.

252 "Along the way": GHW Bush, *All the Best*, 586.

253 "firm and commensurate response": David Von Drehle and R. Jeffrey Smith, "U.S. Strikes Iraq for Plot to Kill Bush," *Washington Post*, June 27, 1993.

253 "It's very difficult": Ibid.

254 "I'm not in the interview": Ibid.

254 "saving the world": GHW Bush, author interview for *Second Acts*.

254 $670 million: "Overview of the Office of George Bush (as of August 26, 2017)," Office of George Bush.

254 the "L-word": Ibid.

255 "Maybe they'll say": GHW Bush, *All the Best*, 586.

CHAPTER 36: JOY AND HEARTACHE

256 "When all those people": Tom Farrey, "Man Builds Ballpark, Ballpark Makes Man," ESPN.com, November 9, 2000.

256 A $130,000 public relations effort: Ibid.

256 "I have to make": Minutaglio, *First Son*, 241.

257 "Ladies and gentlemen, I know": Hollandsworth, "Born to Run," *Texas Monthly*.

257 "For those who mock": Nicholas D. Kristof, "The 2000 Campaign: Breaking into Baseball; Road to Politics Ran Through a Texas Ballpark," *New York Times*, September 24, 2000.

257 By 1996, two years after: Farrey, "Man Builds Ballpark," ESPN.com.

257 "[W]ith the Rangers": Pooley and Gwynne, "Got His Groove," *Time*.

258 amassing a net worth of $2.25 million: Scweizer and Schweizer, *The Bushes*, 419.

258 "You can't win": GW Bush, author interview.

258 "True story": Ibid.

258 "[It's inaccurate] to suggest": J. Bush, author interview.

259 "I was always viewed": GW Bush, author interview.

259 Boasting a 63 percent: Jan Reid, "A Year in, Gov. Richards Was Riding High," *Texas Tribune*, October 3, 2012.

259 "We weren't sure": James Baker, author interview.

260 "[I]t wasn't like I wanted": GW Bush, author interview.

260 "If [Dad] had won": Ibid.

260 "into a *People* magazine story": Sam Howe Verhovek, "The 1994 Campaign: The

Bushes; Two Brothers Share a Quest but Not a Style," *New York Times*, November 5, 1994.

260 "Jeb's my little brother": Ibid.

260 "Former President's Shadow": John King, "Former President's Shadow Looms Over 2 Bush Sons in Governors' Races," Associated Press, October 31, 1993.

261 "a total myth": GW Bush, author interview.

261 "George and I didn't talk": J. Bush, author interview.

261 "I'm not running because": King, "President's Shadow," Associated Press.

261 "I can take care": Schweizer and Schweizer, *The Bushes*, 420.

262 "Once you've had the exposure": Jon Meacham, "Has Jeb's Time Come?," *Time*, July 1, 2013.

262 "I surmise Jeb's and my race": GW Bush, author interview.

262 "I bear no ill will": Maureen Dowd, "New Races (Their Own) for 2 Bush Sons," *New York Times*, November 30, 1993.

262 "This is not a joke": Sam Howe Verhovek, "The 1994 Campaign: Texas; Governor and Her Rival Meet in Debate," October 22, 1994.

262 Deriding him as "Shrub": Ibid.

263 was "constrained by the current": Ibid.

263 "let me tell you": Tom Tyron, "He-Coon Walked into Florida Political Lore," *Herald-Tribune*, October 12, 2014.

263 "You know, he thinks": B. Bush and GHW Bush, author interview.

264 "clear that his dad": Karl Rove, author interview.

264 "The joy is in Texas": Joe Hagan, "Bush in the Wilderness," *New York Magazine*, October 14, 2012.

264 "I must say": Hugh Sidey, "George H. W. Bush: The Time Interview," *Time*, December 19, 2004.

264 "moral and spiritual example": Kevin Phillips, *American Dynasty: Aristocracy, Fortune, and the Politics of Deceit in the House of Bush* (New York: Viking, 2004), 87.

264 "tough moment": Rick Lyman and Mireya Navarro, "George W. and Jeb Bush are Elected Governors in Texas and Florida," *New York Times*, November 4, 1999.

265 "On this day": Schweizer and Schweizer, *The Bushes*, 428.

265 "At first I didn't think": Isaacson, "My Heritage," *Time*.

CHAPTER 37: "CHART YOUR OWN COURSE"

266 "I have never seen": Paul Burka, "The Man Who Isn't There," *Texas Monthly*, February 2004.

266 "batting a thousand": *Texas Monthly* archive, "George W. Bush," *Texas Monthly*, retrieved on May 1, 2015, from: http://www.texasmonthly.com/category/topics/george-w-bush/.

267 "Bullock was not": GW Bush, *Charge to Keep*, 111.

267 "I am genuinely fond": Burka, "Man Who Isn't There," *Texas Monthly*.

267 In 1994, the year before: Anemona Hartocollis, "Most Fourth Graders Fail Albany's New English Test," *New York Times*, May 26, 1999.

267 During the same period: Jeffrey H. Birnbaum, "The Man Who Could Be President," *Fortune*, March 29, 1999.

267 When his father saw: L. Bush, *Spoken from the Heart*, 122.

268 "When somebody says": Hart, "Don't Call," *Texas Monthly*.

268 "out of it": GHW Bush, author interview for *Second Acts*.

268 "just a word": A. M. Rosenthal and Marvin Siegal, eds., *The New York Times: Great Lives of the Twentieth Century* (New York: Times Books, 1988), 194.

269 "Just don't tell": GW Bush, *41*, 256.

269 "The King would have": GHW Bush, *All the Best*, 691.

269 "He stood always": "George Bush Presidential Library Dedication," C-Span, accessed on September 12, 2016, https://www.c-span.org/video/?95003–1/bush-presidential-library-dedication.

270 "knowing the perils": GW Bush, *Charge to Keep*, 218.

270 "The only reason": Ibid.

271 Dear George and Jeb . . . Your mother: GHW Bush, *All the Best*, 615–16.

273 Should Jeb lose in Florida: Ibid., 618–19.

274 "I hardly ever": Rick Lyman and Mireya Navarro, "The 1998 Campaign: The Nation—Party Leaders; George W. and Jeb Bush are Easily Elected in Texas and Florida," *New York Times*, November 4, 1998.

274 "a cork in a raging": Dan Balz, "Bush Derides Talk That He Fears 2000 Run," *Washington Post*, October 15, 1998.

274 "In a dynasty": Lyman and Navarro, "Easily Elected," *New York Times*.

CHAPTER 38: "NO TURNING BACK"

275 "I got some things": Adam Nagourney, "Bush, the Elder, Passes the Political Baton," *New York Times*, June 14, 1999.

275 "I'm coming here": Adam Nagourney, "Bush Iowa Trip Signals the Real Start of the 2000 Race for President," *New York Times*, June 13, 1999.

276 $36 million in financial: Moore, *American President*, 620.

276 "Is compassion beneath us?": Nagourney, "Bush Iowa Trip," *New York Times*.

276 "He doesn't need advice": Nagourney, "Bush, the Elder," *New York Times*.

278 "I know who I am": Burka, "Man Who Isn't There," *Texas Monthly*.

278 "is an unofficial": Jeffrey H. Birnbaum, "The Man Who Could Be President Isn't Just the Spittin' Image of His Dad," *Fortune*, March 29, 1999.

278 "We are the party": Jennifer Steinhauer, "McCain Back at Scene of Collapse in S. Carolina," *New York Times*, October 19, 2007.

278 "chose to sire children": James Mann, *George W. Bush* (New York: Times Books, 2015), 35.

279 "I'm not happy": "1999 Donald Trump Press Conference on Running for Pres-

ident," video retrieved on February 3, 2017, from: https://www.youtube.com /watch?v=hMlz4J4tYbM.

280 "How bad can I be": Pamela Kilian, *Barbara Bush: Matriarch of a Dynasty* (New York: Thomas Dunne, 2002), 214–15.

280 "this boy, this son": Ibid.

280 "This is not going": Schweizer and Schweizer, *The Bushes*, 486.

280 "he's more ideological": Isaacson, "My Heritage," *Time*.

280 "We'll do what": Schweizer and Schweizer, *The Bushes*, 486.

280 "My dad": Isaacson, "My Heritage," *Time*.

281 "Dad said that same": Ibid.

281 "If you need me": Dick Cheney, author interview.

281 "wrap things up": Ibid.

282 "This is a bad idea": Karl Rove, author interview.

282 "warts" including a "misspent youth": Dick Cheney, author interview.

282 "You're my guy": Ibid.

283 "He was very supportive": GW Bush, author interview.

283 "George W. Bush is lucky": GHW Bush, *All the Best*, 639.

CHAPTER 39: GREAT EXPECTATIONS

284 "Please stow your expectations": Nagourney, "Bush Iowa Trip."

284 1 percent lead in the polls: Karl Rove, *Courage and Consequence: My Life as a Conservative in the Fight* (New York: Simon & Schuster, 2010), 177.

284 "Keeping quiet was": "Debacle: What Al Gore's Debate Against George W. Bush Can Teach Hillary Clinton," *New York Times*, September 25, 2016.

285 "a raw, unbridled contempt": Ibid.

285 as wide as ten points: Mann, *George W. Bush*, 37–38.

285 Bush believed he would: GW Bush, author interview.

285 "I've often said": Kellman, "Bush Once Plead Guilty to DUI," *Washington Post*.

286 "Circumstances have changed": David Margolick, "The Path to Florida," *Vanity Fair*, March 19, 2014.

287 "totally cool, calm": J. Bush, author interview.

287 "would you rather win": L. Bush, author interview, Texas Book Festival, October 16, 2010.

287 "We're kicking your name around": James Baker, author interview.

288 "I'll be at the ranch": John Dickerson, "Home on the Range," *Time*, December 18, 2000.

288 "The only way": L. Bush, author interview, Texas Book Festival.

289 "It's a nervous time": Updegrove, *Second Acts*, 378.

289 "I wasn't getting advice": Ibid., 379.

289 "Congratulations, Mr. President": Karl Rove, author interview.

CHAPTER 40: "MR. PRESIDENT"

293 Some 300,000 stood: Joshua Gillin, "Inaugural Crowd Sizes Ranked," Politifact .com, January 20, 2017.

293 While the sting: Andy Card, author interview.

293 "Civility is not a tactic": "President Bush's Inaugural Address," PBS *NewsHour*, January 20, 2001.

294 "President Bush would like you": GHW Bush, *All the Best*, 642.

294 "It was cold and dark": Andy Card, author interview.

294 "again to walk in": GHW Bush, *All the Best*, 642.

294 "Mr. President": Andy Card, author interview.

295 "Dad didn't tell me directly": GW Bush, author interview.

295 "You're not the copresident": Andy Card, author interview.

296 As 43 was quick to point out: GW Bush, author interview.

297 "incomplete state of scientific": Mann, *George W. Bush*, 48.

CHAPTER 41: "SHOCK AND AWE"

298 "soft bigotry": "George W. Bush's Speech to the NAACP," *Washington Post*, July 10, 2000.

298 "Let's show them": Mann, *George W. Bush*, 51.

299 "wartime president": "George W. Bush: The 9/11 Interview," National Geographic, aired August 2011.

299 "Today we've had": P. Baker, *Days of Fire*, 122.

299 "Dad's words must": GW Bush, *Decision Points*, 128.

300 "Is Rumsfeld alive": Karl Rove, author interview, "9/11: A White House View," LBJ Presidential Library, September 3, 2015.

300 The first plane: GW Bush, *Decision Points*, 528.

300 F-16 fighter jets: P. Baker, *Days of Fire*, 126.

300 "immediately started thinking": CNN.com, September 11, 2016.

301 "make sure [Putin]": Ibid.

301 As images of: "George W. Bush: The 9/11 Interview," National Geographic.

301 "Make no mistake": P. Baker, *Days of Fire*, 128.

302 "We don't need some": Ibid.

302 "We're at war": Ibid., 131.

303 "We never talked about": L. Bush, author interview, Texas Book Festival.

303 "Shock and awe": GHW Bush, author interview.

303 "Son, you grounded": B. Bush, author interview.

303 "the sooner [you] get back": GHW Bush, "George H. W. Bush to George W. Bush After 9/11: 'Get Back to Washington Right Now,'" *Newsweek*, November 6, 2014.

304 "The search is underway": "Statement by the President in His Address to the Nation," The White House, September 2001.

CHAPTER 42: MOURNING IN AMERICA

305 But when the elder: Jean Becker, author interview.

306 "more united than ever": Billy Graham, "Billy Graham's 9/11 Message from the Washington National Cathedral," Billy Graham Evangelistic Association, September 6, 2016.

306 "Just three days removed": "Bush's Remarks at Prayer Service," *Washington Post*, September 14, 2001.

306 The memory of the moment: GHW Bush, author interview; and GW Bush, author interview.

306 "Feeling his arm": GW Bush, author interview.

307 "The president really": Condoleezza Rice, author interview.

307 "[Y]ou could smell it": Karen Hughes, author interview, "9/11: A White House View."

307 "It was like you": Karl Rove, author interview, "9/11: A White House View."

308 "I can hear you": Kenneth T. Walsh, "George Bush's 'Bullhorn Moment,'" *U.S. News & World Report*, April 25, 2013.

308 "the most intense experience": Karen Hughes, author interview, "9/11: A White House View."

309 "a reminder of": "George W. Bush: The 9/11 Interview," National Geographic.

309 He left inspired: Ibid.

CHAPTER 43: PRELUDE TO WAR

311 "paper tiger": "George W. Bush: Foreign Affairs," UVA/The Miller Center, https://millercenter.org/president/gwbush/foreign-affairs.

311 "This time we would": Mann, *George W. Bush*, 63.

311 While providing no evidence: P. Baker, *Days of Fire*, 144.

311 "No one will understand": Ibid.

311 "How many times": Ibid.

312 "Islam is peace": "'Islam is Peace,' Says President: Remarks by the President at Islam Center of Washington, DC," The White House, September 2001.

312 "There's an old poster": Toby Harnden, "Bin Laden Is Wanted: Dead or Alive, Says Bush," *Telegraph*, September 18, 2001.

312 43's father offered to fly commercial: Jean Becker, author interview.

313 "the finest, strongest": Frank Pellegrini, "The Bush Speech: How to Rally a Nation," *Time*, September 21, 2001.

313 "They will hand over": Ibid.

314 Afterward, his approval rating: Ibid.

CHAPTER 44: A BROADER BATTLE

315 "I was able": GW Bush, author interview.

315 "I want a plan": Mann, *George W. Bush*, 66.

317 "Behind my presidency": GW Bush, author interview.

318 "amused" by the glib term: Ibid.

318 "will not hesitate": Mann, *George W. Bush*, 75.

318 "the thrill of liberation": GW Bush, *Decision Points*, 204.

319 "True believers cannot": Patrick E. Tyler, "Bush Warns 'Taliban' Will Pay a Price," *New York Times*, October 9, 2001.

CHAPTER 45: 41 AND 43

320 "big, strong, highly respected": GHW Bush, *All the Best*, 496.

320 "You know how tough": GW Bush, *Decision Points*, 243.

321 "[H]e didn't need to": GW Bush, author interview.

321 "It is among the rarest": Fred Kaplan, *John Quincy Adams: American Visionary* (New York: HarperCollins, 2004), 141.

321 "I would definitely": James Baker, author interview.

321 "He always said": Condoleezza Rice, author interview.

322 "George Bush knew what": GW Bush, author interview.

322 If you've been president: Ibid.

323 "He stole a can": Jean Edward Smith, *Bush* (New York: Simon & Schuster, 2016), 658.

323 "few are going to believe": GW Bush, author interview.

323 "more instability in Iraq": Michiko Kakutani, "Love Flows, President to President," *New York Times*, November 11, 2014.

324 "There's no doubt": *World News Tonight* and Terry Moran, "Is Bush's Iraq Stance Rooted in Revenge," ABCNews.go.com, March 18, 2003.

324 "Some Americans have wondered": Ibid.

324 "a virtual go-it-alone": Brent Scowcroft, "Don't Attack Saddam," *Wall Street Journal*, August 15, 2002.

324 "On everything from taxes": Maureen Dowd, "Junior Gets a Spanking," *New York Times*, August 18, 2002.

324 "knew nothing about": Brent Scowcroft, author interview.

324 "did seek [41's] permission": Jean Becker, author interview.

325 "I think I know": Brent Scowcroft, author interview.

325 "gave 41 some heartburn": James Baker, author interview.

325 "The question people will": GW Bush, author interview.

325 "Why did [Scowcroft]": Ibid.

325 "He's in my administration": Andy Card, author interview.

325 "It's an interesting incident": GW Bush, author interview.

325 "how it works inside": Ibid.

326 "They're very close": B. Bush, author interview.

326 "go it alone": James A. Baker III, "The Right Way to Change a Regime," *New York Times*, August 25, 2002.

326 "I think people around": Andy Card, author interview.

326 "What I want to do": Updegrove, *Second Acts*, 231.

326 "some sort of competition": GHW Bush, author interview.

326 "I know that [41] was worried": James Baker, author interview.

CHAPTER 46: TO WHOM MUCH IS GIVEN . . .

331 In a predawn hour: Schweizer and Schweizer, *The Bushes*, 535.

331 "the wrong father to appeal to in terms of strength": Hamilton, "Bush Began to Plan War," *Washington Post*.

331 "was understandable as": James Baker, author interview.

332 The greatest influence: GW Bush, author interview.

332 "I thank God": David Valdez, *George Herbert Walker Bush: A Photographic Profile* (College Station, TX: Texas A&M Press, 2007), 116.

332 "If you believe": GW Bush, author interview.

333 five-year $15 billion commitment: "The Bush Record: President Bush's Global Health Initiatives Are Saving Lives Around the World," The White House, 2008.

333 By 2016, twelve million: email to the author from Freddy Ford, communications director, Office of George W. Bush, June 6, 2017.

333 "Of course, he's": GW Bush, *Decision Points*, 348.

333 "I clearly saw the ideological": Updegrove, "Bush 2.0," *Texas Monthly*.

334 And just as his father's: Schweizer and Schweizer, *The Bushes*, 540.

334 "Living a life": Ibid.

334 "Mr. Secretary, for": P. Baker, *Days of Fire*, 257–58.

335 Dear Dad, At around: GW Bush, *Decision Points*, 223–24.

335 Dear George, Your handwritten: Ibid., 224–25.

336 "My fellow Americans": "Bush Declares War," CNN.com, March 19, 2003.

CHAPTER 47: "REGARDS FROM PRESIDENT BUSH"

337 Acting on a tip that Saddam Hussein: Nancy Gibbs, "We Got Him," *Time*, December 14, 2003.

337 "My name is Saddam Hussein": GW Bush, *Decision Points*, 267.

337 "the father calling": GHW Bush, author interview for *Second Acts*.

337 "It was a touching": Robin Abcarian, "In His Father's Shadow, a Son Charts His Own Course," *Los Angeles Times*, August 29, 2004.

338 "The capture of this": Gibbs, "We Got Him," *Time*.

338 You have borne the burden: GHW Bush, *All the Best*, 661.

338 just 240,000 American troops: Mann, *George W. Bush*, 93.

339 "Major combat operations": "Bush Makes Historic Speech Aboard Warship," CNN.com, May 1, 2003.

339 "The top gun cut": Seth Cline, "The Symbol of George W. Bush's Legacy," *U.S. News & World Report*, May 1, 2013.

339 "He won the war": Heather Digby Parton, "Chris Matthews vs. Obama: How the President Lost One of His Top Supporters," *Salon*, October 3, 2014.

340 "Mission *Not* Accomplished": "Mission *Not* Accomplished," *Time*, October 6, 2003.

340 "Mission *Still* Not Accomplished": Joanne McGeary, "Mission *Still* Not Accomplished," *Time*, September 20, 2004.

340 "It's Worse Than You Think": Scott Johnson and Babak Dehghanpisheh, "It's Worse Than You Think," *Newsweek*, September 12, 2004.

340 an average of 120 Iraqis: "George W. Bush: Foreign Affairs," UVA/The Miller Center.

340 one in five Americans: Todd S. Purdum, "Bush Offers Bold Strokes, but Few Details in Speech," *New York Times*, September 2, 2004.

340 The war's mounting price: Linda J. Bilmes and Joseph E. Stiglitz, "The Iraq War Will Cost Us $3 Trillion, and Much More," *Washington Post*, March 9, 2008.

340 a total of 4,424 American: "Total Deaths KIA Non-hostile Pending WIA OIF U.S. Military," U.S. Department of Defense, https://www.defense.gov/casualty .pdf.

341 "[I]n denying his real": David Greenberg, "Fathers and Sons: George W. Bush and His Forebears," *New Yorker*, July 12, 2004.

341 "I know that [41]": James Baker, author interview.

341 "the most important thing": Susan Glazer, "Does Trump Actually Want to Succeed?" *Politico Magazine*, February 6, 2017.

341 "didn't agree": GW Bush, author interview.

341 "didn't know how": Ibid.

342 "I don't know him": GHW Bush, author interview for *Second Acts*.

342 "I don't think": Dick Cheney, author interview.

342 "He knew the Texas": Elspeth Reeve, "Powell Aide: Cheney Was President 'for All Practical Purposes,'" *Atlantic*, August 30, 2011.

342 "Cheney had his own": Meacham, *Destiny and Power*, 589.

342 "didn't worry": B. Bush and GHW Bush, author interview.

342 "It's true": Ibid.

343 "I never talked": Dick Cheney, author interview.

343 "used to be close": GHW Bush, author interview for *Second Acts*.

343 "I do think": B. Bush and GHW Bush, author interview.

343 "I'm confident they": GW Bush, author interview.

343 "Mom, when you're": B. Bush and GHW Bush, author interview.

344 "I hear the voices": Sheryl Gay Stolberg, "The Decider," *New York Times*, December 24, 2006.

344 "The fact that": GW Bush, author interview.

344 "People around George H. W.": Ibid.

344 "I was content": Ibid.

345 A lesson he taught: Ibid.

345 "shaking up the ticket": Ibid.

CHAPTER 48: THE LAST CAMPAIGN

346 Nearly thirty-eight years: L. Bush, author interview for "Reflections of First Ladies: Barbara Bush and Laura Bush," LBJ Presidential Library, November 15, 2012.

347 "place in the world": L. Bush, *Spoken from the Heart*, 194.

347 "a little jealous": L. Bush, author interview for "Reflections of First Ladies."

347 "Thank you so much": Ibid.

347 "Bushie, you gonna": ABC News, "First Lady Laura Bush on *Good Morning America*," November 27, 2001.

347 "I know he knows it": L. Bush, author interview, Texas Book Festival.

347 "What surprised me": Hugh Sidey, "They're Talking About . . . Our Kid," *Time*, December 19, 2004.

348 "Being in the city": L. Bush, *Spoken from the Heart*, 304.

348 You and Mom have: GW Bush, *Decision Points*, 288–89.

349 "the greatest gift": GW Bush, author interview.

349 "My last campaign": GW Bush, *Decision Points*, 289.

349 "now and then": Purdum, "Bush Offers Bold Strokes," *New York Times*.

349 "This election is": GW Bush, *Decision Points*, 287.

350 "when people are insecure": Purdum, "Bush Offers Bold Strokes," *New York Times*.

350 "How do you ask": "Bush in His Sights," *Sydney Morning Herald*, January 31, 2004.

350 "I'm John Kerry": Associated Press, "I'm John Kerry, and I'm Reporting for Duty!" *Michigan Daily*, July 29, 2004.

350 "flip-flopper": "Bush Outlines 'Where I Will Lead This Country,'" CNN.com, September 3, 2004.

350 "He dishonored his": Howard Kurtz, "Another Swift Boat Ad," *Washington Post*, August 24, 2004.

351 "bad for the system": MSNBC staff, "Bush Calls for Halt to Swift Boat Veterans' Ads," NBCNews.com, August 23, 2004.

351 "punched in the stomach": GW Bush, *Decision Points*, 294.

351 "the most devastated": Condoleezza Rice, author interview.

352 "George, you can't": GW Bush, *Decision Points*, 295.

352 "President Bush decided": Ibid., 296.

353 Prior to their departure: GHW Bush, *All the Best*, 670–71.

353 Will God continue to give: Ibid.

CHAPTER 49: "BROTHER FROM ANOTHER MOTHER"

354 "I want this": William Safire, "Bush's 'Freedom Speech,'" *New York Times*, January 21, 2005.

354 After working closely with Gerson: email to the author from Freddy Ford, communications director, Office of George W. Bush, June 6, 2017.

354 On January 20, 2005: Ibid.

354 "At this second": Ibid.

355 "two bozos": Michael Takiff, "When Presidential Candidates Get Desperate," CNN.com, October 22, 2012.

355 On December 26: Alan Taylor, "Ten Years Since the 2004 Indian Ocean Tsunami," *Atlantic*, December 28, 2014.

355 Initially, the U.S.: Ibid.

355 As the White House: GW Bush, author interview.

356 "Claims no more": GHW Bush, *All the Best*, 673–80.

356 "went out of his way": Ibid.

356 "wasn't 'Let's have'": GW Bush, author interview.

357 "Come on,": Updegrove, *Second Acts*, 235.

357 "We found that": Associated Press, "Bush, Clinton Pal Around on Golf Course," *USA Today*, June 28, 2005.

357 "I think people look": "Bill Clinton and George H. W. Bush: Partners of the Year," *Time*, December 26, 2005.

357 "almost like brothers": Jimmy Carter, author interview for *Second Acts*.

357 a "dividend": GHW Bush, author interview.

357 "Maybe I'm the father": Hugh Sidey, "The Benefits of Being an Ex-President," *Time*, April 23, 2005.

357 "brother from another": GW Bush, author interview.

357 "We all like him": B. Bush, author interview.

357 "[Clinton] would say": GW Bush, author interview.

358 "I talked to Clinton": Ibid.

358 Eight months after: CNN Library, "Hurricane Katrina Statistics Fast Facts," CNN.com, August 23, 2016.

359 "Brownie, you're doing": Marta Jewson and Charles Maldonado, "The Myths of Katrina," *Slate*, August 28, 2016.

359 "recognized deficiencies sooner": Campbell Robertson and Richard Pérez-Peña, "In New Orleans, Bush Lauds School Gains Since Katrina," *Boston Globe*, August 29, 2015.

359 "Our investigation revealed": Eric Lipton, "First Report on Katrina Assails Bush's Response," *New York Times*, February 13, 2007.

359 "the worst moment": GW Bush, *Decision Points*, 326.

360 I am really down about the way the President has been attacked: GHW Bush, *All the Best*, 686–88.

360 "I see you've reunited": GW Bush, *Decision Points*, 325.

360 "That's what they do": "Punchlines," *Time*, September 19, 2005.

361 over $120 million: "Press Release: Announcement of Major Bush-Clinton Katrina Fund Grants," The Clinton Foundation, December 7, 2005.

361 "to have another president": "Opposites Attract," *Guardian*, June 30, 2005.

361 "might become a little": GHW Bush, author interview for *Second Acts*.

361 his vote for Mrs. Clinton: GW Bush, author interview.

CHAPTER 50: "THE FIRST JEWISH PRESIDENT"

362 climbed to their highest: "George W. Bush: Domestic Affairs," UVA/The Miller Center.

362 "By the end": GW Bush, *Decision Points*, 330.

363 from twenty thousand to thirty thousand: "US to Send 30,000 Extra Troops to Afghanistan as War Hits Setbacks," *Guardian*, December 20, 2008.

363 "I disagree with": P. Baker, *Days of Fire*, 499.

363 "[Forty-one] called": James Baker, author interview.

364 "strongly advocated": Ibid.

364 "Cheney?" he asked: Mann, *George W. Bush*, 108.

364 "revolt of the radical": P. Baker, *Days of Fire*, 560.

364 "Peace," he replied: Sidey, "They're Talking About . . . Our Kid," *Time*, December 19, 2004.

364 "Bush the father": Sheryl Gay Stolberg, "Bush's Embrace of Israel Shows Gap with His Father," *New York Times*, August 2, 2006.

365 *We have driveways*: Ibid.

365 "How's the first Jewish": GW Bush, author interview.

365 "[My parents] were concerned": Ibid.

366 "I suspect that some": Ibid.

366 With President Mubarak: Ibid.

366 "unsettled": Ibid.

367 "I think your mother": Jean Becker, author interview.

CHAPTER 51: THE TOUGHEST DECISION

368 Bush read no less: Peter Baker, "Abraham Lincoln, the One President All of Them Want to Be More Like," *New York Times*, April 14, 2015.

368 "everybody's favorite": GHW Bush, author interview.

368 "Always bear in mind": Charles W. Moores, ed., *Lincoln: Addresses and Letters* (New York: American Book Company, 1914), 64.

368 "the toughest and most," GW Bush, *Decision Points*, 355.

368 "bring some troops": Ibid.

369 "We're not winning the war": Peter Baker, "U.S. Not Winning the War in Iraq, Bush Says for 1st Time," *Washington Post*, December 20, 2006.

369 "now allow the U.S.": "Results from the *Los Angeles Times*/Bloomberg Poll: January 13–16, 2007," Study #540, *Los Angeles Times*/Bloomberg.

369 "Thinking about what Lincoln": P. Baker, *Days of Fire*, 527–28.

370 The same month: Thomas R. Mockaitis, ed., *The Iraq War Encyclopedia* (Santa Barvara, CA: ABC-CLIO, 2013), 403.

370 "We are no longer": Peter D. Feaver, "Anatomy of the Surge," *Wall Street Journal*, March 26, 2008.

370 A Pentagon report: Mockaitis, 405.

370 "just right": GW Bush, author interview.

371 "The big mistake": Meacham, *Destiny and Power*, 589.

371 "Well, in the final": GHW Bush, author interview.

371 "The Presidency, even": *Lincoln: Speeches and Writings, 1832–1858* (New York: Library of America, 1989), 253.

CHAPTER 52: LAST DAYS

372 "Is this the worst": GW Bush, *Decision Points*, 440.

372 "September 11, when": L. Bush, *Spoken from the Heart*, 419.

372 "Yes," replied Bernanke: GW Bush, *Decision Points*, 440.

373 "September and October": Tim Worstall, "Ben Bernanke: The 2008 Financial Crisis Was Worse Than the Great Depression," *Forbes*, August 27, 2014.

373 "There will be ample": Mark Landler and Sheryl Gay Stolberg, "Finger-Pointing in Financial Crisis Is Directed at Bush," *New York Times*, September 19, 2008.

373 69 percent of Americans: Peter Baker, "The Final Days," *New York Times Magazine*, August 29, 2008.

373 "To say that Bush": Ibid.

373 "I would like to be": Updegrove, "Bush 2.0," *Texas Monthly*.

374 "If we're looking": GW Bush, *Decision Points*, 444.

374 $700 billion in federal: Mann, *George W. Bush*, 134.

374 Congress's rejection of: "Timeline of Events in the Financial Crisis," *USA Today*, January 28, 2011.

374 "those of Henry Morgenthau": GW Bush, *Decision Points*, 450.

374 By the end of 2008: Joel Havemann, "The Financial Crisis of 2008: Year in Review 2008," *Encyclopaedia Britannica*, https://www.britannica.com/topic/Financial-Crisis-of-2008-The-1484264.

375 "It's my belief": Liz Halloran, "McCain Suspends Campaign, Shocks Republicans," *U.S. News & World Report*, September 24, 2008.

375 "We don't need": Ibid.

375 Even beforehand: Ibid.

375 "One message that": CNN, "Obama, Bush, and Ex-presidents Have 'Historic Moment,'" CNN.com, January 7, 2009.

376 "I felt we were": Jonathan Martin, "George W. on Dick Cheney, 'It's Been Cordial,'" *Politico*, April 2013.

376 "Don't let this": P. Baker, *Days of Fire*, 633.

376 "I understand the way": GW Bush, author interview.

377 Dear Barack, Congratulations on: Meghan Keneally and Lissette Rodriguez, "First on ABC: George W. Bush's Inauguration Day Letter to Barack Obama," ABC.com, January 19, 2017.

377 "Welcoming Barack Obama": GW Bush, author interview.

377 "what he must have": Ibid.

378 "The love of the Bush": L. Bush, *Spoken from the Heart*, 426.

CHAPTER 53: "IT WASN'T ME"

381 "I have had": George H. W. Bush, Essay for Kennebunkport Conversation Trust, Provided by Jean Becker, Office of George H. W. Bush.

382 "It wasn't me": Updegrove, "Bush 2.0," *Texas Monthly*.

382 "By focusing on": Ibid.

382 "He didn't linger": Ibid.

382 "a chapter": Ibid.

383 "Man, Bush, I can't": L. Bush, author interview, Texas Book Festival.

383 "replenish the ol' coffers": "Bush's Great Ambition: Wealthy Boredom," *Guardian*, September 3, 2007.

383 $100,000 to $150,000 in honoraria: Rachel Rose Hartman, "George W. Bush Racks Up $15 Million in Speaking Fees," Yahoo News, May 20, 2011.

383 "It didn't take long": Updegrove, "Bush 2.0," *Texas Monthly*.

383 In his first two: Hartman, "$15 Million in Speaking Fees."

383 "W. could buy": B. Bush, author interview.

384 "I don't view": Mark K. Updegrove, "A Conversation with George W. Bush and Mrs. Bush," *Parade Magazine*, April 20, 2013.

384 "If someone said": George W. Bush, *Portraits in Courage* (New York: Crown, 2017), Foreword.

385 "I kept hoping": Updegrove, "Conversation," Parade.

385 "It's not the cough": GW Bush, *41*, 274.

385 "There's death and": Ibid., 275.

385 "Did I almost": Jean Becker, author interview.

385 she replied, "Eleven": Ibid.

385 George W. believed: GW Bush, author interview.

386 "I'll be there": GW Bush, *41*, 275.

386 "a support group": Judy Keen, "Five Living Presidents Toast New Bush Library," *USA Today*, April 25, 2013.

386   "Too much?": GW Bush, *41*, 275.

386   "strength and resolve": Mark K. Updegrove, "Inside the Presidential Reunion at the Bush Library Dedication," Parade.com, April 26, 2013.

CHAPTER 54: TRUMPED

388   Dear Donald, I did not: Rachel Dicker, "Donald Trump to Hang a Letter from Richard Nixon on the Oval Office Wall," *U.S. News & World Report*, December 12, 2016.

389   "Would President Bush": Jean Becker, author interview.

389   "[Jeb Bush] is going": Michael Scherer, "Jeb Bush Bets on a Sunshine State of Mind," Time.com, December 16, 2014.

389   "I think our politics": Jeb Bush, author interview.

390   "We've had enough Bushes": Associated Press, "'We've Had Enough Bushes in the White House'—Barbara Bush," *Daily Telegraph*, April 26, 2013.

390   "What's the difference": GW Bush, author interview.

390   "Till death do us part": Philip Rucker, "In Drive to Be 45th President, Jeb Bush Faces Legacies of the 41st and 43rd," *Washington Post*, February 14, 2015.

390   GOP "would-be's": GW Bush, author interview.

390   "What do you mean": Jennifer Shutt, "Barbara Bush Reverses on 'Enough Bushes' Line," *Politico*, February 13, 2015.

390   "I thought that": B. Bush, author interview.

391   of $150 million: Eli Stokols, "Inside Jeb Bush's $150 Million Campaign Failure," *Politico*, February 20, 2016.

391   "campaign as I would": Ed O'Keefe, "Jeb Bush Announces Presidential Bid: 'We Will Take Command of Our Future Again,'" *Washington Post*, June 15, 2015.

391   "Our country is in": "Here's Donald Trump's Presidential Announcement Speech," Time.com, June 16, 2015.

391   "derogatory statements": Maria Puente, "NBC to Donald Trump: 'You're Fired,'" *USA Today*, June 29, 2015.

391   Nonetheless, Trump, due: Maria Stainer, "Jeb Bush, Donald Trump Lead GOP Candidates, Still Trail Hillary Clinton: Poll," *Washington Times*, July 1, 2013.

392   "He's not a war hero": Jonathan Martin and Alan Rappeport, "Donald Trump Says John McCain Is No War Hero, Setting Off Another Storm," *New York Times*, July 18, 2015.

392   "had blood coming out": David Catanese, "Trump Testing Fellow GOP Candidates' Messaging," *U.S. News & World Report*, August 10, 2015.

392   "I could stand": Jeremy Diamond, "Trump: I Could 'Shoot Somebody and I Wouldn't Lose Voters,'" CNN.com, January 24, 2016.

392   an estimated $2 billion: Chris Deaton, "Trump Has Earned $2 Billion of Free Media Coverage," *Weekly Standard*, March 15, 2016.

392 "low energy": Brett LoGiurato, "Trump: Here's the Backstory on My 'Low Energy' Takedown of Jeb Bush," *Business Insider*, November 19, 2015.

393 "is a chaos candidate": Harvey Solomon Takoma, "A Chaos President," *New York Times*, December 15, 2015.

393 "Knowing what we": Philip Rucker, "Jeb Bush Now Says, 'I Would Not Have Gone into Iraq,'" *Washington Post*, Mary 14, 2015.

394 "These are tough times": email to the author from Freddy Ford, communications director, Office of George W. Bush, June 15, 2017.

394 "I'm a big W. fan": David A. Graham, "Can George W. Save Jeb," *Atlantic*, February 16, 2016.

394 "They lied. They said": David French, "Donald Trump is No Revolutionary, He's Just a Democrat," *National Review*, February 15, 2016.

394 "In this campaign": Ed O'Keefe, "Jeb Bush Drops Out of 2016 Presidential Campaign," *Washington Post*, February 20, 2016.

395 "I think he would've": GW Bush, author interview.

395 "You can either exploit": Ibid.

395 "Hasn't Donald Trump": NR Symposium, "Conservatives Against Trump," *National Review*, February 15, 2016.

396 "I'm worried that": Eric Bradner, "Bush: 'I'm Worried That I Will Be the Last Republican President," CNN.com, July 19, 2016.

396 "force for good": GW Bush, author interview.

396 "twins of negativity": Updegrove, "Bush 2.0," *Texas Monthly*.

397 *Interesting; won't last*: GW Bush, author interview.

397 "I don't like him": GHW Bush, author interview.

398 "standard": GW Bush, *41*, 105.

398 "Power accompanied by arrogance": GHW Bush, *All the Best*, 179–86.

398 "The question for": GW Bush, author interview.

398 Neither Donald Trump: GW Bush, author interview.

CHAPTER 55: "WHAT ABOUT GEORGE?"

399 "Posterity will scarcely": Harlow Giles Unger, *John Quincy Adams* (Boston: De Capo Press, 2012), 256.

400 approval rating of 34 percent: "Presidential Approval Ratings—Gallup Historical Trends," Gallup, accessed on July 16, 2017, http://www.gallup.com/poll/116677/presidential-approval-ratings-gallup-historical-statistics-trends.aspx.

400 "below average" president: Hugh Sidey, "The Presidency," *Time*, June 24, 2001.

401 "Like the remarkable": Michael D. Shear, "Elder Bush Receives Medal of Freedom," *New York Times*, February 15, 2011.

401 "Qualities once branded": John Solomon, "A Wimp He Wasn't," *Newsweek*, March 20, 2011.

401 Bush's approval ratings: Philip Bump, "George W. Bush Now Polls Better Than

Hillary Clinton and Obama. That Means Less in Context," *Washington Post*, June 3, 2015.

402 "the enemy of the people": David Jackson, "Trump Again Calls Media 'Enemy of the People,'" *USA Today*, February 24, 2017.

402 "indispensable to democracy": Associated Press, "George W. Bush on Trump and Russia: 'We All Need Answers,'" *Los Angeles Times*, February 27, 2017.

402 A Good Man™: Aaron Goldstein, "George W. Bush, Liberals' New Hero," *National Review*, March 6, 2017.

403 "You've become an icon": Jean Becker, author interview.

403 "Let history be": GW Bush, author interview.

403 "I'm glad that": Jean Becker, author interview.

403 "History will ultimately": GW Bush, author interview.

403 "What about George?": Jean Becker, author interview.

EPILOGUE

405 After George W.'s election: Schweizer and Schweizer, *The Bushes*, 510.

405 "just to make sure": Ibid.

406 "I, along with others": Sean Collins Walsh, "George P. Bush Breaks with Family, Endorses Trump," *Austin American-Statesman*, August 8, 2018.

406 "It's time to put": Ibid.

INDEX

ABOUT THE AUTHOR

MARK K. UPDEGROVE is the author of four books on the presidency, including *Indomitable Will: LBJ in the Presidency*. The inaugural CEO of the National Medal of Honor Museum and the former director of the LBJ Presidential Library, Updegrove is a contributor to ABC News and *Good Morning America*, and has written for the *Daily Beast*, the *New York Times*, *National Geographic*, *Parade*, *Politico*, *Texas Monthly*, and *Time*. He lives in Austin, Texas.